Zbigniew Płotnicki

Information-Oriented Logic

Also of interest

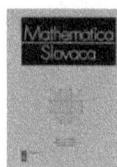

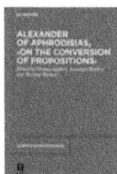

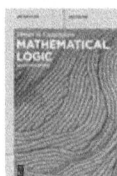

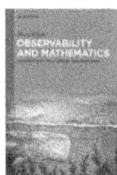

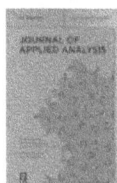

Zbigniew Płotnicki

Information-Oriented Logic

Implications of Natural-Language Sentences and All Logical Paradoxes Solved

DE GRUYTER

Mathematics Subject Classification 2020
03B60, 03B65, 03B70, 03E70

Author
Zbigniew Płotnicki
zbigniew.plotnicki.books@gmail.com

ISBN 978-3-11-144046-0
e-ISBN (PDF) 978-3-11-144138-2
e-ISBN (EPUB) 978-3-11-144156-6

Library of Congress Control Number: 2024945560

Bibliographic information published by the Deutsche Nationalbibliothek
The Deutsche Nationalbibliothek lists this publication in the Deutsche Nationalbibliografie; detailed bibliographic data are available on the internet at http://dnb.dnb.de.

Cover image: ipopba/iStock/Getty Images Plus
Typesetting: Integra Software Services Pvt. Ltd.

www.degruyter.com
Questions about General Product Safety Regulation:
productsafety@degruyterbrill.com

Abstract

Do you want to know a logic that solves all problems of old classical and nonclassical logics? Have you wondered why all previously invented logics are so inept at solving everyday problems for mathematicians? Are you fed up with the fact that these logics can't do the simplest things, e.g.: speak about themselves, define truth decently, solve paradoxes conclusively and reliably, easily solve implications based on the meaning of natural language sentences that are so obvious to us, distinguish a statement from a sentence, give legitimacy to material implication and finally solve the problem of material implication, and lend themselves to subjective sentences such as "This is a beautiful day", etc.? I have been working on this logic for 10 years and it solves all these problems. The fundamental rules of rational thinking are given in this book.

Enjoy!

https://doi.org/10.1515/9783111441382-202

Preface

I marked some dependent clauses using commas to make clearer their structural situation in sentences. For example: "The truth, that she comes to know, nourishes her soul" means the same as "The truth that she comes to know nourishes her soul", "He used to say, that thinking can be an art" means the same as "He used to say that thinking can be an art". Simply, the word "that" preceded by a comma is always a relative pronoun or a conjunction.

The mix of two coordinating conjugations "and/or" have exactly the same meaning as "or", so I will use coordinator "or" in the meaning, that includes the fact, that the statement of an outer sentence can concern all its arguments at the same time. For example: red or sweet = exactly one of three possibilities: red and sweet, red and not sweet, not red and sweet.

A text in square brackets used in citations is a complement information or an additional explanation or my commentary to the information contained in the citation.

In this book, the convention of omitting articles before the names of variables has been adopted to enhance readability.

As you will see, some small fragments of this book are repeated in different places. This is done so that you don't have to leaf through the pages of the book looking for things that are very important in a particular place. This will also be helpful for readers who will be coming back to read after some time or who will not read the book in the traditional linear way.

To minimize redundant text, some sections in this book have little text, so read mathematical expressions in those sections as sentences, the meaning of which is the content of those sections.

In this book, for example, the sentence "For any statements or attributes $x, y, \ldots$: $P(x, y, \ldots)$" means the same as the sentence "$P(x, y, \ldots)$ holds for all $x, y, \ldots$ such that all of them are statements or all of them are attributes", which has the same meaning as the sentence "$P(x, y, \ldots)$ holds for any statements $x, y, \ldots$ and $P(x, y, \ldots)$ holds for any attributes $x, y, \ldots$".

The main idea that guided me in creating this logic was to create a logic that would solve all the problems of previous logics. I noticed that previous logics used sentences leaving out their meaning, which is information. I wanted to create a logic along the lines of arithmetic, which would literally be the "arithmetic" of information and would talk about information instead of sentences. I noticed that a piece of information implicated by another piece of information is literally a part of that piece of information and I discovered new mereology (which is the study of part-whole relationships) to solve that problem, and then I drew all the consequences from that. And that is exactly what I have accomplished in this book. I discovered this new type of approach that allows to develop the complete and the revolutionary-way formal system of information oriented logic.

https://doi.org/10.1515/9783111441382-203

The following citations show us that some great logicians had some faint glimmerings of this idea, which they unfortunately were not able to develop back then:

> *The two . . . logical inferences . . . from the original [set of] propositions . . . give us all the* ***information*** *which it* ***contains*** *respecting the class* ([6] George Boole, 1847, p. 75).

> *it is the office of a conclusion not to present us new truth, but only to bring into explicit form some* ***portion*** *of that truth which was implicitly involved in the premises . . . [some* ***portion*** *of] the particular* ***information conveyed*** *in the premises* ([6] George Boole 1856?, p. 239).

> *Every collective set of premises* ***contains*** *all its valid conclusions; . . . speaking objectively, the assumption of them [the premises] is the assumption of the conclusion; though, ideally speaking, the presence of the premises in the mind is not necessarily the presence of the conclusion* ([7] Augustus De Morgan, 1847, p. 254).

> *All the propositions of pure geometry, which multiply so fast that only a small . . . class . . . among mathematicians . . . know all that has been done . . ., are certainly* ***contained*** *in a very few notions . . . . [The] consequences are virtually* ***contained*** *in the premises* ([7] Augustus De Morgan, 1847, p. 45).

> *The very purpose of syllogism is to deduce a conclusion which will be true when the premises are true. The syllogism enables us to restate in a new form the* ***information*** *. . .* ***contained*** *in the premises, just as a machine may deliver to us in a new form the material . . . put into it* ([8] W. Stanley Jevons, 1870, p. 149).

> *We extract out of the premises all the* ***information*** *. . . useful for the purpose in view – and this is the whole which reasoning accomplishes* ([8] W. Stanley Jevons 1870, p. 15).

> *[To deduce is] . . . to draw . . . propositions as will necessarily be true when the premises are true. By deduction we investigate and unfold the* ***information contained*** *in the premises* ([8] W. Stanley Jevons, 1879, p. 49).

> *These . . . [consequences] . . .* ***contain*** *every particle of* ***information*** *yielded by the original [premise] . . ., or in any way deducible from it* ([9] John Venn, 1881, p. 296).

> *That is, [in making this inference] we have had to let slip* ***a part of the information contained*** *in the data* ([9] John Venn, 1881, p. 362).

> *logicians in overwhelming majority maintain that every conclusion is implicitly* ***contained*** *in the premises* ([9] John Venn, 1889, p. 42).

The most striking and the most important competitive advantage over previous logics is, that this logic solves entirely all elementary problems of previous logics and it is not bragging, because it is the honest truth indeed. First of all, the notion of decidability of a statement solves all logical paradoxes and many other elementary problems. Secondly, new set theory allows to solve all elementary problems in set theory. Additionally, this logic fits into opinions and objective sentences as well. And for dessert, it introduces probabilistic logic for contextual sentences. To see solutions of logical paradoxes and most elementary problems, see Chapter 15.

Ask yourself the following questions:

Do you want to know the logic that solves all the problems of old classical and nonclassical logics?

Have you wondered why all previously invented logics are so inept at solving everyday problems for mathematicians?

Are you fed up with the fact that these logics can't do the simplest things, e.g.:

- speak about themselves,
- define truth decently,
- solve paradoxes conclusively and reliably,
- easily solve implications based on the meaning of natural language sentences that are so obvious to us,
- distinguish a statement from a sentence,
- give legitimacy to material implication and finally solve the problem of material implication,
- lend themselves to subjective sentences such as "This is a beautiful day", etc.?

I have been working on this logic for 10 years and it solves all these problems.

This new logic invented by me can change the foundation of mathematics. This is completely new approach to logic. All previously invented logics use declarative sentences that are symbols, as there were no referents of them, which is like doing arithmetic using only symbols of numbers and not referring to numbers themselves. This logic puts the referent of a sentence, which is a statement (a piece of information) in the center of this theory.

What I did, is taking seriously the fact, that a meaningful declarative sentence (I will call it a **declarence**) is a symbol, that refers to something important – a **piece of information**, exactly the same like numerals (e.g.: "1", "2", etc.) and arithmetic expressions (e.g.: "1 + 1", "1 + 2 ∗ 3", "(1 + 2) ∗ 3", etc.) refer to numbers in arithmetic. A meaningful declarative sentence simply expresses a piece of information (I will call it a **statement** after Peter Strawson [1, 2]) the same as an arithmetic expression expresses the number it refers to. Otherwise arithmetic would be as strange and needlessly difficult as classical logic, e.g.: ("1 ∗ 2e0") + "1.0" ⇔ "3", where plus would be no more an operation on numbers and true equality between symbols would be useless, while very useful true equality between numbers would be unreachable and replaced by limited equivalence. So in this new logic I put a piece of information (statement) in the center of the theory, the same as a number is in the center of arithmetic. That is why I call it information-oriented logic. I simply derived whole new logic from this truth, that a piece of information (statement) is the referent of a meaningful declarative sentence, that is a symbol.

The most important thing is that true logic should always be able to defend itself by its subtle power to prove anything, that is true (which I proved in this book), and this way to convince anybody, who is intelligent enough, without any brute force. Even if someone is ignorant, but intelligent enough to understand logical arguments, then logic can even make him think the way, he do not want to think. With an use of intelligence logic can be understood, so it is the only universal way to convince anyone, that something is true or false.

So why do we have the epistemological crisis? Why is not formal logic suitable to solve even everyday mathematician's practical problems? Why do logicians still think that a sentence, that is only a symbol of a statement, can be true or false? Is a symbol of number 5 positive or is it a symbol of a positive number? Why do not they see clearly the difference between a statement (a piece of information) and a meaningful declarative sentence (a symbol of a piece of information)? Why they think, that the most important thing in reasoning, that is implication, is between symbols and not between pieces of information? Is a symbol of an integer number divisible by another symbol of an integer number? Why do they allow material implication to be ungrounded? Why do they offer insufficient and thin solutions of paradoxes in logic, while paradoxes in a theory always say that the theory is wrong? Why do not they well understand the difference between subjective and objective logic?

I can answer these questions: simply, logic, that would solve all these problems, was not yet known.

Philosopher of language, Peter Strawson made the point [1, 2] that two meaningful declarative sentences (symbols) can make the same statement if they express the same thing (referent) in different ways. I independently came to the conclusion, that information is the referent of a meaningful declarative sentence and I have invented information oriented logic, that can be this logic, that solves all these mentioned above problems. And this book is the complete guide for that logic. In this book I only show you, what there is and was forever in logic, which by its definition is the complete system of correct reasoning and thinking, and should be shown already very long time ago.

This is not just another nonclassical logic as those described in [4, 5], because all those logics are symbol-oriented, while this new logic is information-oriented. That is, why there are not too many references in this book, because no one wrote about this logic yet, since no one knew about this logic yet.

To be perfectly honest after writing this book I found just a few papers on this subject, for example, [11] and it is worth appreciating that anyone paid attention to this problem; however, no real problem was solved there, that would bring us closer to this goal, which is the creation of information-oriented logic. You can be sure that this book contains solid concrete in the form of the complete definition of information-oriented logic and does not contain any unnecessary considerations, because I simply do not want to waste your precious time.

In this logic we consider whole truth as the conjunction of all statements of the true, consistent and complete theory of everything because everything can be covered by this logic. For that reason this logic is not based on a set of standalone axioms (all axioms are used only to define basic notions, e.g.: implication and negation), from which you can derive all provable sentences, but it concerns all pieces of information, the number of which is far greater than the number of all possible sentences in any language. For instance, the set of all statements of sentences in the form "x is an element of set Y" for every element x of set Y can be easily uncountable, for example, for the set of all real numbers. Instead of a set of standalone axioms this logic defines a statement. For example, how a statement can be constructed using other statements. And whatever this definition allows a statement to be, this new logic can process. So it is not ith-order logic, because it allows quantifiers over any set, even over any set of statements.

This logic is defined using only one primitive notion – a conjunction of things – and does not use standalone axioms. It offers the final solution to all paradoxes and many elementary problems of logic (e.g. the problem of Gödel's incompleteness theorem is solved in that logic). You will find in this book also complete theory of things, that, for example, allows to solve natural language implications.

I do not want to bloviate, so this book is very condensed, because the amount of knowledge I want to convey to you is quite big. For the simplicity of a reception the language is as simple as can be and does not use unnecessary difficult words.

For the proof, that Albert Einstein was most likely right about quantum physics see Section 16.4.

Contents

1 Pre-introduction

1.1 Abstract

This is the simplest way to understand basics of new logic.
In this chapter, I will explain shortly the most important things, that I have discovered since I started my intellectual adventure with logic 10 years ago. You will find here true content implication between statements derived from three simple axioms and material implication derived from them, which both build a complete new logic. Implication of simple statements in English is also derived on the basis of this content implication and appropriate examples are given. This logic uses new reduction to absurdity, that allows solving practically all elementary paradoxes of classical logic.

This new logic is fully explained in this book.

1.2 Some comments from positive reviews from Academia Letters

I got the following comment from Jose Maria Frapolli (https://en.wikipedia.org/wiki/María_José_Frápolli):

> Thank you very much for posting this exciting paper. It gives me some hope about the future of logic (and philosophy). I agree with the framework thoroughly. (. . .) this view is original and offers some fresh air (. . .)

I also got the following reviews from reviewers of Academia Letters:

> Yes!! I'm actually having a wider open mind than before with the author. It's worthy to be able to read all the legal information that I need – Hermenegildo Renco,
>
> I believe that material which is insightful, even if an outline of a larger project, merits publication and discussion. – Joseph Smith, University of Adelaide,
>
> I think that it is an interesting approach to logic. Some readers perhaps would like to know it in details. – Francisco Diaz Montilla, Universidad de Panamaa.

1.3 Introduction

For the introduction to this section, see Preface.

https://doi.org/10.1515/9783111441382-001

1.4 Operators

In this chapter, I use the following symbols of operators:

"∼" – logical negation of a statement,

"⇒" – implication **between statements**, and not between sentences, like it is in classical logic (it is **analogous** to "logical consequence" from classical logic),

"⊃" – inclusion of things and thereby also the relation of being a part,

"=" – **true equality**, much stronger than any equivalence from any other logic,

"→" – contextual implication of statements (it is **analogous** to "material implication" from classical logic),

"≡" – the equality of logical values of statements; it will be used for the aesthetic use of logical constants *true, false, undecidable.*

1.5 New logic

This is not a complete definition of new logic, but **the simplest introduction** to it.

Let us start from few definitions:

Let the notion "**thing**" (you can use also notion "object") be defined by the following tautology:

$$(for\ every\ x)(x\ is\ a\ thing)$$

In other words, every**thing** is a thing, so it has the widest class. Every**thing** has its definition, but not always that intended one (e.g. a square circle has its definition, because it has, for example, the following attribute: "it was meant to be a square and a circle at the same time").

Better definition of the notion "thing" is not possible and not necessary, because the word "thing" is built-in the word "everything", so we would get the following problem: "a thing is every**thing** that . . .".

Comment: Notice that one does not say "everything and everyone" when he wants to say about every-thing, that is there, e.g.: "everything is fine", "everything you see", "everything you can imagine" do not turn into "everything and everyone is fine", "everything and everyone you see", "everything and everyone you can imagine", however, it may mean exactly that. It means that **we count people as animate things**. I think the words "everyone", "anyone" when referring to people are more a form of politeness, because of their greater value than the value of inanimate things. For example, when you mean truly "no-thing", you do not say "nothing and no one". A universal quantifier likewise, as far as I know, reads "for all things" rather than "for all things and for anyone". An existential quantifier reads "there is at least one thing" rather than "there is at least one thing or at least someone".

We can always take into consideration any things together and treat them as one thing, the parts of which they are. Such a thing will be **the conjunction of these things**. All common parts of things, that are in a conjunction, are not doubled in the conjunction.

In other words, two or more things taken into consideration as one thing or treated or perceived as one thing are together this one thing that is equal to the conjunction of these things.

We will use the word "**and**" and the small and big symbol "$\wedge$" to express a conjunction of things.

A conjunction of any number of things has the following properties: idempotency (for nonabstract things), commutativity, associativity, and, for example, for sets and statements: distributivity, De Morgan's laws. The only difference is **true equality** in place of logical equivalence in all these properties, e.g. for idempotency:

$$(X\, and\, X) = X$$

After philosopher of language Peter Strawson, a **statement** (in the meaning of a proposition [1, 2, 3]) is, what is expressed by a meaningful declarative sentence. So a meaningful declarative sentence is a symbol of a statement and a statement is the **meaning** of a meaningful declarative sentence. A statement is equal to a piece of **information**, because every statement informs about something and every piece of information states something. The only difference between a statement and a piece of information would be, that an information can be expressed as a sequence of sentences separated by dots, if we did not assume, that a statement can be expressed the same way. So we assume, that a conjunction of statements can be expressed as a sequence of their expressions in any order separated by dots. Here we have to remember, that some sentences can be relative, when they relate to their context, and we have to make them irrelative to write them as a conjunction, e.g.: the statement of the sentence "He went to get coffee. Then he drink it." can be expressed as the following conjunction: "He went to get coffee *c* at time *t* and after time *t* he drank coffee *c*", because now the arguments of the conjunction can be in any order, for example, "After time *t* he drank coffee *c* and he went to get coffee *c* at time *t*".

Comment: The word "statement" does not refer to a speech act. At most, the phrase "to state something" refers to a speech act. We can express a statement, but we cannot express a sentence, because a sentence is already an expression of a language. When we say something, then we have to express a statement in some language in the form of a sentence and utter it, so to say something we need an expression of a statement, so we should not say, that we say a statement, because a statement is not an expression of the language. We can express, make, and state a statement, but we cannot say it. Similarly, we can say a word, for example, "bottle", but we cannot say the meaning of this word, because we cannot say a bottle. And this is how you should start to perceive the difference between a sentence and a statement, if you did not perceive it yet.

A conjunction of two or more statements is also a statement and thanks to this, such a conjunction has a logical value, when it is decidable. A conjunction of two or more sets is the union of these sets, because when you take two or more sets together, then naturally you will get another set that is the union of them and has the appropriate cardinality, e.g.: $\{a, b, c\} = (\{a\} \textit{ and } \{b\} \textit{ and } \{c\})$, where $\{x\}$ for any x is an atomic thing.

Let us define the empty statement $\varnothing_ς$:

$$\varnothing_ς = \bigwedge_{x \in \varnothing} (x \in \varnothing) \equiv true$$

For any statement X:

$$(X \textit{ and } \varnothing_ς) = X$$

So:

$$(X \textit{ or } {\sim}\varnothing_ς) = {\sim}({\sim}X \textit{ and } \varnothing_ς) = {\sim}{\sim}X = X$$

Let us assume the following three quite obvious axioms for any **statements** X, Y and Z, that will partially define implication:

$$(X = Y) = ((X \Rightarrow Y) \textit{ and } (Y \Rightarrow X))$$

$$(X \textit{ and } Y) \Rightarrow X$$

$$(Z \Rightarrow (X \textit{ and } Y)) = ((Z \Rightarrow X) \textit{ and } (Z \Rightarrow Y))$$

Where the fourth optional axiom $(X \Rightarrow X)$ can be derived from the second axiom, because $X = (X \textit{ and } X) \Rightarrow X$.

Let us consider the following statement:

$$(\boldsymbol{X} = (\boldsymbol{Y} \textbf{ and } \boldsymbol{X}))$$

When we use the first axiom, we will get, that it is equal to:

$$((X \Rightarrow (X \textit{ and } Y)) \textit{ and } ((X \textit{ and } Y) \Rightarrow X))$$

When we use the third axiom, we will get, that it is equal to:

$$((\boldsymbol{X} \Rightarrow \boldsymbol{Y}) \textit{ and } (X \Rightarrow X) \textit{ and } ((X \textit{ and } Y) \Rightarrow X))$$

Where "$(X \Rightarrow X)$" and "$((X \textit{ and } Y) \Rightarrow X)$" are tautologies according to respectively the fourth and the second axiom, so their statements are equal to the empty statement (a proof of this you will find further in this book), so we will get that the above statement is equal to:

$$(\boldsymbol{X} \Rightarrow \boldsymbol{Y})$$

So finally, we will get the following equality:

$$(\boldsymbol{X} \Rightarrow \boldsymbol{Y}) = (\boldsymbol{X} = (\boldsymbol{Y}\ \boldsymbol{and}\ \boldsymbol{X}))$$

Mainly from the above equality, whole new logic is derived, so you need only few definitions and these three quite obviously true assumptions about implication to create this new logic based on implication, that is based on the meaning of sentences.

1.6 Statement

So for any statements a and b:

$$(\boldsymbol{a} \Rightarrow \boldsymbol{b}) = (\boldsymbol{a}\ \boldsymbol{implicates}\ \boldsymbol{b}) = (\boldsymbol{a}\ \boldsymbol{is}\ \boldsymbol{sufficient}\ \boldsymbol{condition}\ \boldsymbol{for}\ \boldsymbol{b})$$
$$= (\boldsymbol{b}\ \boldsymbol{is}\ \boldsymbol{necessary}\ \boldsymbol{condition}\ \boldsymbol{for}\ \boldsymbol{a}) = (\boldsymbol{a} = (\boldsymbol{b}\ \boldsymbol{and}\ \boldsymbol{a}))$$

If we treat $a = (b\ and\ a)$ as a recurrence and expand it for all necessary conditions of a, then we will prove very easily, that for any statement a:

$$a = \bigwedge_{a \Rightarrow b} b$$

Which is equivalent to the main definition: $(a \Rightarrow b) = (a = (b\ and\ a))$

That means, that every statement is the conjunction of all its necessary conditions and this conjunction is a sufficient condition for this statement to be true.

Here is the following proof of it:

First of all, this is trivial (see the third axiom) that a statement implicates the conjunction of all its necessary conditions. And it is also trivial (see the second axiom) that the conjunction of all necessary conditions of a statement implicates this statement, because it is one of its necessary conditions. But it can be also justified the following way:

Secondly, a necessary condition of a statement is a condition that has to be fulfilled for the statement to be true. So if all such conditions are fulfilled, then there is nothing more that has to be fulfilled for the statement to be true, so this statement is true.

Q.E.D.

And this is another proof that the main definition is correct.

This new logic is the calculation of statements (pieces of information) and not of their symbols.

A statement is something that fulfills the following four conditions:

$$\bigwedge_{x,y} ((x = y)\ \textit{is an elementary statement})$$

$$\bigwedge_{x \text{ is a statement}} (\sim x \text{ is a statement})$$

A conjunction (“and”) of any number of statements is a statement.
A disjunction (“or”) of any number of statements is a statement.
A universal quantifier is a conjunction of statements.
An existential quantifier is a disjunction of statements.

Implication is a special case of equality and contextual implication (analogy to material implication) is a special case of implication.

And whatever this definition allows a statement to be, this new logic can process. So, this is not *i*-th order logic, because it allows quantifiers over any set, even over any set of statements.

1.7 Inclusion and being a part – new mereology

A **concrete thing** is a thing, the class of which has exactly one member.

So a concrete thing is a thing, that has concretized all details (attributes).

An **abstract thing** is a thing, that is not concrete.

So an abstract thing is a thing, the set of common attributes of all members of the class of which is abstracted from some details (attributes) of all members of its class.

Both these terms (“concrete thing”, “abstract thing”) have these meanings in the whole book.

Inclusion (operators “⊃” and “⊂”) of nonabstract things (and thereby also relation of **being a part**) can be defined the same way as implication:

For any nonabstract things X, Y and Z:

$$(\boldsymbol{X} = \boldsymbol{Y}) = ((\boldsymbol{X} \supset \boldsymbol{Y}) \, \boldsymbol{and} \, (\boldsymbol{Y} \supset \boldsymbol{X}))$$

$$(\boldsymbol{X} \, \boldsymbol{and} \, \boldsymbol{Y}) \supset \boldsymbol{X}$$

$$(\boldsymbol{Z} \supset (\boldsymbol{X} \, \boldsymbol{and} \, \boldsymbol{Y})) = ((\boldsymbol{Z} \supset \boldsymbol{X}) \, \boldsymbol{and} \, (\boldsymbol{Z} \supset \boldsymbol{Y}))$$

So the same as for implication:

$$(\boldsymbol{X} \supset \boldsymbol{Y}) = (\boldsymbol{X} = (\boldsymbol{Y} \, \boldsymbol{and} \, \boldsymbol{X}))$$

So for any statements X and Y:

$$(\boldsymbol{X} \Rightarrow \boldsymbol{Y}) = (\boldsymbol{X} \supset \boldsymbol{Y}) = (X \, \textit{implicates} \, Y) = (X \, \textit{includes} \, Y) = (Y \, \textit{is a part of} \, X)$$
$$= (\boldsymbol{X} = (\boldsymbol{Y} \, \boldsymbol{and} \, \boldsymbol{X}))$$

And for any sets X, Y and Z:

$$(\boldsymbol{X} \supset \boldsymbol{Y}) = (X\,includes\,Y) = (Y\,is\,a\,part\,of\,X) = \left(\boldsymbol{X} = (\boldsymbol{Y}\,\boldsymbol{and}\,\boldsymbol{X})\right)$$

$$(X\,and\,Y) = (X \cup Y)$$

$$\left(Z \supset (X\,and\,Y)\right) = \left((Z \supset X)\,and\,(Z \supset Y)\right)$$

Which is used in Section 1.8 to derive simple implication.

1.8 Simple implication

I will demonstrate a small simplified derivation of some kind of implication that has wide application, because it is very general.

For any different sets x, y and a:

$$((\boldsymbol{a} \supset \mathbf{x}) \Rightarrow (\boldsymbol{a} \supset \boldsymbol{y})) = \Big((a \supset x) = ((a \supset x)\,and\,(a \supset y)) = \left(a \supset (x\,and\,y)\right)\Big)$$

$$= \Big((a \supset x) = \left(a \supset (x\,and\,y)\right)\Big) = \left(x = (x\,and\,y)\right) = (\mathbf{x} \supset \boldsymbol{y})$$

So:

$$((x \subset a) \Rightarrow (y \subset a)) = (y \subset x)$$

At first sight, it can seem not too practical, but nothing could be further from truth. Let us define the following operator:

$$X^{\Lsh} = (the\,set\,of\,all\,past\,Xes)$$

Then from our formula, we get:

$$\Big(\left(X^{\Lsh} \subset A^{\Lsh}\right) \Rightarrow \left(Y^{\Lsh} \subset A^{\Lsh}\right)\Big) = \left(Y^{\Lsh} \subset X^{\Lsh}\right)$$

And we have for any A and B:

$$\left(A^{\Lsh} \subset B^{\Lsh}\right) = (Any\,past\,A\,is\,Some\,past\,B)$$

And now we are ready to evaluate very wide range of implications, for example:

$$A = (Marie\,Sklodowska\,Curie\,won\,a\,Nobel\,Prize)$$

$$= (Some\,past\,Marie\,Sklodowska\,Curie\,won\,a\,Nobel\,Prize)$$

$$= \sim(Any\,past\,Marie\,Curie\,did\,not\,win\,a\,Nobel\,Prize)$$

$$
\begin{aligned}
B &= (\mathit{Some\ woman\ won\ a\ Nobel\ Prize}) \\
&= (\mathit{Some\ past\ woman\ won\ a\ Nobel\ Prize}) \\
&= \sim(\mathit{Any\ past\ woman\ did\ not\ win\ a\ Nobel\ Prize})
\end{aligned}
$$

$$
\begin{aligned}
(A \Rightarrow B) &= (\sim B \Rightarrow \sim A) \\
&= ((\mathit{Any\ past\ woman\ did\ not\ win\ a\ Nobel\ Prize}) \\
&\Rightarrow (\mathit{Any\ past\ Marie\ Sklodowska\ Curie\ did\ not\ win\ a\ Nobel\ Prize})) \\
&= ((\mathit{Any\ past\ woman}\ \textbf{is}\ \mathit{Someone\ past\ who\ did\ not\ win\ a\ Nobel\ Prize}) \\
&\Rightarrow (\mathit{Any\ past\ Marie\ Sklodowska\ Curie}\ \textbf{is}\ \mathit{Someone\ past\ who\ did\ not\ win} \\
&\quad \mathit{a\ Nobel\ Prize}))
\end{aligned}
$$

Let us transform it to:

$$
\begin{aligned}
(A \Rightarrow B) &= (\sim B \Rightarrow \sim A) \\
&= \Big((\mathit{Woman}^{\Lsh} \subset (\mathit{Who\ did\ not\ win\ a\ Nobel\ Prize})^{\Lsh}) \\
&\Rightarrow \Big((\mathit{Marie\ Sklodowska\ Curie})^{\Lsh} \subset (\mathit{Who\ did\ not\ win\ a\ Nobel\ Prize})^{\Lsh} \Big) \Big) \\
&= ((\mathit{Marie\ Sklodowska\ Curie})^{\Lsh} \subset \mathit{Woman}^{\Lsh}) \\
&= (\mathit{Any\ past\ Marie\ Sklodowska\ Curie}\ \textbf{is}\ \mathit{Some\ past\ Woman}) \\
&= (\mathit{Marie\ Sklodowska\ Curie\ was\ always\ a\ woman}) \equiv \boldsymbol{true}
\end{aligned}
$$

So we have evaluated the logical value of our implication. The same way, we can evaluate the logical value of any implication of the same kind (second kind).

Using the previous rule, we have also for any different sets x, y and a (the negation of a set is the complement of the set, so for any sets X and Y: $(X \subset Y) = (\sim Y \subset \sim X)$):

$$((\boldsymbol{a} \subset \mathbf{x}) \Rightarrow (\boldsymbol{a} \subset \boldsymbol{y})) = ((\sim x \subset \sim a) \Rightarrow (\sim y \subset \sim a)) = (\sim y \subset \sim x) = (\mathbf{x} \subset \boldsymbol{y})$$

So:

$$((a \subset x) \Rightarrow (a \subset y)) = (x \subset y)$$

Let us define the following operator:

$$X^{\uparrow} = (\mathit{the\ set\ of\ all\ members\ of\ the\ class\ of\ X})$$

Where we assume, that not only every existing X is a member of the class of X, but also every past, future, potential and impossible X is such a member.

Let us call to mind, that our formulas are true for sets, and we get:

$$\left((A^{\uparrow} \subset X^{\uparrow}) \Rightarrow (A^{\uparrow} \subset Y^{\uparrow})\right) = (X^{\uparrow} \subset Y^{\uparrow})$$

And we have for A and B in singular form:

$$(A^{\uparrow} \subset B^{\uparrow}) = (a\,A\; is\; a\,B)$$

And now we are ready to evaluate very wide range of implications.

For example, the following true implication between false statements:

$$((A\; penguin\; is\; an\; aeroplane) \Rightarrow (A\; penguin\; can\; fly))$$
$$= ((A\; Penguin\; \mathbf{is}\; An\; Aeroplane) \Rightarrow (A\; Penguin\; \mathbf{is}\; A\; Thing\; That\; Can\; Fly))$$

Let us transform it to:

$$\left((Penguin^{\uparrow} \subset Aeroplane^{\uparrow}) \Rightarrow (Penguin^{\uparrow} \subset (Thing\; That\; Can\; Fly)^{\uparrow})\right)$$
$$= (Aeroplane^{\uparrow} \subset (Thing\; That\; Can\; Fly)^{\uparrow})$$
$$= (An\; Aeroplane\; \mathbf{is}\; A\; Thing\; That\; Can\; Fly)$$
$$= (An\; aeroplane\; can\; fly)$$
$$\equiv \boldsymbol{true}$$

So we have evaluated the logical value of our implication. The same way we can evaluate the logical value of any implication of the same kind (first kind).

Now, let us define the following operator:

$$X^{\Rsh} = (the\; set\; of\; all\; future\; Xes)$$

Then from our formula, we get:

$$((A^{\Rsh} \subset X^{\Rsh}) \Rightarrow (A^{\Rsh} \subset Y^{\Rsh})) = (X^{\Rsh} \subset Y^{\Rsh})$$

And we have:

$$\left(((\$A)^{\Rsh} \subset X^{\Rsh}) \Rightarrow ((\$A)^{\Rsh} \subset Y^{\Rsh})\right) = ((\{A\} \subset X^{\Rsh}) \Rightarrow (\{A\} \subset Y^{\Rsh}))$$
$$= ((A \in X^{\Rsh}) \Rightarrow (A \in Y^{\Rsh})) = (X^{\Rsh} \subset Y^{\Rsh})$$

Where $(\$Q)^{\uparrow} = \{Q\}$ for every Q.

So:

$$((A \in X^{\Rsh}) \Rightarrow (A \in Y^{\Rsh})) = (X^{\Rsh} \subset Y^{\Rsh})$$

And we have for any A and B:

$$(A^{\Rsh} \subset B^{\Rsh}) = (\mathit{Any\ future\ A\ is\ Some\ future\ B})$$

$$(A \in B^{\Rsh}) = (\mathit{The\ future\ A\ is\ Some\ future\ B})$$

And now again, we are ready to evaluate very wide range of implications.

For example, the following true implication between statements about the predictable future (for the unpredictable future we would have to consider also potential things):

$$\begin{aligned}&((\mathit{Some\ force\ will\ act\ on\ the\ body}) \Rightarrow (\mathit{The\ body\ will\ react}))\\ &\quad = ((\mathit{The\ Future\ Body}\ \textbf{\textit{is}}\ \mathit{A\ Thing\ Some\ Force\ Will\ Act\ On})\\ &\quad \Rightarrow (\mathit{The\ Future\ Body}\ \textbf{\textit{is}}\ \mathit{A\ Thing\ That\ Will\ React}))\end{aligned}$$

Let us transform it to:

$$\begin{aligned}&\Big(((\mathit{The\ Future\ Body}) \in (\mathit{Thing\ Some\ Force\ Will\ Act\ On})^{\Rsh})\\ &\quad \Rightarrow ((\mathit{The\ Future\ Body}) \in (\mathit{Thing\ That\ Will\ React})^{\Rsh})\Big)\\ &\quad = ((\mathit{Thing\ Some\ Force\ Will\ Act\ On})^{\Rsh} \subset (\mathit{Thing\ That\ Will\ React})^{\Rsh})\\ &\quad = (\mathit{A\ Thing\ Some\ Force\ Will\ Act\ On}\ \textbf{\textit{is}}\ \mathit{A\ Thing\ That\ Will\ React})\\ &\quad = (\mathit{A\ thing,\ that\ some\ force\ will\ act\ on,\ will\ react}) \equiv \textbf{\textit{true}}\end{aligned}$$

So we have evaluated the logical value of our implication, which is a paraphrase of the Newton's third law of motion. The same way, we can evaluate the logical value of any implication of the same kind (third kind).

1.9 Other examples of implications

Here are some examples in the present tense of implications of the first kind:

$$((\mathit{Rose\ is\ Flowering\ Plant}) \Rightarrow (\mathit{Rose\ is\ Plant})) = (\mathit{Flowering\ Plant\ is\ Plant}) \equiv \mathit{true}$$

$$\begin{aligned}&((\mathit{Sky\ is\ Being\ That\ Is\ Blue}) \Rightarrow (\mathit{Sky\ is\ Being\ That\ Has\ A\ Color}))\\ &= (\mathit{Being\ That\ is\ Blue\ is\ Being\ That\ Has\ A\ Color}) \equiv \mathit{true}\end{aligned}$$

$$\begin{aligned}&((\mathit{Pterosaur\ is\ Being\ That\ Can\ Fly}) \Rightarrow (\mathit{Pterosaur\ is\ Being\ That\ Do\ Not\ Only\ Walk}))\\ &= (\mathit{Being\ That\ Can\ Fly\ is\ Being\ That\ Do\ Not\ Only\ Walk}) \equiv \mathit{true}\end{aligned}$$

And here are some examples in the present tense of implications of the second kind:

$$((\textit{Rose is Flowering Plant}) \Rightarrow (\textit{French Rose is Flowering Plant}))$$
$$= (\textit{French Rose is Rose}) \equiv \textit{true}$$

$$((\textit{Flightless Bird is Being That Cannot Fly}) \Rightarrow (\textit{Penguine is Being That Cannot Fly}))$$
$$= (\textit{Penguine is Flightless Bird}) \equiv \textit{true}$$

$$((\textit{Star is Being That Is Heavier Than Its Planet})$$
$$\Rightarrow (\textit{The Sun is Being That Is Heavier Than Its Planet}))$$
$$= (\textit{The Sun is Star}) \equiv \textit{true}$$

1.10 Grounded material implication

Truth (*TR*) is simply the conjunction of all true statements. So a statement is true if and only if it is a part of truth.

Let us define **contextual implication** (the analogy to **material implication**) for any statements a and b the following way:

$$(\boldsymbol{a} \rightarrow \boldsymbol{b}) = ((\boldsymbol{a} \textbf{ and } \boldsymbol{TR}) \Rightarrow \boldsymbol{b})$$

because material implication is nothing else than the statement of the sentence "If a is true, then b is true", which is implication in the context of truth.

Then we have:

$$(\boldsymbol{a} \rightarrow \boldsymbol{b}) = ((a \textit{ and } TR) \Rightarrow b) = ((a \textit{ and } TR) = (b \textit{ and } a \textit{ and } TR))$$
$$= \left((a \textit{ and } TR) = ((b \textit{ and } a \textit{ and } TR) \textit{ or } {\sim}\varnothing_{\varsigma})\right)$$
$$= \left((a \textit{ and } TR) = ((b \textit{ and } a \textit{ and } TR) \textit{ or } {\sim}(TR \textit{ or } {\sim}TR \textit{ or } {\sim}a))\right)$$
$$= \left((a \textit{ and } TR) = ((b \textit{ and } a \textit{ and } TR) \textit{ or } ({\sim}TR \textit{ and } a \textit{ and } TR))\right)$$
$$= \left((a \textit{ and } TR) = ((b \textit{ or } {\sim}TR) \textit{ and } a \textit{ and } TR)\right)$$
$$= ((\boldsymbol{a} \textbf{ and } \boldsymbol{TR}) \Rightarrow (\boldsymbol{b} \textbf{ or } {\sim}\boldsymbol{TR}))$$

Where "*TR or ~TR or ~a*" is, of course, a tautology, because "*~(~TR and TR and TR is decidable)*" is a tautology and (*TR is decidable*) is a part of *TR*. And the statement of a tautology is equal to the empty statement.

So:

$$(\boldsymbol{a} \rightarrow \boldsymbol{b}) = ((a\, and\, TR) \Rightarrow (b\, or \sim TR)) = (\sim(b\, or \sim TR) \Rightarrow \sim(a\, and\, TR))$$
$$= ((\sim b\, and\, TR) \Rightarrow (\sim a\, or \sim TR)) = (\sim \boldsymbol{b} \rightarrow \sim \boldsymbol{a})$$

And if b is true then $(a \rightarrow b)$ is true, because then b is a part of TR. And if b is false and a is true, then $(a \rightarrow b)$ is false, because b is not a part of $(a\, and\, TR)$, that is then equal to TR. And if b is false and a is false, then $\sim a$ is true (so we have the first case), so $(\sim b \rightarrow \sim a)$ is true, so $(a \rightarrow b)$ is true.

Summing up, $(a \rightarrow b) = (b\, or \sim a)$ for decidable statements a and b, which is like classical material implication, but for statements, so this is not the same as material implication from classical logic.

Remember, that material implication in classical logic is completely ungrounded.

1.11 Age-old mathematicians dream

I strongly encourage you to read this book. The whole new logic is explained here step by step in a transparent and precise way. As you probably know, it was age-old mathematicians dream to understand true implication based on the meaning of sentences. It cost me 10 years to accomplish that and finally here it is.

2 Introduction

Here, you will find new fundaments of mathematics: complete true logic defined by definitions and complete true set theory defined by definitions.

Although they are fundamental things, they are very advanced, so need a great concentration.

2.1 Main differences between new logic and classical logic

2.1.1 Introduction

In this section, I will explain briefly the main differences between new logic and classical logic. Some of these topics will be explained in more detail further in this introductory part. Further, in this book, you will also find complete explanation of these topics.

The main difference between this new logic and classical logic is that this new logic is derived from the definition of implication, so it is defined by a set of axioms about implication instead of a set of axioms about material implication. And material implication (contextual implication) is derived from implication, so it is not ungrounded as in classical logic. So it is done completely oppositely to what is done in classical logic.

2.1.2 Theory as a set of statements versus theory as a set of sentences

We can have at most countable number of sentences in any classical language, while the number of all statements is uncountable (e.g.: for every real number x, we have the following true statements: $(x\ \textit{is a real number})$, $(x\ \textit{is a number})$, $(x\ \textit{is a being})$, $(x\ \textit{is a thing})$, etc.). So the notion of a classical theory, as the set of all "true sentences" about something, is insufficient, and that is one of the main reasons why classical logic is handicapped and why we all need this new logic. **So classical logic cannot even cover fully real number theory and anything that includes it**, simply because all true statements about real numbers, as in my above example, cannot be expressed at the same time in any language, but they should be covered by a reliable theory of real numbers. So a theory must be the set of all true statements about some topic, instead of the set of all true sentences.

2.1.3 A statement (a piece of information) versus a sentence (a symbol of a piece of information)

A meaningful declarative sentence is only a statement expressed in some language. So a meaningful declarative sentence is only a symbol of a statement. It is admittedly a

https://doi.org/10.1515/9783111441382-002

complex symbol, but it is still just a symbol, that has its referent. Every statement is equal to some piece of information and every piece of information is equal to some statement. Simply, both these terms ("piece of information" and "statement") have the same meaning.

Comment: Remember that not every part of data must be a piece of information. For example, when we have some data expressed as "55", then out of its context, it is not understandable as a declarative sentence (though it is understandable as a symbol of a natural number), so only its meaning, that come from its context (e.g. a row in a table of prices of products in a shop), that determines the structure of the outer data (e.g.: a price list), is some piece of information (that can be expressed as "55$ is the price of a pen drive in this shop").

That is why a decidable statement is a truth-bearer, while a sentence is not, the same as a symbol of a nonzero number is not positive and is not negative, while the number itself is positive or is negative. So it is a mistake to say that some sentence is true or false, because only its statement can be true or false. It can only be a mental shortcut because a sentence does not change in different models (in different contexts for different speakers), while in classical logic, its logical value changes in different models, if it depends on the model (on the context or the speaker), because the thing that changes is the meaning of the sentence, that comes from different interpretation of the same sentence. And the meaning is, of course, equal to a statement (a piece of information).

Important: Remember that when we say "the statement of a/the sentence", we always mean the statement determined by the defined (current is always default) context and the defined (current is always default) speaker, while when we say "a statement of a/the sentence", we mean one of all statements, each of which is determined by its context and its speaker.

For example, the statement of the sentence "This is my dog" can be true or false, depending on the context in which it is said and the speaker, by whom it is said, so this sentence cannot be just true or just false. So, correctly we can only say that in given model (in given context for given speaker), the statement of a sentence is true or false, or that in given model (in given context for given speaker), a sentence expressed some truth or some untruth (true or false statement). Remember that the logical value of a decidable statement does not ever change, because the statement of a sentence incorporates all details that its logical value depends on and that are default in the sentence and comes from the context, in which the sentence is said, and from the speaker by which the sentence is said, e.g.: (You, Paris at location l_1 tomorrow at time t_1)^"It is raining" = (You, Paris at location l_1 tomorrow at time t_1)^"It is raining **here and now**" = ^"It will be raining in Paris at location l_1 tomorrow at time t_1", and since in mathematics we deal only with theorems that are true everywhere and always and for everyone, e.g.: (You, here and now)^"$1+1=2$" = (She, in London at location l_2 last Sunday at time t_2)^"**everywhere and always for everyone** $(1+1=2)$" = ^"$1+1=2$".

Where:

$(S,C)^{\wedge}X$ is the meaning of standalone symbol X, when said by speaker S in context C. So it is the referent of symbol X. So it is the statement of sentence X, if X is a declarence.

$$^{\wedge}X = (current\ speaker, current\ context)^{\wedge}X$$

The same way, this is a mistake to say that some sentence implicates another sentence (the same as, for example, a symbol of a natural number has not a divisor, while the number has its divisors), because only a statement that is a piece of information and not only a symbol of a piece of information can implicate other statement.

So this new logic is the calculation of statements (pieces of information) and not of their symbols that are called sometimes formulas to hide the fact that they are only symbols of statements. So this new logic is information-oriented logic, while classical logic is symbol-oriented logic, so it does not take into account something so important in logic as information.

For more details, see Section 5.18.

2.1.4 New implication

When we can infer one piece of information (second) from the other piece of information (first), then the first piece of information, from which we infer, has to contain this second (inferred) piece of information (so the second piece of information must be a part of the first piece of information), because otherwise we would infer the second piece of information not just from the first piece of information only, but from other additional information, so we would have not the right to say that we inferred the second piece of information from the first piece of information only, which only is true implication.

From English Wikipedia (10-10-2018) from the term *Logical consequence*:

> *[In classical logic] a sentence is said to be a logical consequence of a set of sentences, for a given language, if and only if, using only logic (i.e. without regard to any personal interpretations of the sentences), the sentence must be true if every sentence in the set is true.*

In other words, in classical logic, implication concerns sentences, while they are only symbols of statements. A symbol of a statement cannot implicate anything, similarly like the symbol of a number cannot have a divisor. So, only a statement can implicate a statement. So, this is already a mistake. Moreover, some statements cannot be expressed as sentences, but they also have their implications.

Additionally, x implicates y if and only if piece of information x is enough to infer piece of information y, which means that piece of information x includes piece of information y, so we do not need to know anything more than piece of information x to infer piece of information y. So piece of information x is enough to infer piece of information y if and only if x states y among alternatively other things. And when piece of information y is a part of piece of information x, then x states y among alternatively other things, so piece of information x is enough to infer piece of information y. But in logical consequence from classical logic, we would need to know, that in every model sentence b is true if sentence a is true, to confirm logical consequence, so sentence a would not be enough to infer sentence b. So, logical consequence from classical logic is not true implication.

For more details, see Sections 5.18 and 5.35 and Chapters 6 and 9.

New implication allows to also define very easily "material implication". For more details, see Section 5.38.

2.1.5 Reduction to absurdity (lat. reductio ad absurdum)

When you assume that a statement is, for example, true, which leads to a contradiction, then you cannot conclude automatically and truthfully, that the statement is false, if you have shown neither that the assumption, that it is false, does not lead to a contradiction, nor that the statement is decidable. Indeed, a contradiction to the assumption that a statement has some logical value, always means that it has the opposite logical value, but despite this, we cannot decide what value it has, when it is undecidable, because when it is undecidable and, for example, the assumption, that it is true, leads to a contradiction, then the statement, that such a statement is not true (is false), would be undecidable also (for more details see Section 5.23). We also cannot say that it has not defined logical value, because such a statement would be also undecidable.

If a statement is undecidable, then it is still true/false (1), if it is not false/not true (2), because both these statements are negations to each other:

$$(x\ is\ not\ true) = \sim(x\ is\ true) = \sim x = (x\ is\ false)$$

$$(x\ is\ not\ false) = \sim(x\ is\ false) = \sim\sim x = x = (x\ is\ true)$$

It is easy to prove that given statement is decidable (most statements in usage are decidable), and in such a case, a contradiction to the assumption that it is true, implicates, that it is decidably false, and a contradiction to the assumption, that it is false, implicates, that it is decidably true. And when the assumption, that a statement is true, and the assumption, that the statement is false, both lead to contradictions, then the statement is undecidable. To see the simplest example of such a situation, see Section 5.16. Symbol *undecidable* is introduced to inform, that a statement is undecidable. Remember, that, for example, the negation of the undecidable statement of the sentence “The statement of this sentence is false” cannot be expressed as “The statement of this sentence is true”, and cannot be expressed as “The statement of this sentence is not false”, but can be expressed as “The negation of the statement of this sentence is true” and “The statement of this sentence is false”, because this statement is equal to its negation.

So be careful and do not forget that.

For example, for the statement from Liar paradox:

$$x = (x \, is \, false) = {\sim}x$$

So:

$$x = {\sim}x$$

If we assume, that x is decidable, which means, that assumption x or assumption ${\sim}x$ does not lead to a contradiction, then we get a contradiction:

If we assume, that x, then ${\sim}x$, we get a contradiction to the assumption, that x is true.

If we assume, that ${\sim}x$, then x, we get a contradiction to the assumption, that x is false.

So finally we have a contradiction to the main assumption, so x is undecidable.

Q.E.D.

So, I have understood the property, that joins all statements, that are not just either true or false and I call them undecidable statements. It allowed me to solve probably all paradoxes of classical logic, like, for example, Liar paradox, Quine’s paradox, Yablo’s paradox, and so on. All these you will find in this book.

For more details, see Section 5.23.

2.1.6 Speaker and context versus model

Instead of a model, I use the **speaker**, by which a sentence is said, and the **context**, in which the sentence is said, to determine, what information (statement) is expressed by the sentence and whether it is true information (true statement) or not (false statement). The ordered pair of a speaker and a context will be called in this book a **situation** (of a speech act). It is much simpler, more natural and more useful than model theory.

Remember also, that either way the classical definition of a model is not enough, because the true interpretation of a meaningful declarative sentence is a piece of information (statement), which is something unknown for classical logic. Remember also, that we can have at most countable number of sentences in any classical language, while the number of all statements is uncountable.

For more details, see Chapter 8.

2.1.7 Definitions-based definition of new logic

This new logic is defined differently than a formal system in classical logic. First of all, there are no standalone axioms, because all axioms are divided into separate definitions of basic notions. And except the notions such as a symbol and its referent, a sentence and its statement, that are necessary to understand any text, there is only one primitive notion, which is a conjunction, from which all other notions are derived. A statement is defined by the rules it can be constructed with from other statements. And truth is the conjunction of all true statements of the theory of everything.

2.2 What among other things you will find here

2.2.1 Introduction

I will explain shortly the most important things, that I have discovered since I started my intellectual adventure with logic 10 years ago.

For information about the symbols used in this section, see Section 2.4.

2.2.2 It is quite easy

The notion **thing** fulfills the following assumption:

$$\bigwedge_{X}(X \textit{ is a thing})$$

That can be easily proved:

$$t = \bigwedge_{X}(X \textit{ is a thing}) = (\textit{Everything is a thing}) = (\textit{Every thing is a thing})$$
$$= (\textit{A thing is a thing}) \equiv \textit{true}$$

We can see clearly, that statement t fulfills:

$$t = (\boldsymbol{A\ thing\ is\ a\ thing}) = (\mathit{Every\ thing\ is\ something}) = (\boldsymbol{Everything\ is\ something})$$
$$= (\mathit{Nothing\ is\ not\ something}) = (\mathit{Nothing\ is\ not\ a\ thing}) = (\boldsymbol{Nothing\ is\ nothing})$$
$$= (\mathit{Any\ thing\ is\ something}) = (\boldsymbol{Anything\ is\ something})$$
$$= \bigwedge_{X}\bigvee_{Y}(\boldsymbol{X\ is\ Y}) \Leftarrow \bigwedge_{X}\bigvee_{Y}(X = Y)$$

In other words, every**thing** is a thing, so it has the widest class. Every**thing** has its definition, but not always that intended one (e.g. a square circle has its definition, because it has, for example, the following attribute: "it was meant to be a square and a circle at the same time").

Better definition of the notion "thing" is not possible and not necessary, because the word "thing" is built in the word "everything", so we would get the following problem: "a thing is every**thing** that . . .".

We can always take into consideration many things together and treat them as one thing, the parts of which they are. Such a thing will be **the conjunction of these things**. All common parts of things, that are in a conjunction, are not doubled in the conjunction.

In other words, two or more things taken into consideration as one thing or treated or perceived as one thing, are together this one thing that is equal to the conjunction of these things.

We will use the word "and" and the big and small symbol "$\wedge$" to express a conjunction of things.

An **empty thing** is a thing, that is the empty conjunction of things, e.g. the empty set.

An **atomic thing** is a nonempty thing, that is not a conjunction of two different nonempty things.

A **complex thing** is a nonempty thing, that is not atomic.

A **concrete thing** is a thing, the class of which has exactly one member.

So a concrete thing is a thing, that has **concretized** all details (attributes).

An **abstract thing** is a thing, that is not concrete.

So an abstract thing is a thing, the set of common attributes of all members of the class of which is **abstracted** from some details (attributes) of all members of its class.

A **plain thing** is a thing, that is a conjunction of atomic parts.

A conjunction of any number of things has the following properties: idempotency (only for concrete things), commutativity, associativity, distributivity (only for plain things and attributes and statements), De Morgan's laws (only for plain things and attributes and statements). The only difference is true equality in place of logical equivalence, e.g. for idempotency: $(X\ and\ X) = X$.

For more details, see Section 5.3.

After philosopher of language Peter Strawson, a **statement** (in the meaning of a proposition [1, 2]) is, what is expressed by a meaningful declarative sentence. So, a meaningful declarative sentence is a symbol of a statement and a statement is the **meaning** of a meaningful declarative sentence. A statement is equal to a piece of **information**, because every statement informs about something and every piece of information states something.

A conjunction of two or more statements is also a statement and thanks to this, such a conjunction has a logical value, when it is decidable. A conjunction of two or more sets is the union of these sets, because when you take two or more sets together, then naturally you will get another set that is the union of them and has the appropriate cardinality, e.g.: $\{a, b, c\} = (\{a\} \textit{ and } \{b\} \textit{ and } \{c\})$, where $\{x\}$ for any x is an atomic thing.

Let us define the empty statement $\varnothing_\varsigma$:

$$\varnothing_\varsigma = \bigwedge_{x \in \varnothing} (x \in \varnothing) \equiv \textit{true}$$

For any statement X:

$$(X \textit{ and } \varnothing_\varsigma) = X$$

So:

$$(X \textit{ or } \sim\varnothing_\varsigma) = \sim(\sim X \textit{ and } \varnothing_\varsigma) = \sim\sim X = X$$

For any concrete things X and Y:

$$\textit{isTaken}(X \textit{ and } Y) = (\textit{isTaken}(X) \textit{ and } \textit{isTaken}(Y))$$

X equals Y if and only if X is a part of thing Y and Y is a part of thing X.

We have, that for any X and Y:

$$(X \textit{ and } Y) = (\textit{the conjunction of } X \textit{ and } Y) = (\textit{the union of } X \textit{ and } Y) = (X \cup Y)$$

$$(X \textit{ is a part of } Y) = (X \textit{ is a subthing of } Y) = (Y \textit{ is a superthing of } X)$$

$$= (Y \textit{ includes } X) = (X \subset Y)$$

$$((X \subset Y) \textit{ and } (X \supset Y)) = (X = Y)$$

For any concrete thing X:

$$(X \textit{ and } X) = X$$

For example, the conjunction of two concrete sets X and Y is the conjunction of all atomic parts of both sets X and Y.

For example, for the following concrete sets X and Y:

$$X = \{1, 2, 3\} = (\{1\} \textit{ and } \{2\} \textit{ and } \{3\})$$

$$Y = \{2, 4\} = (\{2\} \textit{ and } \{4\})$$

Where every single element subset of a set is indivisible, so every single element subset of a set is an atomic part of the set.

We have:

$$(X\,and\,Y) = (X \cup Y) = (\{1\}\,and\,\{2\}\,and\,\{3\}\,and\,\{2\}\,and\,\{4\})$$
$$= (\{1\}\,and\,\{2\}\,and\,\{3\}\,and\,\{4\}) = \{1, 2, 3, 4\}$$
$$(X \cap Y) = \{2\}$$

Now let us show that the statement of every tautology is the empty statement.

For any model m and any two tautologies q and p, we have:

$$true \equiv (\varnothing = \varnothing)$$
$$= (\{n{:}(n \in M)\,and \sim [n]^q\} = \{n{:}(n \in M)\,and \sim [n]^p\})$$
$$\Rightarrow ((m \in \{n{:}(n \in M)\,and \sim [n]^q\}) = (m \in \{n{:}(n \in M)\,and \sim [n]^p\}))$$
$$= (\sim [m]^q = \sim [m]^p) = ([m]^q = [m]^p)$$

Where M is the set of all models and $[a]^X$ is the statement of sentence x in model a.

The same for any logically possible situation s and any two tautologies q and p, we have:

$$true \equiv (\varnothing = \varnothing)$$
$$= (\{t{:}(t \in S)\,and \sim t^q\} = \{t{:}(t \in S)\,and \sim t^p\})$$
$$\Rightarrow ((s \in \{t{:}(t \in S)\,and \sim t^q\}) = (s \in \{t{:}(t \in S)\,and \sim t^p\}))$$
$$= (\sim s^q = \sim s^p) = (s^q = s^p)$$

Where S is the set of all logically possible situations and a^x is the statement of sentence x in logically possible situation a.

So the statements of any two tautologies are equal and the empty statement is a tautological statement, so the statement of any tautology is equal to the empty statement. It is so, because the statement of every tautology, by the definition of a tautology, must be **unconditionally true** just like the empty statement.

The definition of implication:

The conjunction of the following four quite obvious conditions for any concrete statements X, Y and Z is the necessary and sufficient condition for relation $\Rightarrow$ to be implication:

$$(X = Y) = ((X \Rightarrow Y)\,and\,(Y \Rightarrow X))$$
$$(X\,and\,Y) \Rightarrow X$$
$$(Z \Rightarrow (X\,and\,Y)) = ((Z \Rightarrow X)\,and\,(Z \Rightarrow Y))$$
$$((X\,or\,Y) \Rightarrow Z) = ((X \Rightarrow Z)\,and\,(Y \Rightarrow Z))$$

Where the fifth optional condition $(X \Rightarrow X)$ can be derived from the second condition, because $X = (X \, and \, X) \Rightarrow X$.

So:

$$(\text{“}\Rightarrow\text{” is the symbol of implication})$$

$$= \bigwedge_{X,Y,Z \text{ are concrete statements}} and \begin{pmatrix} (X = Y) = ((X \Rightarrow Y) \, and \, (Y \Rightarrow X)), \\ (X \, and \, Y) \Rightarrow X, \\ (Z \Rightarrow (X \, and \, Y)) = ((Z \Rightarrow X) \, and \, (Z \Rightarrow Y)), \\ ((X \, or \, Y) \Rightarrow Z) = ((X \Rightarrow Z) \, and \, (Y \Rightarrow Z)) \end{pmatrix}$$

Where the last axiom can be replaced by “$(X \Rightarrow Y) = (\sim Y \Rightarrow \sim X)$”, because they are equivalent.

From the above definition of implication and the definitions of conjunction (the only primitive notion), disjunction and negation, whole new logic is derived in this book, so you need only those few definitions and four quite obviously true assumptions about implication to create new logic based on implication, that is based on the meaning of sentences.

Let us consider the following statement:

$$(\boldsymbol{X} = (\boldsymbol{Y} \, \textbf{and} \, \boldsymbol{X}))$$

When we use the first condition, we will get that it is equal to:

$$((X \Rightarrow (X \, and \, Y)) \, and \, ((X \, and \, Y) \Rightarrow X))$$

When we use the third condition, we will get that it is equal to:

$$((\boldsymbol{X} \Rightarrow \boldsymbol{Y}) \, and \, (X \Rightarrow X) \, and \, ((X \, and \, Y) \Rightarrow X))$$

Where “$(X \Rightarrow X)$” and “$((X \, and \, Y) \Rightarrow X)$” are tautologies according to respectively the fifth (optional) and the second condition, so their statements are equal to the empty statement, so we will get that the above statement is equal to:

$$(\boldsymbol{X} \Rightarrow \boldsymbol{Y})$$

So, finally we will get the following equality:

$$(\boldsymbol{X} \Rightarrow \boldsymbol{Y}) = (\boldsymbol{X} = (\boldsymbol{Y} \, \textbf{and} \, \boldsymbol{X}))$$

So, as you can see, the first three conditions (the fifth optional condition can be derived from the second condition) are sufficient for the above equality to be true.

Here is a proof that these conditions are also necessary for the above equality to be true:

$$(\boldsymbol{Y} \Leftarrow (\boldsymbol{X} \, \textbf{and} \, \boldsymbol{Y})) = ((X \, and \, Y) = (Y \, and \, (X \, and \, Y))) \equiv \boldsymbol{true}$$

$$(\boldsymbol{Y} \Leftarrow \boldsymbol{Y}) = (Y \Leftarrow (Y \, and \, Y)) \equiv \boldsymbol{true}$$

$$((\boldsymbol{X} \Leftarrow \boldsymbol{Y})\ \boldsymbol{and}\ (\boldsymbol{Y} \Leftarrow \boldsymbol{X})) = \big((Y = (X\ and\ Y))\ and\ (X = (Y\ and\ X))\big)$$
$$= (X = (X\ and\ Y) = Y) \Rightarrow (\boldsymbol{X} = \boldsymbol{Y}) \Rightarrow ((\boldsymbol{X} \Leftarrow \boldsymbol{Y})\ \boldsymbol{and}\ (\boldsymbol{Y} \Leftarrow \boldsymbol{X}))$$

$$((\boldsymbol{X} \Leftarrow \boldsymbol{Z})\ \boldsymbol{and}\ (\boldsymbol{Y} \Leftarrow \boldsymbol{Z})) = \big((Z = (X\ and\ Z))\ and\ (Z = (Y\ and\ Z))\big) \Rightarrow \big(Z = (X\ and\ (Y\ and\ Z))\big)$$
$$= \big(Z = ((X\ and\ Y)\ and\ Z)\big) = ((\boldsymbol{X}\ \boldsymbol{and}\ \boldsymbol{Y}) \Leftarrow \boldsymbol{Z})$$

$$((\boldsymbol{X}\ \boldsymbol{and}\ \boldsymbol{Y}) \Leftarrow \boldsymbol{Z})$$
$$= \big(Z = ((X\ and\ Y)\ and\ Z)\big) \Rightarrow \Big(\big(Z = ((X\ and\ Y)\ and\ Z)\big)\ and\ \big((X\ and\ Z)$$
$$= (X\ and\ (X\ and\ Y)\ and\ Z)\big)\ and\ \big((Y\ and\ Z) = (Y\ and\ (X\ and\ Y)\ and\ Z)\big)\Big)$$
$$= \Big(\big(Z = ((X\ and\ Y)\ and\ Z)\big)\ and\ \big((X\ and\ Z) = ((Y\ and\ X)\ and\ Z)\big)\ and\ \big((Y\ and\ Z)$$
$$= ((X\ and\ Y)\ and\ Z)\big)\Big) \Rightarrow \big(((X\ and\ Z) = Z)\ and\ ((Y\ and\ Z) = Z)\big)$$
$$= ((\boldsymbol{X} \Leftarrow \boldsymbol{Z})\ \boldsymbol{and}\ (\boldsymbol{Y} \Leftarrow \boldsymbol{Z}))$$

Q.E.D.

And inclusion of any concrete things has its definition that is a subdefinition of the definition of implication, so implication is inclusion of statements, so for any concrete statements X and Y:

$$(X \subset Y) = (X \Leftarrow Y)$$

The definition of inclusion:

The conjunction of the following four quite obvious conditions for any concrete things X, Y and Z is the necessary and sufficient condition for relation $\subset$ to be inclusion of things:

$$(X = Y) = ((X \subset Y)\ and\ (Y \subset X))$$
$$Y \subset Y$$
$$Y \subset (X\ and\ Y)$$
$$((X\ and\ Y) \subset Z) = ((X \subset Z)\ and\ (Y \subset Z))$$

Then, using these conditions, we have:

$$(\boldsymbol{Y} = (\boldsymbol{X}\ \boldsymbol{and}\ \boldsymbol{Y})) = \big(((X\ and\ Y) \subset Y)\ and\ (Y \subset (X\ and\ Y))\big)$$
$$= \big((\boldsymbol{X} \subset \boldsymbol{Y})\ and\ (Y \subset Y)\ and\ (Y \subset (X\ and\ Y))\big)$$

Where "$(Y \subset Y)$" and "$\left(Y \subset (X\,and\,Y)\right)$" are tautological statements, so:

$$(X \subset Y) = \left(Y = (X\,and\,Y)\right)$$

So, for any concrete statements X and Y:

$$(X \subset Y) = (X \Leftarrow Y) = \left(Y = (X\,and\,Y)\right)$$

Q.E.D.

It may seem like there is a cycle in definitions, because the definition of equality uses the relation of inclusion and vice versa, but this is not true, because we do not need equality at all, since we can replace it always by the conjunction of two-way inclusion. So, remember that every equality is always only a shortcut for two-way inclusion of two things. As I proved above, the same we can replace inclusion by equality, so you can choose to use one of them and do not use the other.

You can treat the statements used in the definition of implication as axioms, but remember that the set of all statements is greater than the set of all sentences in any language (for example, the set of statements $\{(x\,is\,irrational\,number): x \in R\}$ will be always greater than the set of all possible sentences in English) and that a proof can have even an uncountable number of steps in this logic (for more details, see Section 7.1).

I have used definitions, each of which can be composed of a set of axioms, instead of a set of standalone axioms, because when you use standalone axioms, it always means that your definitions are incomplete or inexistent, so there is no need for standalone axioms, when your definitions are complete. In this book, I define whole definitions-based logic and definitions-based set theory using only one primitive notion, which is a conjunction.

2.2.3 Being a part and implication

Firstly, I have noticed that the problem of true implication based on the meaning is analogous to the problem of being a part of a thing. I have simply understood, that a piece of information, that can be inferred from other piece of information, is simply a part of this piece of information. In other words, a statement implicates only everything, that it includes in itself, so it implicates, what it states among alternatively other statements, and states among alternatively other statements, what it implicates. Please, see, that for both problems, we have the same solution. For any concrete things a and b:

$$(\boldsymbol{a} \supset \boldsymbol{b}) = (a\,includes\,b) = (b\,is\,a\,part\,of\,a) = \left(\boldsymbol{a} = (\boldsymbol{b}\,\boldsymbol{and}\,\boldsymbol{a})\right)$$

$$(\boldsymbol{a} = \boldsymbol{b}) = \left((\boldsymbol{a} \supset \boldsymbol{b})\,\boldsymbol{and}\,(\boldsymbol{b} \supset \boldsymbol{a})\right)$$

And for any concrete statements a and b:

$$(\boldsymbol{a} \Rightarrow \boldsymbol{b}) = (a\ implicates\ b) = (a\ is\ sufficient\ condition\ for\ b)$$
$$= (b\ is\ necessary\ condition\ for\ a) = (\boldsymbol{a} \supset \boldsymbol{b}) = (\boldsymbol{a} = (\boldsymbol{b}\ \boldsymbol{and}\ \boldsymbol{a}))$$

IMPORTANT: The above defined implication is new implication based on the meaning of sentences and has nothing to do with implication from classical logic. In this book, the symbol "⇒" is always the symbol of new implication. And everywhere in this book, the symbol of equality for statements is the symbol of equality of statements, that can be derived from implication:

$$((\boldsymbol{a} \Rightarrow \boldsymbol{b})\ \boldsymbol{and}\ (\boldsymbol{a} \Leftarrow \boldsymbol{b})) = ((b \subset a)\ and\ (a \subset b)) = (\boldsymbol{a} = \boldsymbol{b})$$

Where, as you can see, I generalized all operations on sets to operations on any things. I used here a conjunction of things, that corresponds to a union of sets, to define what for a concrete thing is being a part of other concrete thing. So, concrete thing b is a part of concrete thing a if and only if we can represent a as the conjunction of b and a itself.

In case of statements, we can also interpret implication as the inclusion of one statement in other statement, but we can also say, that one statement implicates other statement, if and only if it states this statement and alternatively something else.

In this book, I explain, what concrete things are. In short, concrete things are things, that are not abstract, so their classes have only one member. They have some properties, that allowed me to keep the above formulas simple.

So, the above formulas are correct for concrete sets also. A concrete set has defined all elements. For example, a set of ten numbers is an abstract set, because his class has many members, and the set $\{a, b, c\} = (\{a\}\ and\ \{b\}\ and\ \{c\})$ is a concrete set. As you can see, every set can be divided into atomic parts, that are nonempty parts, that cannot be divided into smaller nonempty parts. The same is for all plain things, so in their case, the validity of formulas is easy to be understood by analogy to sets. All finer points are described in this book.

If we treat $a = (b\ and\ a)$ as a recurrence and expand it for all necessary conditions of a, then we will prove very easily, that for any concrete statement a:

$$\boldsymbol{a} = \bigwedge_{a \Rightarrow b} \boldsymbol{b}$$

Which is equivalent to the main definition: $(a \Rightarrow b) = (a = (b\ and\ a))$.

That means, that every statement is the conjunction of all its necessary conditions and this conjunction is a sufficient condition for this statement to be true. It may seem a bit strange, that nobody noticed it until now, but it requires noticing, that every statement not only states, that it is true, but also states, that it is true only if all conditions, that are defined in it, are fulfilled. Therefore, every such condition is necessary and all taken together are sufficient for the statement to be true. Simply, a necessary condition of a statement is a condition, that has to be fulfilled for the statement to be true. So, if all such conditions are fulfilled, then there is nothing more that has to be fulfilled for the statement to be true, so this statement is true.

So also for any statement a and for $q = {\sim}a$:

$$\boldsymbol{a} = {\sim}q = {\sim}\bigwedge_{q \Rightarrow b} b = \bigvee_{q \Rightarrow b} {\sim}b = \bigvee_{q \Rightarrow {\sim}b} b = \bigvee_{b \Rightarrow {\sim}q} b = \bigvee_{b \Rightarrow a} \boldsymbol{b}$$

For more details, see Sections 5.18 and 5.35 and Chapter 9.

2.2.4 Simple implication

I will demonstrate a simplified derivation of some kind of implication, that has wide application, because it is very general.

For any different concrete things x, y and a:

$$((\boldsymbol{a} \supset \mathbf{x}) \Rightarrow (\boldsymbol{a} \supset \boldsymbol{y})) = \left((a \supset x) = ((a \supset x)\ and\ (a \supset y)) = (a \supset (x\ and y))\right)$$

$$= \left((a \supset x) = (a \supset (x\ and y))\right) = (x = (x\ and y)) = (\mathbf{x} \supset \boldsymbol{y})$$

So:

$$((x \subset a) \Rightarrow (y \subset a)) = (y \subset x)$$

So, using the above rule (assume, that the negation of a thing is the analogy to complement of a set, $(X \subset Y) = ({\sim}Y \subset {\sim}X)$):

$$((\boldsymbol{a} \subset \mathbf{x}) \Rightarrow (\boldsymbol{a} \subset \boldsymbol{y})) = (({\sim}x \subset {\sim}a) \Rightarrow ({\sim}y \subset {\sim}a)) = ({\sim}y \subset {\sim}x) = (\mathbf{x} \subset \boldsymbol{y})$$

So:

$$((a \subset x) \Rightarrow (a \subset y)) = (x \subset y)$$

At first sight, it can seem not too practical, but nothing could be further from truth. Let us define the following operator:

$$X^{\uparrow} = (\textit{the set of all members of the class of } X)$$

Where we assume, that not only every existing concrete X is a member of the class of X, but also every past, future, potential and impossible concrete X is such a member.

Let us call to mind, that our formulas are true also for sets, and we get:

$$\left((A^{\uparrow} \subset X^{\uparrow}) \Rightarrow (A^{\uparrow} \subset Y^{\uparrow})\right) = (X^{\uparrow} \subset Y^{\uparrow})$$

And we have for A and B in singular form:

$$\left(A^{\uparrow} \subset B^{\uparrow}\right) = (a\,A\,is\,a\,B)$$

And now we are ready to evaluate very wide range of implications.

For example, the following false implication between a true statement and a false statement:

$$\begin{aligned}\big((A\,mole\,has\,eyes) &\Rightarrow (A\,mole\,can\,see\,well)\big)\\ &= \big((A\,Mole\,\textbf{is}\,A\,Thing\,That\,Has\,Eyes)\\ &\Rightarrow (A\,Mole\,\textbf{is}\,A\,Thing\,That\,Can\,See\,Well)\big)\end{aligned}$$

Let us transform it to:

$$\begin{aligned}&\left(\left(Mole^{\uparrow} \subset (Thing\,That\,Has\,Eyes)^{\uparrow}\right) \Rightarrow \left(Mole^{\uparrow} \subset (Thing\,That\,Can\,See\,Well)^{\uparrow}\right)\right)\\ &= \left((Thing\,That\,Has\,Eyes)^{\uparrow} \subset (Thing\,That\,Can\,See\,Well)^{\uparrow}\right)\\ &= (A\,Thing\,That\,Has\,Eyes\,\textbf{is}\,A\,Thing\,That\,Can\,See\,Well)\\ &= (A\,thing,\ that\,has\,eyes,\ can\,see\,well) \equiv \textbf{\textit{false}}\end{aligned}$$

So, we have evaluated the logical value of our implication. The same way, we can evaluate the logical value of implication for any statements of the same kind (in this book, I present how wide is this class).

For more details, see Chapter 6, especially Section 6.3.

Other examples:

$$\big((Rose\,is\,Flowering\,Plant) \Rightarrow (Rose\,is\,Plant)\big) = (Flowering\,Plant\,is\,Plant) \equiv true$$

$$\begin{aligned}\big((Sky\,is\,Being\,That\,Is\,Blue) &\Rightarrow (Sky\,is\,Being\,That\,Has\,A\,Color)\big)\\ &= (Being\,That\,is\,Blue\,is\,Being\,That\,Has\,A\,Color) \equiv true\end{aligned}$$

$$\begin{aligned}\big((Pterosaur\,is\,Being\,That\,Can\,Fly) &\Rightarrow (Pterosaur\,is\,Being\,That\,Do\,Not\,Only\,Walk)\big)\\ &= (Being\,That\,Can\,Fly\,is\,Being\,That\,Do\,Not\,Only\,Walk) \equiv true\end{aligned}$$

$$\begin{aligned}\big((Rose\,is\,Flowering\,Plant) &\Rightarrow (French\,Rose\,is\,Flowering\,Plant)\big)\\ &= (French\,Rose\,is\,Rose) \equiv true\end{aligned}$$

$$((\mathit{Flightless\ Bird\ is\ Being\ That\ Cannot\ Fly}) \Rightarrow (\mathit{Penguine\ is\ Being\ That\ Cannot\ Fly}))$$
$$= (\mathit{Penguine\ is\ Flightless\ Bird}) \equiv \mathit{true}$$
$$((\mathit{Star\ is\ Being\ That\ Is\ Heavier\ Than\ Its\ Planet})$$
$$\Rightarrow (\mathit{The\ Sun\ is\ Being\ That\ Is\ Heavier\ Than\ Its\ Planet}))$$
$$= (\mathit{The\ Sun\ is\ Star}) \equiv \mathit{true}$$

2.2.5 Material implication derived from new implication

I will also mention here, that I derived the logical values of material implication from new implication. As you know, they were just assumed without any proof until now. And material implication itself was not well understood. I defined material implication the following way:

$$(a \rightarrow b) = ((a\ \mathit{and}\ \mathit{truth}) \Rightarrow b)$$

because material implication is nothing else than a phrase "If a is true, then b is true", that is no more and no less than implication in the context of whole truth. *Truth* is simply the conjunction of all true statements. The derivation of logical values of material implication from the above definition is quite simple and I have done it in this book.

For more details, see Section 5.38.

2.2.6 Undecidable statements

I have understood the property, that joins all statements, that are not just either true or false and I call them undecidable statements. It allowed me to solve all paradoxes of classical logic, like for example, Liar paradox, Quine's paradox, Yablo's paradox, and so on. All these, you will find in this book.

For more details, see Section 5.23 and many subsections of Chapter 15.

2.3 More about speaker and context versus model

Remember, that you do not need to understand, what is a speaker and a context and a model, to understand elementary new logic.

In this book, **model (interpretations)** from model theory is replaced by **definitions**, that define strict terms, a **speaker**, that said given sentence, and a **context**, in which this sentence is said, because the meanings of all symbols depend only on definitions, that in case of strict terms (e.g.: "Socrates", "France", "DNA", "sinus function", etc.) and strict references (e.g.: "Leonardo da Vinci's", "Alfred Hitchcock's", etc.) do not depend on anything else and in case of non-strict terms (e.g.: "high"/"low", "hot"/"cold", "good"/"bad", etc.) and non-strict references (e.g.: "here", "she", "your", "it", "the Dog", etc.) depend only on the speaker or the context. So, of course, in case of non-strict terms and non-strict references for every speaker in every context, we also have their appropriate definitions. Such an approach to the matter is much more understandable and intuitive than models and, most of all, correct.

A speaker could be treated as a part of the context, so the context could be always enough to interpret any sentence, but I singled out a speaker as the one, that determines the definitions of all non-strict terms, and in this book, I will treat a context as only everything that surrounds (not only physically) the act of speaking of a speaker, so the information who speaks is not included in the context.

Of course, a speaker does not have to be a human and it does not have to speak. It can be, for example, a computer, that uses fuzzy logic, and generates output in the form of sentences.

Remember, that, of course, if you really need to, then you can use models terminology, but then it will be much more difficult for you to understand the difference between subjective and objective logic.

For more details, see Chapter 8.

2.4 Symbols

2.4.1 Operators

In this book, I use the following symbols of operators:

"∼" – logical special negation of a statement, attribute, complement of a plain thing,

"≀" – natural general negation ("not") of anything; for more details see Section 9.2.4,

"⇒" – implication **between statements**, and not between sentences, like it is in classical logic (it is **analogous** to "logical consequence" from classical logic); for more details, see Section 5.35,

"⊃" – inclusion of things and thereby also the relation of being a part,

"=" – **true equality**, much stronger than any equivalence from any other logic,

"→" – contextual implication of statements (it is **analogous** to "material implication" from classical logic); for more details, see Section 5.38,

"↔" – contextual equality of statements (it is **analogous** to "material equivalence"); for more details, see Section 5.38.10,

"⇌" and "≡" – the equality of logical values of statements; it will be used for the aesthetic use of logical constants *true, false, undecidable*; for more details, see Section 5.44,

"⇛" – deterministic implication; for more details, see Section 6.1,

"⇉" – nondeterministic implication; for more details see Section 6.1,
"⇚⇛" – deterministic equality of statements; for more details, see Section 6.1,
"⇇⇉" – nondeterministic equality of statements; for more details, see Section 6.1.

2.4.2 Nothing, things, sets and conjunctions

At the beginning, you do not have to know all these symbols. They will be introduced in this book, when necessary.

Nothing:

$$\wr\omega = \Phi = (nothing)$$

Things:

$$\omega = (Thing)$$

$$\alpha = (Atomic\ Thing)$$

$$\beta = (Being)$$

$$\eta = (Nobeing)$$

$$\rho = (Past\ Being)$$

$$\varphi = (Future\ Being)$$

$$\psi = (Set)$$

$$\gamma = (Tuple)$$

$$\zeta = (Declarence)$$

$$\varsigma = (Statement)$$

$$\nu = (Decidable\ Statement)$$

$$\mu = (Undecidable\ Statement)$$

$$\tau = (Tautological\ Statement)$$

$$\xi = (Attribute)$$

$$\chi = (Shape)$$

$$\kappa = (Physical\ Shape)$$

Zeroes (empty parts):

$$\psi_{\varnothing} = \varnothing_{\psi} = \varnothing = (The\ Empty\ Set)$$

$$\gamma_{\varnothing} = \varnothing_{\gamma} = \varnothing = (The\ Empty\ Tuple)$$

$$\varsigma_{\varnothing} = \varnothing_{\varsigma} = (\textit{The Empty Statement})$$

$$\xi_{\varnothing} = \varnothing_{\xi} = (\textit{The Empty Attribute})$$

$$\chi_{\varnothing} = \varnothing_{\chi} = (\textit{The Empty Shape})$$

$$\kappa_{\varnothing} = \varnothing_{\kappa} = (\textit{The Empty Physcial Shape})$$

Sets:

$$\varnothing = \{\Phi\} = (\textit{the concrete empty set})$$

$$\Omega = \omega^{\mho} = (\textit{the concrete set of everything})$$

$$\mathcal{B} = (\textit{the concrete set of all concrete or abstract beings})$$

$$\mathcal{E} = \begin{pmatrix} \textit{the concrete set of all concrete or abstract} \\ \textit{present, past, future beings} \end{pmatrix}$$

$$\omega^{\uparrow} = (\textit{the concrete set of all concrete things})$$

$$\Psi = (\textit{the concrete set of all concrete sets})$$

$$\Sigma = (\textit{the concrete set of all concrete statements})$$

Conjunctions:

$$\Upsilon = @\Omega = (\textit{totality})$$

$$\mathcal{R} = @\mathcal{B} = (\textit{reality})$$

$$\Lambda_{A}^{\mho} = \left(\textit{the totality for negation} \sim_{A}^{\mho}\right)$$

2.5 Default quantifiers

In this book, free variables from list x used in any relation R, that is written in separate line, have a default universal quantifier:

$$R(x) = \bigwedge_{x} R(x) = \left(\textit{for every } x \textit{ this is true that } R(x)\right)$$

For example:

You should read:

$$(x+y) = (y+x)$$

As:

$$\bigwedge_{x}\bigwedge_{y}\left((x+y) = (y+x)\right)$$

3 Two important terms

In this book I will use the following two important terms:

a.) **kind** – in the sense of a type, a class,

b.) **representation of the kind of** K – the set, that contains every thing, that was consistently meant to be concrete K – the set of all representatives of the kind of K – the set of all members of the class of K – the set of all examples of the type of K.

In formal sentences singular and plural kinds will be usually marked by beginning significant words with big letters, e.g.:

$$\textit{CelestialBody} \longrightarrow \textit{CelestialBodies}$$

$$\textit{Race of Horses} \longrightarrow \textit{Races of Horses}$$

Remember, that if X is a bound variable, then in the restriction of its quantifier "concrete X"/"abstract X" means a concrete/an abstract thing X, but otherwise "concrete X" means a member of the class of X and "abstract X" means a thing, the class of which is a subclass of (the class of) X. For example: "For any concrete X: X is a thing" means, that every concrete thing is a thing, but "For any X: for any x, that is concrete X: x is a thing" means that any current member of the class of any thing is a thing.

Operator $K^{\uparrow}$ gives the representation of the kind of K, so it gives the set, that contains only every past concrete, every present concrete, every future concrete, every potential concrete and every impossible (but consistently meant to be) concrete K. Operator $S_{\downarrow}$ is the inversion of operator $K^{\uparrow}$, because it gives the thing, the kind of which has representation S. Operator $S_{\downarrow}$ gives the result for every set S.

Operator $K^{\Uparrow}$ gives the set, that contains only every concrete K, that **is** K. So if K is a being, then $K^{\Uparrow}$ contains only every existing concrete K.

Operator $K^{\circlearrowleft}$ gives the set of the things of all subclasses of the class of K, so it gives the set, that contains only every past concrete, every present concrete, every future concrete, every potential concrete and every impossible (but consistently meant to be) concrete K and every past abstract, every present abstract, every future abstract, every potential abstract and every impossible (but consistently meant to be) abstract K. Operator $S_{\circlearrowright}$ is the inversion of operator $K^{\circlearrowleft}$. Operator $S_{\circlearrowright}$ gives the result only for such a set S, that is equal to $K^{\circlearrowleft}$ for some K.

Operator $K^{\circlearrowleft\circlearrowright}$ gives the set, that contains only every concrete K, that **is** K, and every abstract K, that **is** K. So if K is a being, then $K^{\Uparrow}$ contains only every existing concrete K and every existing abstract K.

For more details see Section 10.9.

https://doi.org/10.1515/9783111441382-003

4 Priorities of operators

I do not set any priorities to operators of the same arity in logical expressions, so this is, why I use brackets so often. Order of not associative operations is defined by grouping left to right, if not explicitly stated otherwise. Unary operators have a higher priority than multi-ary (binary, ternary, etc.) operators. Prefix unary operators have a higher priority than suffix unary operators, so e.g.: $(\sim x \textit{ is false}) = \big((\sim x) \textit{ is false}\big) \neq \sim(x \textit{ is false})$. And although prefix unary operators (e.g.: $\wr X$) have a higher priority than suffix unary operators (e.g.: $X@$), upper and lower index unary operators (e.g.: $X^{\uparrow}$) have even higher priority, so $\wr X^{\uparrow} = \wr\big(X^{\uparrow}\big) \neq (\wr X)^{\uparrow}$. Quantifiers have the same priority as prefix unary operators. Multiple equality and exclusive disjunction (xor) should be treated as a single multi-ary operator.

And for $R_i \in \{=, \Rightarrow, \Leftarrow, \leftrightarrow, \rightarrow, \leftarrow, \equiv\}$:

$$(A_1\, R_1\, A_2\, R_2\, A_3 \ldots A_n\, R_n\, A_{n+1}) = \big((A_1\, R_1\, A_2)\ \textit{and}\ (B_2\, R_2\, A_3)\ \textit{and} \ldots \textit{and}\ (A_n\, R_n\, A_{n+1})\big)$$

https://doi.org/10.1515/9783111441382-004

5 Definition

5.1 Introduction

5.1.1 Standalone symbols

There are two important in logic kinds of standalone symbols: names (symbols of things, where, for example, “the smallest prime number greater than 5” is the complex name of 7) and declarences (only symbols of statements), where, of course, a statement is a thing, so it can have a name. A name is a direct symbol, because its meaning is its referent, but a declarence is an indirect symbol, because its meaning (a piece of information) is not its realization (some state of affairs). Every symbol of a thing has to fulfill only the condition, that it refers to a thing, so it has its referent, the definition of which gives the meaning to this symbol. Remember, that the following symbols are not standalone: “nothing”, “Φ”, “neither of things”, that are used in declarences as the same quantifier. Non-standalone symbols serve only as modifiers of the meaning in standalone symbols, e.g.: adjectives serve as attributes of objects and adverbs serve as attributes of attributes and as attributes of relations.

A reference means just that one thing points to something, which thereby is the referent of that thing. Every reference has only exactly one referent.

So there are basically two types of standalone symbols: names and declarences. A name has its direct referent, but a declarence is a symbol of a statement, realizations of which are something else. So a declarence has this intermediate being (statement) between itself and the states of affairs it refers to. The meaning of a symbol is always its referent, so the meaning of a declarence is a statement. On top of that, the states of affairs about which a declarence says are realizations of the statement of that declarence. So a statement is the direct referent of a declarence and realizations are like “direct referents” of a statement, so they are also like “indirect referents” of a declarence.

$(S,C)^X$ the same as $(S,C)^\wedge X$ is the meaning of standalone symbol X, when said in situation (S,C), which means: said by speaker S in context C. So it is the referent of symbol X. So it is the statement of sentence X, if X is a declarence.

$$(C)^X = (C)^\wedge X = (\mathit{current\ speaker}, C)^X$$

$$(S,\_)^X = (S,\_)^\wedge X = (S, \mathit{current\ context})^X$$

$$(\_,\_)^X = {}^\wedge X = (\mathit{current\ speaker}, \mathit{current\ context})^X$$

For example, ^“This is the beginning of the endless journey to the great unknown” is the statement of this sentence said by current speaker in current context.

$[m]^X$ is the meaning of standalone symbol X in model m. A model is always completely determined by a situation.

https://doi.org/10.1515/9783111441382-005

The symbol "|" is a symbol of the symbols concatenation operator. For example: ("abc" | "def") = "abcdef".

5.1.2 A thing and nothing, a being and a nobeing

A **thing** (ω) has to fulfill only the condition, that it has a definition (that does not have to be expressed), so its symbol has the referent and its kind has at least one representative, because a definition always defines something. If the attempt to define it failed, then it does not exist, so it is not a being, so it is a nobeing, and otherwise, if the attempt was not to define it as a nobeing, then it exists, so it is a being. Attributes of a nobeing come from two facts: that it is a nobeing, so it has all the attributes of a nobeing, and that it was meant to be, what the failed attempt to define it claims it to be, e.g.: a square circle was meant to be a square and a circle at the same time, which is impossible, and that is why it is a nobeing. As a result every thing, that has failed attempt to define it, has its not contrary definition. So every**thing** has a unique definition and every**thing**, that exists, is a being and every**thing**, that does not exist, is a nobeing. Every**thing**, that can exist, is a possible thing and can be a being and is a nobeing, when it does not exist, and every**thing**, that cannot ever exist, is an impossible thing and is always a nobeing, so is never a being.

For example, a square circle, the same as a circle, is something, that was meant to be a circle, but a square circle is not a circle and never will be one.

Every being exists at the present time, so you always define a being as if the time it exists was right now. For example, the complete attempt to define a dog, that is sleeping at the time it exists, will have the attribute, that states, that it is sleeping (at the time it exists, which is the time it is a being, which is the time it is itself). Thanks to this we have always only one stripped attempt (for the definition of a stripped attempt see Section 5.1.3) of a definition for each thing. So if it is a past, future or potential being, then the attempt to define it fails, because the time it was meant to exist (and to be a being and to be itself), contradicts the current time, so at the present time it is a nobeing, that respectively was sleeping, will be sleeping or can be sleeping in the future.

The meaning of **nothing** (Φ) is the opposite of a thing, so nothing is not a thing, which means that everything is a thing. The word "nothing" in a declarence has the same meaning as the phrase "neither of things", that serves in a declarence as a quantifier and not as an object or a subject. All you can say using the word "nothing" follow from the fact, that ((Nothing does not X) = (Everything X)) and ((Nothing is not X) = (Everything is X)), etc., and in general:

$$\left(Everything\, R(\_)\right) = \bigwedge_{a \in \Omega} R(a) = \sim \bigvee_{a \in \Omega} \sim R(a) = \left(Nothing \sim R(\_)\right)$$

So, nothing is not a thing. So, nothing is the opposite of a thing, thus "nothing" means in a declarence the same as the phrase "neither of things". Nothing has no definition, nothing does not have any attributes and cannot be treated as the subject of a sentence and cannot be treated as an object of a sentence. So, nothing cannot be referenced by the word "it", because everything can be referenced by the word "it". And "nothing" is not a symbol of anything, because the symbol "nothing" does not have any referent. Although you can say something using the word "nothing", 'nothing' (when we use apostrophes in place of quotation marks, we will mean the intentional counterfactual use of the word "nothing" as a symbol of a thing) has no definition, because all statements of sentences, that would use 'nothing' as a subject, would be true and false at the same time (including the simplest one: "Nothing is (not) a being"), if 'nothing' had a definition. This is this way, because then 'nothing' would be something, so 'nothing' could be referenced by the word "it", so 'nothing' could be treated as the subject of a sentence, e.g.:

$$false \equiv (Everything\ is\ not\ a\ being) = (Nothing\ is\ a\ being) = X$$

$$true \equiv \sim(Everything\ is\ a\ being) = \sim(Nothing\ is\ not\ a\ being) = \sim(It\ is\ not\ a\ being)$$
$$= (It\ is\ a\ being) = (Nothing\ is\ a\ being) = X$$

So 'nothing' would have self-contrary definition. Notice, that the above statement is not even undecidable, because the logical value of an undecidable statement is undecidable. So 'nothing' should be a nobeing (the result of a failed attempt to define a being), but then we would have the same problem with the statement of the sentence "Nothing was meant to be a being", that would be true and false at the same time, so it would be also self-contradicting, and so on to infinity. In other words, we could not define 'nothing' even as a nobeing, so we could not define 'nothing' at all. So we have a contradiction. So 'nothing' cannot be defined, so the word "nothing" (understood as the replacement for the phrase "neither of things") cannot be a symbol of a thing. For more details see Section 5.1.5.

Beside it, if nothing has a definition, then everything does not have a definition, because both these sentences mean the same, since the word "nothing" means in a sentence the same as "neither of things". And that is simply not true, because every**thing** has a definition. The word "nothing" has a meaning only as a part of a sentence, but, of course, it does not mean that nothing has a meaning and it does not mean that nothing has a definition. Not every word has an independent meaning and not every word has a definition. Only standalone symbols have their independent meanings and definitions. For example, the meaning of the word "bright" is not a thing, because it is not a noun, so it is only a modifier (it is a modifier as a word, but it does not mean, that bright is a modifier) of the meaning of nouns, that symbolize bright things, so you cannot correctly say, that bright is a property, because a property is a thing, but the meaning of the word "brightness" is a thing, and you can correctly say, that brightness is a property. The word "bright" is not a noun, so it cannot be an object and cannot be the subject of a sentence and that is indirectly, why it cannot be defined, because we simply cannot say anything about 'bright', for example, we cannot say that **it** is the meaning of the word "bright", though we can say something about brightness. At the beginning it may all sound difficult, but in time everything will come to its place and settle in your mind and then you will understand it. Just remember, that you just cannot treat the word "nothing" as a symbol of an object, since it is only a quantifier in a sentence.

Of course, nothing can be defined as the opposite of a thing, but it only means, that there cannot be (anything, that is) the opposite of a thing, so it does not mean, that nothing can have a definition, because complete opposite is true, since everything can have a definition (= ^"Nothing cannot have a definition").

The **set of nothing** (the empty set) has symbol ∅.

The **set of everything** has symbol **Ω**.

5.1.3 Totality, equality, sameness and equality of identities

The conjunction of all **things** is called **totality ("ϒ")** and includes all past, present, future (predictable, necessary future consequences) and potential realities.

Totality changes in every moment, because admittedly in every moment some possible things become impossible, some potential beings become either future or present, some nobeings become beings and some beings become nobeings, etc., but all **stripped attempts** to define things (i.e. complete attempts abstracted from attributes, that came from undecidable statements about the unpredictable future, when it became predictable – past, present or future) do not ever change, so, stripped totality, is always the same thing, so the stripped set of everything is always the same set, whereas totality constantly changes.

So we have two kinds of "equality":

$$(a=b)=(a\ and\ b\ are\ equal)=(a\ and\ b\ are\ the\ same)=\sim(a\ and\ b\ are\ different)$$
$$=(a\ and\ b\ have\ the\ same\ definition)=\big(attr(a)=attr(b)\big)$$

$$(a\parallel b)=(a\ and\ b\ are\ the\ same\ thing)=\sim(a\ and\ b\ are\ different\ things)$$
$$=(a\ and\ b\ have\ equal\ stripped\ attempts\ to\ define\ them)$$

$$(a=b)\Rightarrow(a\parallel b)$$

$$\bigvee_{a,b}\sim\big((a\parallel b)\Rightarrow(a=b)\big)$$

So for two things A and B, that were meant to be concrete sets:

$$(A=B)\Rightarrow\left(\bigwedge_{\substack{it\ was\ meant\ that\\ x\in A}}\ \bigvee_{\substack{it\ was\ meant\ that\\ y\in B}}(x=y)\ and\ \bigwedge_{\substack{it\ was\ meant\ that\\ y\in B}}\ \bigvee_{\substack{it\ was\ meant\ that\\ x\in A}}(x=y)\right)$$

$$(A \parallel B) \Rightarrow \left(\bigwedge_{\substack{\text{it was meant that} \\ x \in A}} \bigvee_{\substack{\text{it was meant that} \\ y \in B}} (x \parallel y) \text{ and } \bigwedge_{\substack{\text{it was meant that} \\ y \in B}} \bigvee_{\substack{\text{it was meant that} \\ x \in A}} (x \parallel y) \right)$$

And we have:

$$\bigcup_{x \subset Attribute^{\Uparrow}} \{(\textit{thing that has attempted set of attributes } x)\} = \Omega$$

For example, the statement of the following sentence can be true: "The apple, that I was eating yesterday at high noon, is [in the strict sense] the same thing as the apple (or: the same apple), about which I am thinking right now, but it is not the same, because it no longer exists." But remember, that in the strict sense, an apple, that exists at some moment, no longer exists in another moment, even if in a not strict sense we would be rather inclined to agree, that it still exists and it is the same apple. For more details see Section 10.1.

For any things *a* and *b*:

$$(a \doteqdot b) = (\textit{a and b have equal identities})$$

Two things have the same identity if and only if they are perceived by someone as the same thing or they are equal when abstracted from their existence in time, so identity is always subjective for physical things and is objectively defined for all nonphysical things.

So all nonphysical things and all identifiers of things have objective identities.

And all physical things can have subjective identities and we deal with such a subjective identity if and only if for some speaker some current at different moments physical things have the same identity.

All self-explainable symbols composed only of symbols of full things (this term is explained below) have defined identities, because definitions of such things come from the definitions of terms, that they are explained with, e.g.: "1" (explained by Number One), "My Cat" (explained by My + Cat), "Something That Runs" (explained by Runs), "The United States of America" (explained by America + States + United).

For more details about the equality of identities see Section 10.5.

A thing *X*, all members of the class of which are only all things with the same identity, is a **full thing**, and, as for every thing, then every thing, that was consistently meant to be *X* (for more details see Section 10.9), is a member of the class of *X*. So a full thing is the identity $(\vdots Z)$ of every member *Z* of its class. And we will use the prefix operator $\downarrow$ before any thing to refer to its **current thing**, so only every thing, that **is** *Y*, is a member of the class of $\downarrow Y$.

So:

$$(\downarrow X) = (X^{\Uparrow})_{\downarrow}$$

For example, if I have only one my dog, then the thing *My Dog* is a full thing, the members of the class of which are all things with the same identity, consistently meant to be my dog. And the thing ↓My Dog is a concrete current thing, the only member of the class of which is a thing, that is currently my dog. So:

$$(\downarrow \mathit{My\,Dog}) = \left((\mathit{My\,Dog})^{\Uparrow}\right)_{\downarrow}$$

$$(\downarrow \mathit{My\,Dog})^{\uparrow} = (\downarrow \mathit{My\,Dog})^{\Uparrow} = (\mathit{My\,Dog})^{\Uparrow} \subset (\mathit{My\,Dog})^{\uparrow} \subset (\mathit{Dog})^{\uparrow}$$

And we assume, that all numbers written using digits are current things:

$$1 = \downarrow (\mathit{Number\ One})$$

$$2 = \downarrow (\mathit{Number\ Two})$$

And so on.

Remember that, for example, some past dogs (where past beings are nobeings) **are** currently My Past Dogs, because we have that for some *X*: (My Past Dog *X* **is** My Past Dog) while (My Past Dog *X* **was** My Dog). And remember, that the class of Past/Future/Potential/Impossible *Y* has only members, that were consistently meant to be *Y*, but now are past/future/potential/impossible.

The rest of this section is for advanced readers.

Even completely artificial thing *X* the class of which has members only from set *Y*:

$$X = \left(\bigcup_{x \in Y} \{x\}\right)_{\downarrow}$$

can have its current thing, if only its class has members, that are beings or are logically impossible, but are not double nobeings. All its members are consistently meant to be *X*, but not all have to be *X*. And only every member that is *X*, is an element of $X^{\Uparrow}$.

But, of course, not every thing is a full thing.

Any ***X*** will be the thing, the members of the class of which are all things consistently meant to be *X*, which include all members of the classes of things that are at different moments $\downarrow X$ as well as all potential things, that could become *X* at some moments, and all impossible things, that were such potential *X* in the past.

The thing *My Dog* is for me a full thing, a members of the class of which are, for example, my dog, that existed yesterday, and my dog, that exists now, and my dog, that will exist in the future, so the class of it has many members, but assuming, that I have exactly one dog, ↓My Dog is only this current being, that exists now, that is a concrete being, so the class of it has it as the only member. So the members of the class of full thing *X* are only all things, that for a speaker has the same identity as $\downarrow X$, including past, future, potential and impossible things.

We have for any things *a* and *b*:

$$(a = b) \Rightarrow (a \parallel b) \Rightarrow (a \doteqdot b)$$

But we often assume, that for any two things a and b:

$$(a=b)=(a \parallel b)$$

because in such a case we assume, that we make the whole reasoning (e.g.: a proof) at a single moment. We can do it in abstract calculations, because then we do not care about physical attributes of things, that were not meant to be physical beings.

If we limit our reasoning only to nonphysical properties of beings, then we can completely abstract nonphysical beings from physical totality, since all their nonphysical attributes do not ever change. Then we can even assume, that for any nonphysical things a and b:

$$(a=b)=(a \parallel b)=(a \div b)$$

Numbers are not concrete things, because, for instance, the number one yesterday had different physical attributes, than it has now (e.g. now we are saying about it, which makes a physical attribute of the number one), so their classes have many members, but they are eternal, so at any time their classes have an existing member and the symbols of concrete numbers we use always refer to this one current concrete being.

Any two different nonphysical things have the same identity only if their stripped attempts to define them differ only in the times of their existence.

5.1.4 Reality, possible and impossible things

The common kind of all elements of any set of things is a **class**.

All things divide into possible and impossible things. Possible things are beings, unless they do not exist, and impossible things are always nobeings. We will name all things, that were not ever able to exist and cannot exist now and will not be ever able to exist, as never possible things (logically impossible things; always impossible things; things, that cannot ever exist) and these things, that at least were able to exist, as ever possible things (logically possible things; sometimes possible things; things, that can ever exist). All past, present, future and potential beings will be called possible things and all past, present and future beings will be called forever possible things. All past beings are possible in the past and all present are possible now and all future and potential are possible in the future. Consider, that, for example, a thing, that at some point was a potential being, but currently is impossible, is now ever possible (logically possible) thing, though it is no longer possible.

And for any thing X we have, that:

$$(X\ will\ be\ possible) \Rightarrow (X\ is\ possible) \Rightarrow (X\ was\ always\ possible)$$

Only those things, that were at some point potential beings, but now are impossible, are alternatively possible and impossible at the same time. So for any other thing X:

$$(X\ is\ possible)=(X\ is\ alternatively\ possible)$$

Something is alternatively possible if and only if it is possible under counterfactual conditions, that could be true, if undetermined (free will) actions took a turn. So if something is not alternatively possible, it was never possible and is not possible now and will be never possible regardless of everything – regardless of the course of all undetermined actions. So an alternatively possible thing is nothing else than a possible thing in some alternatively possible world, since, of course, any alternatively possible world can differ from the actual world only in the course of undetermined actions at some points in time and their consequences. Where there is not a big difference between an alternatively possible world and any other alternatively possible thing, because an alternatively possible world is just another alternatively possible thing, that is either a potential world or it was at some point a potential world and now it is either the actual world or an impossible world.

For every thing X:

$$(X\ is\ possible) \Rightarrow (X\ is\ alternatively\ possible) \Rightarrow (X\ is\ logically\ possible)$$

If a successful attempt to define X can be nonphysical, then:

$$\begin{aligned}(X\ is\ logically\ possible) &= (X\ is\ alternatively\ possible) = (X\ is\ possible)\\ &= (X\ is\ a\ being) = (X\ exists)\end{aligned}$$

Possible things are divided into past beings, that are things, that were beings in the past, but are no longer beings, so are defined as nobeings, future beings, that are things, that will be beings in the predictable future, but are not beings yet, so are also defined as nobeings, potential beings, that can be beings in the future, but are not beings yet, so are defined as nobeings too, and just beings, that are things, that exist at the present time. Nonphysical attributes of beings, that are nonphysical, are unchangeable and these beings are eternal, so they exist always, where thing X is eternal if and only if its symbol always refers to some being, so the kind of identity of that thing $(\vdots X)$ has always some existing representative.

If free will exists, then the future is not fully predictable due to the existence of unpredictable beings, that have free will and can act freely in this reality. Such beings are unpredictable, because their will can be unknowable even for the idealized perfect mind. So, if free will exists, then there are potential beings, that can become in time either impossible things or either future beings or just beings. And future beings become in time beings and beings become at every moment past beings. If free will does not exist, then there are no potential beings, because every thing, that is a potential being if free will exists, is either an impossible thing or a future being if free will does not exist. An impossible thing cannot become possible and only a potential being can become impossible.

The set of all things, that are beings, ever were beings or for sure will be beings, that is the set of all **past, present and future beings**, has symbol $\boldsymbol{\varepsilon}$.

Remember, that past physical beings are neither physical nor beings already, and future physical beings the same as potential physical beings are neither physical nor beings yet.

The set of all **beings** has symbol $\mathcal{B}$.

The conjunction of all **beings** is equal to **reality ("$\mathcal{R}$")**. So every being is a part of reality.

5.1.5 The attempt and the result of a definition

If an attempt to define something is incorrect, then nothing is defined as a result.

A thing is a being if and only if the attempt to define it was an attempt to define a being and was successful.

A thing is logically impossible, if and only if its primary attributes (the intended attributes of the attempt to define it as a being) are contrary to themselves (for example, when they are contrary to the definition of some term used in this attempt).

For more details see Section 6.2.

In this book, I use the word "self-contradictory" and the word "contradict" in their strict meaning not to confuse them with respectively the word "self-contrary" and the phrase "is contrary to", so only the Liar statement ($x = \sim x$) is truly self-contradictory. For the definitions of these words see Section 5.15 and Section 5.12. These definitions of a contradiction and a contrariety are based on the square of opposition.

Since the word "contradict" does not have its equivalent in a verb form for the term "is contrary to", I propose creating the new verb "contrare" and its new form "contraring", which I do not use so far in this book to make it more reader-friendly.

However, to be more reader-friendly I still use the word "contradiction" for the result of a proof by a contradiction, although such a proof should be strictly called "proof by a contrariety", and its result should be called "contrariety". Indeed, to be strict, every occurrence of the word "contradiction," with the exception of this section and Section 5.15, should be replaced by the word "contrariety". For a proof of this fact see Section 7.1.

The class of a logically possible thing is not narrower than the class of an alternatively possible thing, but it is rather an open question whether it is a wider class.

If an attempt to define something is contrary to itself or contrary to (truth of) reality, then this thing is a nobeing, because no being is defined by this attempt. So it has a definition, because it is a nobeing, that has a definition, but not that definition intended by the attempt to define it. Everything, that is not a nobeing, is a being and everything, that is a being, is not a nobeing, because they are opposite to each other.

For example: *X* was meant to be a president and it was a president in year *Y* and it is no more a president, so it is a **past** president, a **past** being and a nobeing at the present time.

This paragraph is only for very advanced readers: If the attempt to define something is contrary to itself and reality, then such a thing will be called a **double nobeing**, e.g. a past or future or potential or impossible square circle is not a circle and is not a square, but it also is not even a square circle, because the time of its presence as a square circle is not the current time and, of course, due to the presence of a square circle in time it can have, for example, different physical attributes (e.g. someone will say something about it or not) than a present square circle, not saying anything about the different time of presence. Of course, it is present now but not as a square circle – only as something, that was meant to be a square circle. So it can have also the double presence time: as something, that was meant to be a square circle, but is not a square circle (always a continuum of moments), because the time it was meant to exist is not now, and as a square circle (always exactly one moment, which is now), when the time it was meant to exist is now and it is not an impossible square circle (then it is only present as a square circle, but does not exist, because the attempt to define it is contrary to itself). This is quite complicated, but you do not have to understand it at the beginning.

Remember, that if something was consistently meant to be a being, then it can be past or present or future or potential or even impossible (that at some point was potential, but never became present) being right now.

Important note: A physical/nonphysical attribute, definition, attempt of definition will be mental shortcuts for an attribute, a definition, an attempt of definition, respectively, that reference/does not reference physical totality, where a reference to a part of physical totality is a reference to physical totality. The definition of a physical thing must reference directly or indirectly a concrete part of physical totality, which is always (a part of) time or (a part of) space or a contingent thing.

A **statement** has symbol **ς** and the set of all **statements** has symbol **Σ**.

A **being** (**"*β*"**) – anything, that exists; anything, that is not a nobeing; every part of current reality.

A **nobeing** (**"*η*"**) – opposite to being; anything, that does not exist; anything, that is not a being; everything, that is not a part of current reality.

The rest of this section is for very advanced readers

So we have for every X:

$$(X \text{ is logically impossible}) = \left(\bigwedge_{a \in primaryAttr(X)} a \Rightarrow \sim \bigwedge_{a \in primaryAttr(X)} a \right)$$

For more details see Section 6.2.

A **protostatement** is an elementary statement or the negation of an elementary statement, so it is equality or inequality of any things. For more details see Section 5.19.

Every true disjunction of protostatements is implicated by the fact obtained from an attribute of an object of each true protostatement in this disjunction, because such a true protostatement implicates any disjunction of itself and any other statement. And all other true statements can be decomposed to the conjunction of true disjunctions of protostatements. So:

$$truth = \bigwedge_{x} \bigwedge_{a \in attr(x)} attrToMonoStat(a, x)$$

Remember, that for any basic relation $R(x)$ for any tuple (a) of arguments that fits into tuple (x):

$$\sim R(a) = (!R)(a)$$

By the way, the statement of a basic declarence is a disjunction of elementary statements (in the form $(x = y)$ for any things x and y), because for any basic relation $R(x)$ for any tuple (a) of arguments that fits into tuple (x) and for any statement r we have:

$$(r = R(a)) = \left(((a), r) \in R\right) = \bigvee_{((b),q) \in R} \left(((b), q) = ((a), r)\right)$$

$$R(a) = \left(((a), true) \in R\right) = \bigvee_{((b),true) \in R} ((b) = (a)) = \bigvee_{q \text{ is a true decidable statement}} \bigvee_{((b),q) \in R} ((b) = (a))$$

For more details see Section 5.19 and Section 10.12.

We have for any X:

$$(X \text{ is a being}) = (primaryAttr(X) = attr(X)) = \bigwedge_{a \in primaryAttr(X)} attrToMonoStat(a,X)$$

$$= \bigwedge_{x} \sim \left(\bigwedge_{a \in primaryAttr(X)} a \Rightarrow \sim \bigwedge_{a \in attr(x)} a \right)$$

$$= \sim \bigvee_{x} \left(\bigwedge_{a \in primaryAttr(X)} a \Rightarrow \sim \bigwedge_{a \in attr(x)} a \right)$$

$$\Rightarrow \sim \left(\bigwedge_{a \in primaryAttr(X)} attrToMonoStat(a,X) \right.$$

$$\left. \Rightarrow \bigvee_{x} \sim \bigwedge_{a \in attr(x)} attrToMonoStat(a,x) = \sim \bigwedge_{x} \bigwedge_{a \in attr(x)} attrToMonoStat(a,x) \right)$$

$$= \sim \left(\bigwedge_{f \in primaryFacts(X)} f \Rightarrow \sim truth = untruth \right)$$

So if the attempt to define a being is contrary to *truth*, then we get a nobeing as its result:

$$\left(\bigwedge_{f \in primaryFacts(X)} f \Rightarrow \sim truth \right) \Rightarrow (X \text{ is a nobeing})$$

5.1.6 Presence and existence

We have to distinguish between presence and existence. Presence concerns all things and we will use phrases “there is/was/will be”, “there are/were/will be” to state it, e.g.: “there is a nobeing”, “there is a being”, etc. This is also expressed in the following relations: $((a)\, X\, is\backslash are\, (a)\, Y)$, $((a)\, X\, was\backslash were\, (a)\, Y)$, $((a)\, X\, will\ be\, (a)\, Y)$. Existence concerns only all beings and we will use verb “exist” to state it, e.g.: “there exists a being”, “a being exists”, “a nobeing does not exist”, etc. And we have:

$$(there\ is\ [present]\ X) = (there\ is\ X)$$

$$(there\ is\ past\ X) = (there\ was\ X)$$

$$(there\ is\ future\ X) = (there\ will\ be\ X)$$

$$(X\ is\ there) = (there\ is\ X) = (X\ can\ be\ there) = (there\ can\ be\ X)$$

$$(X\ exists) = (X\ is\ a\ being) \Rightarrow (X\ is\ a\ present\ thing) = (there\ is\ X)$$

$$(X\ existed) = (X\ was\ a\ being) \Rightarrow (X\ is\ a\ past\ thing) = (there\ was\ X)$$

$$(X\ will\ exist) = (X\ will\ be\ a\ being) \Rightarrow (X\ is\ a\ future\ thing) = (there\ will\ be\ X)$$

$$(X\ exists) \Rightarrow (X\ is\ able\ to\ exist) = (X\ can\ exist)$$

$$(X\ existed) \Rightarrow (X\ was\ able\ to\ exist)$$

$$(X\ will\ exist) \Rightarrow (X\ will\ be\ able\ to\ exist)$$

$$(X\ is\ present\ and\ nonphysical) = ((X\ exists) = (X\ can\ exist))$$

$$\begin{pmatrix} X\ is\ present\ and\ nonphysical \\ and\ is\ logically\ possible \end{pmatrix} = ((X\ exists) = (X\ is\ there))$$

For example, there is a square circle, but any square circle cannot exist.

The only necessary and sufficient condition for a thing to be is, that it is defined.

The only necessary and sufficient condition for a thing to exist is, that it is successfully defined as a being, which means, that it is a part of reality, that includes physical reality.

All things are present now. Some of them (past beings) was/were beings (existed in the past), some of them (present beings) are beings now (exists now, at the present time), some of them (future beings) still waits to become beings (will exist in the future), some of them (potential beings) can become beings in the future (are able to exist in the future), and some of them (impossible beings and impossible things in general) will never be beings (will never exist).

5.2 Definition of a set

Concrete set X is a being, for which $(Y \in X)$ is decidable for any thing Y and that has no other nonempty parts than its subsets.

For any Z, that is not a concrete set, $(Y \in Z)$ is not meaningful for any Y.

$$(Y \in X) = (\textit{thing } Y \textit{ is selected to concrete set } X) = (\textit{thing } Y \textit{ is an element of concrete set } X)$$
$$= (\textit{thing } Y \textit{ belongs to concrete set } X)$$

A **selection** of things cannot be distinguished from the set of these things, because it would have exactly the same attributes.

So a set is simply a selection of some things out of all things.

A **conjunction** is a thing, that consists only of things from some selection taken together and treated jointly as one thing. Every set determines the conjunction of its elements. The same conjunction of non-atomic things can be expressed as a conjunction of different arguments.

So a conjunction is a being if and only if it is a conjunction of beings.

Every thing can be an argument of a conjunction.

A conjunction is a being if and only if all its parts are beings.

A **set** has symbol ψ.

The set of all **concrete sets** has symbol $\boldsymbol{\Psi}$.

A concrete set is simply some selection of things. It is not a being constituted by these things taken together, because it is not the conjunction of these things, since it has marked out elements. A conjunction of two sets is a set, so a set is the conjunction of single element sets, that contain things, that are its elements, but the conjunction of all elements of any set is not this set.

Every thing is the conjunction of all its parts, and they together form this thing.

A thing selected to a set is not a part of it, but such a thing is only referenced and pointed out by the set. Only other set can be a part of a set, including a single element set.

A set, that has specified all its elements, is a **concrete set.**

$$(\textit{concrete set } X \textit{ exists}) = \Big(\textit{for every thing } x \; \big((x \in X) \textit{ is decidable}\big)\Big)$$

where a statement is decidable if and only if it is either true or false. For more details see Sections 5.10 and 5.23.

The **empty set** and the **set of everything** exist, because they are not nobeings, since they are, what they were meant to be.

Nothing belongs to the empty set, because everything does not belong to the empty set. And nothing does not belong to the set of everything, because everything belongs to the set of everything. But it is false, that nothing belongs or does not belong to any other set, because something belongs to any other set and something does not belong to any other set.

It is slightly modified so-called “naïve” set theory with one simple assumption:

$$\bigwedge_{S\ is\ a\ set}\ \bigwedge_{x\ is\ a\ thing}((x \in S)\ is\ decidable)$$

And I solved for this set theory all paradoxes of “naïve” set theory. For more details see Sections 15.6, 15.7, 15.8 and 15.10 in Chapter 15.

5.3 Conjunction and being a part

Before you move on to the more advanced topics of the new logic, first become more familiar with its foundations, which are conjunction and the relation of being a part.

We can always take into consideration any things together and treat them as one thing, the parts of which they are. Such a thing will be **the conjunction of these things**.

In other words, two or more things taken into consideration as parts of one thing or treated or perceived as parts of one thing are together this one thing that is equal to the conjunction of these things.

We will use the word “and” and the symbol “∧” to express a conjunction of things.

A conjunction of things has the following properties:

a.) idempotency:

$$\bigwedge_{X \subset \Omega}\ \bigwedge_{x\ is\ a\ concrete\ thing}\left(x = \bigwedge_{e \in X} x\right)$$

b.) commutativity and associativity:

$$\bigwedge_{X \subset \Omega, Y \subset (\Omega - X)}\left(\bigwedge_{a \in (X \cup Y)} a = \left(\bigwedge_{a \in X} a\ and \bigwedge_{a \in Y} a\right)\right)$$

c.) distributivity:

$$\bigwedge_{x \in \Omega, X \subset \Omega}\left(\left(x\ and \overset{\oplus}{\bigvee_{a \in X}} a\right) = \overset{\oplus}{\bigvee_{a \in X}}(x\ and\ a)\right)$$

$$\bigwedge_{\substack{x \in Statement^{\Uparrow},\\ X \subset Statement^{\Uparrow}}}\left(\left(x\ and \bigvee_{a \in X} a\right) = \bigvee_{a \in X}(x\ and\ a)\right)$$

$$\bigwedge_{\substack{x \in Statement^{\Uparrow},\\ X \subset Statement^{\Uparrow}}}\left(\left(x\ or \bigwedge_{a \in X} a\right) = \bigwedge_{a \in X}(x\ or\ a)\right)$$

$$\bigwedge_{\substack{x \in (Plain\ Thing)^{\Uparrow},\\ X \subset (Plain\ Thing)^{\Uparrow}}}\left(\left(x\ and \bigcap_{a \in X} a\right) = \bigcap_{a \in X}(x\ and\ a)\right)$$

$$\bigwedge_{\substack{x \in (Plain\ Thing)^{\Uparrow}, \\ X \subset (Plain\ Thing)^{\Uparrow}}} \left(\left(x \cap \bigwedge_{a \in X} a \right) = \bigwedge_{a \in X} (x \cap a) \right)$$

d.) De Morgan's laws:

$$\bigwedge_{X \subset (Plain\ Thing)^{\Uparrow}} \left(\sim \bigwedge_{a \in X} a = \bigcap_{a \in X} \sim a \right)$$

$$\bigwedge_{X \subset (Plain\ Thing)^{\Uparrow}} \left(\sim \bigcap_{a \in X} a = \bigwedge_{a \in X} \sim a \right)$$

$$\bigwedge_{X \subset Statement^{\Uparrow}} \left(\sim \bigwedge_{a \in X} a = \bigvee_{a \in X} \sim a \right)$$

$$\bigwedge_{X \subset Statement^{\Uparrow}} \left(\sim \bigvee_{a \in X} a = \bigwedge_{a \in X} \sim a \right)$$

For any concrete things X and Y:

$$isTaken(X\ and\ Y) = \big(isTaken(X)\ and\ isTaken(Y)\big)$$

X equals Y if and only if X is a part of thing Y and Y is a part of thing X.

We have, that for any X and Y:

$$(the\ conjunction\ of\ X\ and\ Y) = (the\ union\ of\ X\ and\ Y) = (X \cup Y)$$

$$(the\ intersection\ of\ X\ and\ Y) = (X \cap Y)$$

$$(X\ is\ a\ part\ of\ Y) = (X\ is\ a\ subthing\ of\ Y) = (Y\ is\ a\ superthing\ of\ X) = (Y\ includes\ X)$$
$$= (X \subset Y)$$

$$(X \cup Y) = (X\ and\ Y)$$

$$\big((X \subset Y)\ and\ (X \supset Y)\big) = (X = Y)$$

And:

$$(X\ isA\ Y) = (a\ X\ is\ a\ Y \backslash X\ are\ Y) = \left(X^{\uparrow} \subset Y^{\uparrow}\right)$$

A **concrete thing** is a thing, that is not abstract, so the class of this thing has only one member.

For any concrete things X and Y:

$$(X \subset Y) = \big(Y = (X\ and\ Y)\big)$$

An **atomic thing (α)** Y is a thing, that has itself as the only nonempty part of itself, so cannot be $(X\, and\, Y)$ for X being nonempty something other than this thing. Not every abstract thing has atomic parts other than itself, so every abstract thing, that is not a conjunction of nonempty things, is atomic.

So the empty parts and their conjunctions are not atomic.

Things, that are neither atomic nor empty, are **complex**.

A **plain thing** is a thing, that is a conjunction of atomic parts. For example, concrete sets and concrete shapes are plain things. Concrete statements and concrete attributes are not plain things.

For any concrete plain things X and Y:

$$\left(X \breve{\cap} Y\right) = (X \cap Y)$$

And for any concrete attributes or statements X and Y:

$$\left(X \breve{\cap} Y\right) = (X\, or\, Y)$$

The **empty parts (zeroes)** are as follows:

$$\psi_\varnothing = \varnothing_\psi = \varnothing = (The\, Empty\, Set)$$

$$\gamma_\varnothing = \varnothing_\gamma = \varnothing = (The\, Empty\, Tuple)$$

$$\varsigma_\varnothing = \varnothing_\varsigma = (The\, Empty\, Statement)$$

$$\xi_\varnothing = \varnothing_\xi = (The\, Empty\, Attribute)$$

$$\chi_\varnothing = \varnothing_\chi = (The\, Empty\, Shape)$$

$$\kappa_\varnothing = \varnothing_\kappa = (The\, Empty\, Physcial\, Shape)$$

For any $K \in \{\psi, \gamma, \varsigma, \xi, \chi, \kappa\}$:

$$\Lambda_K = \sim\varnothing_K$$

$$(X\, isA\, K) = (X^\uparrow \subset K^\uparrow) \Rightarrow (\varnothing_X = \varnothing_K)$$

For any $X \in t^\uparrow$ for $t \in \{\psi, \gamma, \chi, \kappa\}$:

$$(X - X) = (X \cap \sim X) = (X \cap \varnothing_t) = \varnothing_t$$

$$(X \cap \Lambda_t) = (X - \varnothing_t) = X$$

For any $X \in t^\uparrow$ for $t \in \{\varsigma, \xi\}$:

$$(X\, or \sim X\, or\, X\, is\, undecidable) = (X\, or\, \varnothing_t) = \varnothing_t$$

$$(X\, or\, \Lambda_t) = (X - \varnothing_t) = X$$

For any $X \in t^{\uparrow}$ for $t \in \{\psi, \gamma, \varsigma, \xi, \chi, \kappa\}$:

$$(X \, and \, \varnothing_t) = (X - \varnothing_t) = X$$

$$(\varnothing_t \, and \, \varnothing_t) = \varnothing_t$$

$$(\Lambda_t \, and \, X) = \Lambda_t$$

$$(\Lambda_t - X) = {\sim}X$$

For any X:

$$(\Lambda \, and \, X) = \Lambda$$

The following operator ("$\pitchfork$") will be defined only for any atomic thing x and any naturally or artificially plain thing Y such that $(x \cup Y) \subset \Lambda_A^{\mho}$ and $x \in A$:

$$(x \pitchfork Y) = (x \subset Y) = (Atomic \, thing \, x \, is \, a \, part \, of \, plain \, thing \, Y)$$

And we have:

$${\sim}(x \pitchfork Y) = \left(x \pitchfork {\sim}_A^{\mho} Y\right)$$

For any things x and y:

$$(\mathbf{x} \subset \mathbf{y}) = \bigvee_{p \subset y} (p = x) = {\sim} \bigwedge_{p \subset y} (\mathbf{p} \neq \mathbf{x})$$

So for any things x and y:

$$(\mathbf{x} \subset \mathbf{y}) = \bigwedge_{q \subset x} (q \subset y) = \bigwedge_{q \subset x} \bigvee_{r \subset y} (q = r) = \bigwedge_{q \subset x} {\sim} \bigwedge_{r \subset y} (q \neq r) = {\sim} \bigvee_{\mathbf{q} \subset x} \bigwedge_{\mathbf{r} \subset y} (\mathbf{q} \neq \mathbf{r})$$

And for any plain things x and y:

$$(\mathbf{x} \subset \mathbf{y}) = \bigwedge_{q \pitchfork x} (q \pitchfork y) = \bigwedge_{q \pitchfork x} \bigvee_{r \pitchfork y} (q = r) = \bigwedge_{q \pitchfork x} {\sim} \bigwedge_{r \pitchfork y} (q \neq r) = {\sim} \bigvee_{\mathbf{q} \pitchfork x} \bigwedge_{\mathbf{r} \pitchfork y} (\mathbf{q} \neq \mathbf{r})$$

And the relation $\subset$ is transitive relation, so for any things X, Y and Z:

$$((X \subset Y) \, and \, (Y \subset Z)) \Rightarrow (X \subset Z)$$

If you want to get the **identifier** of thing X then use operator $\$X$:

$$(\$X)^{\Uparrow} = \{X\} \neq X^{\Uparrow} = \{FirstX, \ldots, LastX\}$$

For example:

$$(\$Dog)^{\Uparrow} = \{Dog\} \neq Dog^{\Uparrow} = \{FirstDog, \ldots, LastDog\}$$

And we have the following two operators:

$$\widehat{@}_K X = @_K X = \bigwedge_{e \in X, e \in K\uparrow} e = \left(\begin{array}{c} \textit{the conjunction of all elements of set } X, \\ \textit{that are members of the class of } K \end{array} \right)$$

$$X\widehat{@}_K = X @_K = \bigcup_{p \subset X, p \in K\uparrow} p = \left(\begin{array}{c} \textit{the set of all parts of } X, \\ \textit{that are members of the class of } K \end{array} \right)$$

$$\breve{@}_K X = \bigvee_{e \in X, e \in K\uparrow} e = \left(\begin{array}{c} \textit{the disjunction of all elements of set } X, \\ \textit{that are members of the class of } K \end{array} \right)$$

$$X\breve{@}_K = \bigcup_{p \supset X, p \in K\uparrow} p = \left(\begin{array}{c} \textit{the set of all superthings of } X, \\ \textit{that are members of the class of } K \end{array} \right)$$

$$\widehat{@}X = \widehat{@}_\omega X$$

$$X\widehat{@} = X\widehat{@}_\omega$$

$$\breve{@}X = \breve{@}_\omega X$$

$$X\breve{@} = X\breve{@}_\omega$$

For any thing X, that is not a concrete set (this is the only allowed exception from the general rules, because any other interpretation would have no sense):

$$\widehat{@}_K X \widehat{@}_L = \widehat{@}_K \left(X \widehat{@}_L \right)$$

$$\widehat{@} X \widehat{@} = \widehat{@} \left(X \widehat{@} \right) = X$$

$$\breve{@}_k X \breve{@}_L = \breve{@}_K \left(X \breve{@}_L \right)$$

$$\breve{@} X \breve{@} = \breve{@} \left(X \breve{@} \right) = X$$

For any concrete set X:

$$\widehat{@}_K X \widehat{@}_L = \left(\widehat{@}_K X \right) \widehat{@}_L$$

$$\widehat{@} X \widehat{@} = \left(\widehat{@} X \right) \widehat{@} \supset X$$

$$\breve{@}_k X \breve{@}_L = \left(\breve{@}_K X \right) \breve{@}_L$$

$$\breve{@} X \breve{@} = \left(\breve{@} X \right) \breve{@} \supset X$$

$$X\widehat{@} = (\textit{the power set of set } X)$$

5.4 Definitions-based and almost noncircular definition of logic, set theory and theory of things – for advanced readers

This section is for advanced readers.

In this section is the definitions-based (because it contains only definitions of terms; they use axioms but only to define these terms) definition of New Logic, Set Theory and Theory of Things based only on one very easily and intuitively understandable primitive notion: conjunction/union (of course, a universal quantifier is a conjunction). Of course, some terms must be defined before you will read and interpret this definition. It will be a symbol and its referent, a sentence and its statement and a conjunction, which is the minimal subset of logic, that is necessary before you will read and interpret logic, because without these terms you cannot even read and interpret anything, that is written down. But remember, that this definition defines logic, the minimal necessary subset of which is used in the language, in which it is written, and this is inevitable. So this subset of logic is undefinable at all any other way than in this definition of logic itself.

The symbol "Φ" is the alternative for the symbol "nothing".

So we have the following pre-assumptions about statements:

There are only two logical values: true and untrue (false).

The following true statement can be derived from the definition of implication:

$$(x\ is\ decidable\ and\ x) = (assumption\ x\ does\ not\ lead\ to\ a\ contradiction)$$

from which we have (for a proof of it see Section 5.13):

$$(x\ is\ decidable) = \sim(both\ assumptions\ x\ and \sim x\ lead\ to\ contradictions)$$

And the following new law in place of the law of excluded middle:

$$x\ or \sim x\ or\ x\ is\ undecidable$$

"and", "or", "xor" are words of the language (for multi-ary relations, that can take statements as their arguments).

A statement is something, that fulfills the following seven conditions:

$$\bigwedge_{x\ is\ a\ statement} (\sim x\ is\ a\ statement)$$

$$\bigwedge_{x\ is\ a\ statement} (x = (x\ is\ true) = (x\ is\ not\ false))$$

$$\bigwedge_{x\ is\ a\ statement} (\sim x = (x\ is\ not\ true) = (x\ is\ false))$$

$$\bigwedge_{x,y} ((x = y)\ is\ an\ elementary\ statement)$$

A conjunction ("and") of any number of statements is a statement.

A disjunction ("or") of any number of statements is a statement.

An exclusive disjunction (“xor”) of any number of statements is a statement.

A universal quantifier is a conjunction of statements.

An existential quantifier is a disjunction of statements.

Comment: Implication is a special case of equality and contextual implication (analogy to material implication) is a special case of implication.

A statement is equal to a piece of information.

Comment: Whatever this definition allows a statement to be, this new logic can process. So this is not *i*-th order logic, because it allows quantifiers over any set, even over any set of statements.

And we have the following definition of new logic:

Comment: Below it may seem like there is a cycle in definitions, because the definition of equality uses the relation of inclusion and vice versa, but this is not true, because we do not need equality at all, since we can replace it always by the conjunction of two-way inclusion. So remember, that every equality is always only a shortcut for two-way inclusion of two things. As I will prove below, the same we can replace inclusion by equality, so you can choose to use one of them and do not use the other. The same is with relation *is*.

For any X and Y:

$$(X \subset Y) = (X \; is \; a \; part \; of \; Y) = (Y \; includes \; X)$$

$$(X \cup Y) = (X \; and \; Y) = (the \; union \; of \; X \; and \; Y) = (the \; conjunction \; of \; X \; and \; Y)$$

Introduction

$$\bigwedge_X (X \; is \; a \; thing) = (Every \; thing \; is \; some \; thing) \equiv true$$

The conjunction of the following four quite obvious conditions for any concrete things X, Y and Z is the necessary and sufficient condition for relation $\subset$ to be inclusion:

$$(X = Y) = ((X \subset Y) \; and \; (Y \subset X))$$

$$(Y \subset Y)$$

$$(Y \subset (X \; and \; Y))$$

$$((X \; and \; Y) \subset Z) = ((X \subset Z) \; and \; (Y \subset Z))$$

Then:

$$(\boldsymbol{Y} = (\boldsymbol{X} \; \textbf{and} \; \boldsymbol{Y})) = (((X \; and \; Y) \subset Y) \; and \; (Y \subset (X \; and \; Y)))$$

$$= ((\boldsymbol{X} \subset \boldsymbol{Y}) \; and \; (Y \subset Y) \; and \; (Y \subset (X \; and \; Y)))$$

The part for logic

For any concrete statements X, Y and Z and . . .:

$$(X\,xor\,Y) = (\sim(X\,and\,Y)\,and \sim(\sim X\,and \sim Y))$$

$$(X\,xor\,Y\,xor\,Z) = \begin{pmatrix} \sim(X\,and\,Y\,and\,Z) \\ and \sim(\sim X\,and\,Y\,and\,Z) \\ and \sim(X\,and \sim Y\,and\,Z) \\ and \sim(X\,and\,Y\,and \sim Z) \\ and \sim(\sim X\,and \sim Y\,and \sim Z) \end{pmatrix}$$

and so on.

$$(X\,or\,Y) = \begin{pmatrix} (X\,and \sim Y)\,xor\,(\sim X\,and\,Y)\,xor\,(X\,and\,Y) \\ or\,((X\,is\,undecidable)\,and\,Y)\,or\,(X\,and\,(Y\,is\,undecidable)) \end{pmatrix}$$

The negation (~) of a concrete statement is something, that fulfills the following two conditions for any concrete statements X and Y:

$$\sim\sim X = X$$

$$\sim(X\,and\,Y) = \begin{pmatrix} (X\,and \sim Y)\,xor\,(\sim X\,and\,Y)\,xor\,(\sim X\,and \sim Y) \\ or\,((X\,is\,undecidable)\,and \sim Y)\,or\,(\sim X\,and\,(Y\,is\,undecidable)) \end{pmatrix}$$
$$= \begin{pmatrix} (\sim\sim X\,and \sim Y)\,xor\,(\sim X\,and \sim\sim Y)\,xor\,(\sim X\,and \sim Y) \\ or\,((X\,is\,undecidable)\,and \sim Y)\,or\,(\sim X\,and\,(Y\,is\,undecidable)) \end{pmatrix} = (\sim X\,or \sim Y)$$

Then for any concrete statements X and Y:

$$(X - Y) = (X -_{and} Y) = \sim(\sim X\,and\,Y) = (X\,or \sim Y)$$

$$(X -_{or} Y) = \sim(\sim X\,or\,Y) = (X\,and \sim Y)$$

And for any concrete statement X:

$$(X\,is\,a\,statement) = \bigwedge_{Y \subset X, X \neq Y} (Y\,is\,a\,statement)$$

For any concrete statements a and b:

$$(a \Rightarrow b) = (a\,implicates\,b) = (b \subset a)$$

And:

$$\varnothing_\varsigma = \bigwedge_{x \in \varnothing} (x \in \varnothing) \equiv true$$

$$\sim\varnothing_\varsigma = \Lambda_\varsigma \equiv false$$

$$\sim X = \sim(\varnothing_\varsigma\; and\, X) = (\sim\varnothing_\varsigma\; or \sim X) = (\sim\varnothing_\varsigma - X) = (\Lambda_\varsigma - X)$$

For any statement x:

$$(X\; and\; \varnothing_\varsigma) = X$$

So:

$$(X\; or\; \Lambda_\varsigma) = \sim(\sim X\; and\, \varnothing_\varsigma) = \sim\sim X = X$$

For any model m and any two tautologies q and p we have:

$$true \equiv (\varnothing = \varnothing) = (\{n: (n \in M)\; and \sim[n]^q\} = \{n: (n \in M)\; and \sim[n]^p\})$$
$$\Rightarrow \big((m \in \{n: (n \in M)\; and \sim[n]^q\}) = (m \in \{n: (n \in M)\; and \sim[n]^p\})\big)$$
$$= (\sim[m]^q = \sim[m]^p) = ([m]^q = [m]^p)$$

where M is the set of all models and $[a]^x$ is the statement of sentence x in model a.

Similarly for any logically possible situation s and any two tautologies q and p we have:

$$true \equiv (\varnothing = \varnothing) = (\{t: (t \in S)\; and \sim t^q\} = \{t: (t \in S)\; and \sim t^p\})$$
$$\Rightarrow \big((s \in \{t: (t \in S)\; and \sim t^q\}) = (s \in \{t: (t \in S)\; and \sim t^p\})\big) = (\sim s^q = \sim s^p) = (s^q = s^p)$$

where S is the set of all logically possible situations and a^x is the statement of sentence x in logically possible situation a.

So the statements of any two tautologies are equal and the empty statement is a tautological statement, so the statement of any tautology is equal to the empty statement. It is so, because the statement of every tautology by the definition of a tautology must be **unconditionally true** just like the empty statement.

And we have:

$$(\boldsymbol{Y} = (\boldsymbol{X}\; \mathbf{and}\; \boldsymbol{Y})) = \big((\boldsymbol{X} \subset \boldsymbol{Y})\; and\, (Y \subset Y)\; and\, (Y \subset (X\; and\, Y))\big)$$

where "$(Y \subset Y)$" and "$(Y \subset (X\; and\, Y))$" are tautologies, so their statements are equal to the empty statement.

So we have:

$$(\boldsymbol{X} \subset \boldsymbol{Y}) = (\boldsymbol{Y} = (\boldsymbol{X}\; \mathbf{and}\; \boldsymbol{Y}))$$

And then we have, that:

$$(a\; and\, (a \Rightarrow b)) \Rightarrow b$$

Here is the proof:

$$(a \text{ and } (a \Rightarrow b)) = \left(a \text{ and } (a = (b \text{ and } a))\right) = \left(b \text{ and } a \text{ and } (a = (b \text{ and } a))\right) \Rightarrow b$$

Q.E.D.

And for any concrete things *a*, *b* and *x*:

$$((\boldsymbol{a} \subset \boldsymbol{x}) \supset (\boldsymbol{b} \subset \boldsymbol{x})) = \left((a \subset x) = ((a \subset x) \text{ and } (b \subset x)) = ((a \cup b) \subset x)\right)$$

$$= \left((a \subset x) = ((a \cup b) \subset x)\right) = (a = (a \cup b)) = (\boldsymbol{b} \subset \boldsymbol{a})$$

For more details see Section 6.3.1.

And we have additional axiom:

For any concrete things with negation *a*, *b* and *x*:

$$\left(x \subset (a \check{\cap} b)\right) = ((x \subset b) \text{ and } (x \subset a))$$

So for any concrete things with negation *a*, *b* and *x*:

$$(\boldsymbol{a} \subset \boldsymbol{b}) = ((x \subset b) \subset (x \subset a)) = \left(((x \subset b) \text{ and } (x \subset a)) = (x \subset a)\right)$$

$$= \left(\left(x \subset (a \check{\cap} b)\right) = (x \subset a)\right) = \left((\boldsymbol{a} \check{\cap} \boldsymbol{b}) = \boldsymbol{a}\right)$$

So for any concrete statements *a* and *b* we have:

$$(a \Rightarrow b) = (a = (a \text{ and } b)) = (\sim a = (\sim a \text{ or } \sim b)) = (\sim b \Rightarrow \sim a)$$

So:

$$(a \Rightarrow b) = (\sim b \Rightarrow \sim a)$$

And we have, that:

$$(\sim b \text{ and } (a \Rightarrow b)) \Rightarrow \sim a$$

Here is the proof:

$$(\sim b \text{ and } (a \Rightarrow b)) = (\sim b \text{ and } (\sim b \Rightarrow \sim a)) = \left(\sim b \text{ and } (\sim b = (\sim a \text{ and } \sim b))\right)$$

$$= \left(\sim a \text{ and } \sim b \text{ and } (\sim b = (\sim a \text{ and } \sim b))\right) \Rightarrow \sim a$$

Q.E.D.

And we have:

$$truth = \bigwedge_{X \text{ is a decidable statement and } X} X$$

$$(a \rightarrow b) = \begin{pmatrix} a \text{ implicates } b \\ \text{in the context of truth} \end{pmatrix} = ((a \text{ and } truth) \Rightarrow b)$$

Truth is simply the conjunction of all true statements. So a statement is true if and only if it is a part of truth.

So we define **contextual implication** (the analogy to **material implication**) for any statements *a* and *b* the following way:

$$(\boldsymbol{a} \rightarrow \boldsymbol{b}) = ((\boldsymbol{a} \textbf{ and } \boldsymbol{truth}) \Rightarrow \boldsymbol{b})$$

because material implication is nothing else than the statement of the sentence "If *a* is true, then *b* is true", which is implication in the context of truth.

Then we have:

$$\begin{aligned}(\boldsymbol{a} \rightarrow \boldsymbol{b}) &= ((a \text{ and } truth) \Rightarrow b) = ((a \text{ and } truth) = (b \text{ and } a \text{ and } truth)) \\ &= \left((a \text{ and } truth) = ((b \text{ and } a \text{ and } truth) \text{ or } \sim\varnothing_{\varsigma})\right) \\ &= \left((a \text{ and } truth) = ((b \text{ and } a \text{ and } truth) \text{ or } \sim(truth \text{ or } \sim truth \text{ or } \sim a))\right) \\ &= \left((a \text{ and } truth) = ((b \text{ and } a \text{ and } truth) \text{ or } (\sim truth \text{ and } a \text{ and } truth))\right) \\ &= \left((a \text{ and } truth) = ((b \text{ or } \sim truth) \text{ and } a \text{ and } truth)\right) \\ &= ((\boldsymbol{a} \textbf{ and } \boldsymbol{truth}) \Rightarrow (\boldsymbol{b} \textbf{ or } \sim\boldsymbol{truth}))\end{aligned}$$

where "*truth or ~truth or ~a*" is, of course, a tautology, because "*~(~truth and truth and truth is decidable)*" is a tautology and (*truth is decidable*) is a part of *truth*. And the statement of a tautology is equal to the empty statement.

So:

$$\begin{aligned}(\boldsymbol{a} \rightarrow \boldsymbol{b}) &= ((a \text{ and } truth) \Rightarrow (b \text{ or } \sim truth)) = (\sim(b \text{ or } \sim truth) \Rightarrow \sim(a \text{ and } truth)) \\ &= ((\sim b \text{ and } truth) \Rightarrow (\sim a \text{ or } \sim truth)) = (\sim\boldsymbol{b} \rightarrow \sim\boldsymbol{a})\end{aligned}$$

So:

$$(a \rightarrow b) = (\sim b \rightarrow \sim a)$$

The part for set theory

$$\bigwedge_{X}(|X| = (The\ Cardinality\ of\ X))$$

Comment: The cardinality concerns every plain thing and it is equal to the number of its atomic parts. For non-plain things it is undefined.

A set is something, that fulfills the following four conditions:

$\varnothing$ is the empty set.

$$\bigwedge_{X}(\{X\}\ is\ a\ set)$$

$$\bigwedge_{Y \subset \{X\}}((Y = \{X\})\ xor\ (Y = \varnothing))$$

Comment: The above statement states, that $\{X\}$ is an atomic thing for every X.

$$(X\ is\ a\ set) = \bigwedge_{Y \subset X, X \neq Y}(Y\ is\ a\ set)$$

$$0 = |\varnothing|$$

$$\bigwedge_{X}(|\{X\}| = 1)$$

For any X:

$$(X\ is\ an\ Element\ of\ Concrete\ Set\ Y) = (X \in Y)$$

$$((X \in Y)\ and\ (Y\ is\ a\ set)) = ((\{X\} \subset Y)\ and\ (Y\ is\ a\ set))$$

For any concrete sets X, Y, Z and Q:

$$(X \cap Y) = (the\ intersection\ of\ X\ and\ Y)$$

$$((X \cap Y) = Z) = \Big(((Q \subset X)\ and\ (Q \subset Y)) = (Q \subset Z)\Big)$$

For any concrete sets A, B:

$$(A \cap B) = \{x\colon ((x \in A)\ and\ (x \in B))\}$$

$$(A \cup B) = (A\ and\ B) = \{x\colon ((x \in A)\ or\ (x \in B))\}$$

$$(A - B) = (A -_{\cup} B) = (A \cap {\sim}B) = \{x\colon ((x \in A) -_{or} (x \in B))\}$$
$$= \{x\colon ((x \in A)\ and {\sim}(x \in B))\}$$

$$(A -_{\cap} B) = (A \cup \sim B) = \{x: ((x \in A) - (x \in B))\} = \{x: ((x \in A) -_{and} (x \in B))\}$$
$$= \{x: ((x \in A)\ or \sim(x \in B))\}$$

So for any concrete set X:

$$X = \bigwedge_{Y \in X} \{Y\}$$

$$(concrete\ set\ X\ exists) = \bigwedge_{Y} ((Y \in X)\ is\ decidable)$$

For any thing $R(Y)$ and any concrete statement $S(Y)$ decidable for every Y:

$$(X = \{R(Y) : S(Y)\}) = \left(\left(R(Y) \in X\right) = S(Y)\right)$$

The part for theory of things

For any X:

$$X\& = (the\ singular\ form\ of\ X,\ if\ exists)$$
$$\&X = (the\ plural\ form\ of\ X)$$
$$X^{\uparrow} = (the\ set\ of\ all\ members\ of\ the\ class\ of\ X)$$
$$X_{\downarrow} = (the\ thing\ Y\ for\ which\ Y^{\uparrow} = X)$$

For any concrete X and any Y:

$$(X\ is\backslash are\ a\ Y) = \left((X \in Y^{\uparrow})\ and\ (X\ is\ a\ being\ or\ it\ is\ logically\ impossible,\ but\ it\ is\ not\ a\ double\ nobeing)\right)$$

For any X and Y:

$$(a\ X\ is\ a\ Y\backslash X\ are\ Y) = (X^{\uparrow} \subset Y^{\uparrow})$$

The part for additional symbols

$$\omega = (Thing)$$
$$\psi = (Set)$$
$$\Omega = \omega^{\circlearrowleft} = (the\ concrete\ set\ of\ everything) = \bigwedge_{X} \{X\} = \{X : true\}$$
$$\Psi = \psi^{\Uparrow} = (the\ concrete\ set\ of\ all\ concrete\ sets) = \bigwedge_{X \in \psi^{\Uparrow}} \{X\} = \{X : X\ is\ a\ set\}$$

5.5 Extensional definition

A list of given elements can be written as symbols of these elements separated by comma symbol. A finite set of given elements can be written as a list of these elements surrounded by the curly brackets: "{", "}".

Of course, a list the same as any other thing can be the value of a variable, but then its commas are not the commas of the outer expression. In this book If I want the commas of a list to be the commas of the outer expression, I declare a tuple in the form (x) and use its list x in the expression, for example, $f(x)$ and if I want to use list x in such a situation as a variable, then I surround it by angle brackets.

Since a set is a conjunction of beings marked out as elements, we can also write it as the conjunction of bracketed symbols of these things and then a conjunction of sets is just a set, that is the union of these sets, so e.g.: $\{a, b, c\} = (\{a\}\ \mathit{and}\ \{b\}\ \mathit{and}\ \{c\})$.

Because for any concrete sets A and B we have, that:

$$(\mathit{the\ union\ of}\ A\ \mathit{and}\ B) = (A \cup B) = (A\ \mathit{and}\ B)$$

The brackets, that separate marked out things, are necessary to not confuse different sets, that otherwise would be equal, e.g. to not confuse:

$$\{a\ \mathit{and}\ b, c, d\} = (\{a\ \mathit{and}\ b\}\ \mathit{and}\ \{c\}\ \mathit{and}\ \{d\})$$

with:

$$\{a, b, c, d\} = (\{a\}\ \mathit{and}\ \{b\}\ \mathit{and}\ \{c\}\ \mathit{and}\ \{d\})$$

Remember, that $\{x\}$ is an atomic thing for any thing x.

So we have:

$$\begin{aligned}(e \in S) &= ((\{e\} \subset S)\ \mathit{and}\ (S \in \Psi)) = ((\{e\} \pitchfork S)\ \mathit{and}\ (S \in \Psi)) \\ &= \Big((S = (\{e\}\ \mathit{and}\ S))\ \mathit{and}\ (S \in \Psi)\Big) = \Big((S = (\{e\}\ \mathit{and}\ S))\ \mathit{and}\ (S \subset \Omega)\Big)\end{aligned}$$

5.6 Other operations on things

Firstly, we have a relative complement (a subtraction) operator ("–"):

For any concrete statements a and b:

$$(a - b) = (a\ \mathit{or} \sim b)$$

For any concrete artificially or naturally plain things A and B such that $(A \cup B) \subset \Lambda_A^{\mho}$:

$$(A -_A^{\mho} B) = (A \cap \sim_A^{\mho} B)$$

Where $\sim_A^{\mho} X$ is the complement of X in universe $\mho$ for the set A of atoms. For more details see Section 9.2.4.

So for any concrete sets A and B:

$$(A - B) = (A \cap \sim B)$$

Secondly, we have an exclusive disjunction operator (“*xor*”):

I will treat the longest sequence of consecutive operators *xor* at the same level of expression as the single multi-ary generalization of exclusive disjunction operator, that is true if and only if exactly one its arguments is true, applied to things, e.g.:

$$\overset{\oplus}{\bigvee_{x \in X}} x$$

Which means: exactly one element of set X.

For any relation $R(x)$:

$$R\left(\overset{\oplus}{\bigvee_{x \in X}} x\right) = \overset{\oplus}{\bigvee_{x \in X}} R(x) = \bigvee_{x \in X}\left(R(x)\ and \bigwedge_{y \in (X-\{x\})} \sim R(y)\right) \Rightarrow \bigvee_{x \in X} R(x)$$

$$\overset{\oplus}{\bigvee_{e \in X}} R(e) \Rightarrow \bigwedge_{Y \subset X}\left(\sim \bigwedge_{e \in Y} \sim R(e) = \bigvee_{e \in Y} R(e) = \sim \bigvee_{e \in \sim_X Y} R(e) = \bigwedge_{e \in \sim_X Y} \sim R(e)\right)$$

So:

$$(x\ and\ (a_1\ xor \ldots xor\ a_n)) = ((x\ and\ a_1)\ xor \ldots xor\ (x\ and\ a_n))$$

$$(x\ or\ (a_1\ xor \ldots xor\ a_n)) = ((x\ or\ a_1)\ xor \ldots xor\ (x\ or\ a_n))$$

$$(x\ xor\ (a_1\ xor \ldots xor\ a_n)) = ((x\ xor\ a_1)\ xor \ldots xor\ (x\ xor\ a_n))$$

The rest of this section is for advanced readers.

And for the same reason we have the following rules:

$$\bigwedge_{Y \in X} \overset{\oplus}{\bigvee_{y \in Y}} y = \overset{\oplus}{\bigvee_{Z \in \left(\times_{Y \in X}(Y)\right)}} \bigwedge_{v \in \cup_{\{i, \{z\}\} \in Z}\{z\}} v$$

$$\bigvee_{Y \in X} \overset{\oplus}{\bigvee_{y \in Y}} y = \overset{\oplus}{\bigvee_{Z \in \left(\times_{Y \in X}(Y)\right)}} \bigvee_{v \in \cup_{\{i, \{z\}\} \in Z}\{z\}} v$$

And:

$$(exactly\ one\ member\ of\ the\ class\ of\ X) = \overset{\oplus}{\bigvee_{x \in X^{\uparrow}}} x$$

$$(exactly\ one\ X) = \overset{\oplus}{\bigvee_{x \in X^{\Uparrow}}} x$$

So:

$$R(exactly\ one\ member\ of\ the\ class\ of\ X) = R\left(\bigvee_{x \in X^{\uparrow}}^{\oplus} x\right) = \bigvee_{x \in X^{\uparrow}}^{\oplus} R(x)$$

$$R(exactly\ one\ X) = R\left(\bigvee_{x \in X^{\Uparrow}}^{\oplus} x\right) = \bigvee_{x \in X^{\Uparrow}}^{\oplus} R(x)$$

If you want to use a disjunction of statements, then you have to use the following formula:

$$(A\ or\ B) = \left(\begin{array}{c}(A\ and \sim B)\ xor\ (\sim A\ and\ B)\ xor\ (A\ and\ B)\\ or\ (A\ is\ undecidable\ and\ B)\ or\ (A\ and\ B\ is\ undecidable)\end{array}\right) = \sim(\sim A\ and \sim B)$$

And for any concrete statements a, b and c we have:

$$(a\ xor\ b) = ((a\ or\ b)\ and \sim(a\ and\ b)) = (\sim(\sim a\ and \sim b)\ and\ (\sim a\ or \sim b)) = (\sim a\ xor \sim b)$$

$$(a\ xor\ b) = (b\ xor\ a)$$

$$(a\ xor\ b) = ((a\ and \sim b)\ xor\ (\sim a\ and\ b)) = \sim((a\ and\ b)\ xor\ (\sim a\ and \sim b))$$

$$(a\ xor\ b\ xor\ c) = ((a\ and \sim b\ and \sim c)\ xor\ (\sim a\ and\ b\ and \sim c)\ xor\ (\sim a\ and \sim b\ and\ c))$$

and so on.

Remember, that:

$$(a\ xor\ (b\ xor\ c)) \neq (a\ xor\ b\ xor\ c)$$

$$\begin{aligned}(a\ xor\ \varnothing_t) &= ((a\ or\ \varnothing_t)\ and \sim(a\ and\ \varnothing_t)) = (\sim(\sim a\ and\ \Lambda_t)\ and \sim a) = (\sim\Lambda_t\ and \sim a)\\ &= (\varnothing_t\ and \sim a) = \sim a\end{aligned}$$

$$(a\ xor\ \Lambda_t) = ((a\ or\ \Lambda_t)\ and \sim(a\ and\ \Lambda_t)) = (a\ and \sim\Lambda_t) = (a\ and\ \varnothing) = a$$

$$\begin{aligned}(a\ xor\ b\ xor\ \Lambda_t) &= ((a\ and \sim b\ and \sim\Lambda_t)\ or\ (\sim a\ and\ b\ and \sim\Lambda_t)\ or\ (\sim a\ and \sim b\ and\ \Lambda_t))\\ &= ((a\ and \sim b\ and\ \varnothing_t)\ or\ (\sim a\ and\ b\ and\ \varnothing_t)\ or\ (\sim a\ and \sim b\ and\ \Lambda_t))\\ &= ((a\ and \sim b)\ or\ (\sim a\ and\ b)\ or\ \Lambda_t) = ((a\ and \sim b)\ or\ (\sim a\ and\ b)) = (a\ xor\ b)\end{aligned}$$

and so on.

$$\begin{aligned}(a\ xor\ b\ xor\ \varnothing_t) &= ((a\ and \sim b\ and \sim\varnothing_t)\ or\ (\sim a\ and\ b\ and \sim\varnothing_t)\ or\ (\sim a\ and \sim b\ and\ \varnothing_t))\\ &= ((a\ and \sim b\ and\ \Lambda_t)\ or\ (\sim a\ and\ b\ and\ \Lambda_t)\ or\ (\sim a\ and \sim b\ and\ \varnothing_t))\\ &= (\Lambda_t\ or\ \Lambda_t\ or\ (\sim a\ and \sim b)) = (\sim a\ and \sim b)\end{aligned}$$

and so on.

$$(a\,xor \ldots xor\,a) = a$$

And:

$$\big((a\,xor\,{\sim}a)\;or\;a\;is\;undecidable\big) = \varnothing_t$$

$$\Big(\big((a\,xor\,{\sim}a)\;or\;a\;is\;undecidable\big)\;and\;\big((b\,xor\,{\sim}b)\;or\;b\;is\;undecidable\big)\Big) = \varnothing_t$$

and so on.

And we have:

$$\left(\begin{array}{c}((a\;and\,{\sim}b)\;xor\;({\sim}a\;and\;b)\;xor\;(a\;and\;b)\;xor\;({\sim}a\;and\,{\sim}b))\\ or\;a\;is\;undecidable\;or\;b\;is\;undecidable\end{array}\right) = \varnothing_t$$

$$(a\;or\;b) = \left(\begin{array}{c}((a\;and\,{\sim}b)\;xor\;({\sim}a\;and\;b)\;xor\;(a\;and\;b))\\ or\;(a\;is\;undecidable\;and\;b)\;or\;(a\;and\;b\;is\;undecidable)\end{array}\right) = {\sim}({\sim}a\;and\,{\sim}b)$$

$$(a\,xor\,b) = \big((a\;and\,{\sim}b)\;xor\;({\sim}a\;and\;b)\big) = {\sim}\big((a\;and\;b)\;xor\;({\sim}a\;and\,{\sim}b)\big)$$

$$(a\;and\,{\sim}b) = {\sim}\left(\begin{array}{c}({\sim}a\;and\;b)\;xor\;(a\;and\;b)\;xor\;({\sim}a\;and\,{\sim}b)\\ or\;(a\;is\;undecidable\;and\;b)\;or\;({\sim}a\;and\;b\;is\;undecidable)\end{array}\right) = {\sim}({\sim}a\;or\;b)$$

$$\Lambda_t = {\sim}\left(\begin{array}{c}((a\;and\,{\sim}b)\;xor\;({\sim}a\;and\;b)\;xor\;(a\;and\;b)\;xor\;({\sim}a\;and\,{\sim}b))\\ or\;a\;is\;undecidable\;or\;b\;is\;undecidable\end{array}\right)$$

and so on.

Below for simplicity we assume, that undecidable statements do not exist:

$$\begin{aligned}\big((a\,xor\,b)\;and\;({\sim}a\,xor\,c)\big) &= \big((a\;and\,{\sim}a)\;xor\;(a\;and\;c)\;xor\;(b\;and\,{\sim}a)\;xor\;(b\;and\;c)\big)\\ &= \big({\sim}(a\;or\,{\sim}a)\;xor\;(a\;and\;c)\;xor\;(b\;and\,{\sim}a)\;xor\;(b\;and\;c)\big)\\ &= \big({\sim}\varnothing_t\;xor\;(a\;and\;c)\;xor\;(b\;and\,{\sim}a)\;xor\;(b\;and\;c)\big)\\ &= \big(\Lambda_t\;xor\;(a\;and\;c)\;xor\;(b\;and\,{\sim}a)\;xor\;(b\;and\;c)\big)\\ &= \big((a\;and\;c)\;xor\;(b\;and\,{\sim}a)\;xor\;(b\;and\;c)\big)\end{aligned}$$

$$\begin{aligned}\big((a\,xor\,b)\;or\;({\sim}a\,xor\,c)\big) &= \big((a\;or\,{\sim}a)\;xor\;(a\;or\;c)\;xor\;(b\;or\,{\sim}a)\;xor\;(b\;or\;c)\big)\\ &= \big(\varnothing_t\;xor\;(a\;or\;c)\;xor\;(b\;or\,{\sim}a)\;xor\;(b\;or\;c)\big)\\ &= \big({\sim}(a\;or\;c)\;and\,{\sim}(b\;or\,{\sim}a)\;and\,{\sim}(b\;or\;c)\big) =\\ &= \big(({\sim}a\;and\,{\sim}c)\;and\;({\sim}b\;and\;a)\;and\;({\sim}b\;and\,{\sim}c)\big)\\ &= (a\;and\,{\sim}a\;and\,{\sim}b\;and\,{\sim}c) = (\Lambda_t\;and\,{\sim}b\;and\,{\sim}c) = \Lambda_t\end{aligned}$$

5.7 New logic as a lattice and partial order by implication – for advanced readers

This section is for advanced readers.
This logic is a lattice (uncountable, complemented, complete, bounded but infinite, distributive) of concrete statements, in which the set of all concrete statements is partially ordered by implication. The set theory is also a lattice (uncountable, complemented, complete, bounded but infinite, distributive) of concrete sets, in which the set of all concrete sets is partially ordered by inclusion. And the theory of inclusion of any concrete things is a join-semilattice, the set of all concrete things is partially ordered by inclusion.

But remember, that in these lattices a conjunction of statements ("*and*", big and small "∧") and a union (big and small "∪") of sets are joins and a disjunction of statements ("*or*", big and small "∨") and an intersection of (big and small "∩") sets are meets.

Here is the proof:

1. $(\boldsymbol{x}\ \boldsymbol{and}\ \boldsymbol{x}) = x$
   Idempotency of a conjunction of concrete things
2. $(\boldsymbol{x}\ \boldsymbol{or}\ \boldsymbol{x}) = \sim(\sim\boldsymbol{x}\ \boldsymbol{and} \sim\boldsymbol{x}) = \sim\sim\boldsymbol{x} = \boldsymbol{x}$
3. $(\boldsymbol{x}\ \boldsymbol{and}\ (\boldsymbol{y}\ \boldsymbol{and}\ \boldsymbol{z})) = ((\boldsymbol{x}\ \boldsymbol{and}\ \boldsymbol{y})\ \boldsymbol{and}\ \boldsymbol{z})$
   Associativity of a conjunction of things
4. $\left(\boldsymbol{x} \breve{\cap} \left(\boldsymbol{y} \breve{\cap} \boldsymbol{z}\right)\right) = \left(\left(\boldsymbol{x} \breve{\cap} \boldsymbol{y}\right) \breve{\cap} \boldsymbol{z}\right)$

$$\left(x \breve{\cap} \left(y \breve{\cap} z\right)\right) = \sim\left(\sim x\ and \sim\left(y \breve{\cap} z\right)\right) = \sim(\sim x\ and\ (\sim y\ and \sim z))$$

$$= \sim((\sim x\ and \sim y)\ and \sim z)$$

$$= \sim\left(\sim\left(x \breve{\cap} y\right)\ and \sim z\right) = \left(\left(x \breve{\cap} y\right) \breve{\cap} z\right)$$

5. $(\boldsymbol{x}\ \boldsymbol{and}\ \boldsymbol{y}) = (\boldsymbol{y}\ \boldsymbol{and}\ \boldsymbol{x})$
   Commutativity of a conjunction of things
6. $\left(\boldsymbol{x} \breve{\cap} \boldsymbol{y}\right) = \left(\boldsymbol{y} \breve{\cap} \boldsymbol{x}\right)$

$$\left(x \breve{\cap} y\right) = \sim(\sim x\ and \sim y) = \sim(\sim y\ and \sim x) = \left(y \breve{\cap} x\right)$$

7. $\left(\boldsymbol{x}\ \boldsymbol{and}\ \left(\boldsymbol{y} \breve{\cap} \boldsymbol{x}\right)\right) = x$

$$\left(\left(x\ and \left(y \breve{\cap} x\right)\right) = x\right) = \left(x \supset \left(y \breve{\cap} x\right)\right) = ((\sim y\ and \sim x) \supset \sim x) = \varnothing_{\varsigma}$$

8. $\left(\boldsymbol{x} \breve{\cap} (\boldsymbol{y}\ \boldsymbol{and}\ \boldsymbol{x})\right) = \boldsymbol{x}$

$$\left(\left(x \breve{\cap} (y\ and\ x)\right) = x\right) = ((y\ and\ x) \supset x) = \varnothing_\varsigma$$

So it fulfills all axioms of a lattice.

It is complete, because for infinite join we have a universal quantifier and for infinite meet we have an existential quantifier.

And for every concrete statement x:

$$\sim\varnothing_\varsigma = \Lambda_\varsigma \supset x \supset \varnothing_\varsigma$$

so this new logic is bounded.

And for every concrete set x:

$$\sim\varnothing = \Omega \supset x \supset \varnothing$$

so the set theory is bounded.

And for any things:

$$\Upsilon \supset x$$

The set theory is complemented.

This new logic is complemented, but for different conditions:

$$(x\ or \sim x\ or\ x\ is\ undecidable) = \varnothing_\varsigma$$

$$(x\ and \sim x\ and\ x\ is\ decidable) = \sim(x\ or \sim x\ or\ x\ is\ undecidable) = \sim\varnothing_\varsigma = \Lambda_\varsigma$$

The definition of a union ("∪") and an intersection ("∩") and a subtraction ("–") of any sets A and B:

$$(\boldsymbol{A} \cap \boldsymbol{B}) = \{\boldsymbol{x}: ((\boldsymbol{x} \in \boldsymbol{A})\ \boldsymbol{and}\ (\boldsymbol{x} \in \boldsymbol{B}))\}$$

$$(\boldsymbol{A} \cup \boldsymbol{B}) = (\boldsymbol{A}\ \boldsymbol{and}\ \boldsymbol{B}) = \{\boldsymbol{x}: ((\boldsymbol{x} \in \boldsymbol{A})\ \boldsymbol{or}\ (\boldsymbol{x} \in \boldsymbol{B}))\}$$

$$(\boldsymbol{A} - \boldsymbol{B}) = (\boldsymbol{A} -_{\cup} \boldsymbol{B}) = (A \cap \sim B) = \{x: ((x \in A) -_{or} (x \in B))\} = \{\boldsymbol{x}: ((\boldsymbol{x} \in \boldsymbol{A})\ \boldsymbol{and} \sim(\boldsymbol{x} \in \boldsymbol{B}))\}$$

$$(\boldsymbol{A} -_{\cap} \boldsymbol{B}) = (A \cup \sim B) = \{x: ((x \in A) - (x \in B))\} = \{x: ((x \in A) -_{and} (x \in B))\}$$
$$= \{\boldsymbol{x}: ((\boldsymbol{x} \in \boldsymbol{A})\ \boldsymbol{or} \sim(\boldsymbol{x} \in \boldsymbol{B}))\}$$

De Morgan's laws:

For any concrete statements a and b:

$$\sim(\boldsymbol{a}\ \boldsymbol{and}\ \boldsymbol{b}) = \begin{pmatrix} ((\sim a\ and\ b)\ xor\ (a\ and \sim b)\ xor\ (\sim a\ and \sim b)) \\ or\ (\sim a\ and\ b\ is\ undecidable)\ or\ (a\ is\ undecidable\ and \sim b) \end{pmatrix} = (\sim\boldsymbol{a}\ \boldsymbol{or} \sim\boldsymbol{b})$$

Q.E.D.
For any concrete sets A and B we have:

$$(\sim\boldsymbol{A} \cap \sim\boldsymbol{B}) = \{x: (x \in \sim A) \; and \; (x \in \sim B)\} = \{x: \sim(\sim(x \in \sim A) \; or \sim(x \in \sim B))\}$$
$$= \{x: \sim((x \in A) \; or \; (x \in B))\} = \sim\{x: (x \in A) \; or \; (x \in B)\} = \sim(\boldsymbol{A} \cup \boldsymbol{B})$$

Q.E.D.
Distributivity:
For any concrete statements a, b and x:

$$(\boldsymbol{x} \; \boldsymbol{and} \; (\boldsymbol{a} \; \boldsymbol{or} \; \boldsymbol{b}))$$
$$= \left(x \; and \left(\begin{array}{c} ((a \; and \sim b) \; xor \; (\sim a \; and \; b) \; xor \; (a \; and \; b)) \\ or \; (a \; and \; b \; is \; undecidable) \; or \; (a \; is \; undecidable \; and \; b) \end{array}\right)\right)$$
$$= \left(\begin{array}{c} (x \; and \; ((a \; and \sim b) \; xor \; (\sim a \; and \; b) \; xor \; (a \; and \; b))) \\ or \; (x \; and \; ((a \; and \; b \; is \; undecidable) \; or \; (a \; is \; undecidable \; and \; b))) \end{array}\right)$$
$$= \left(\begin{array}{c} ((x \; and \; a \; and \sim b) \; xor \; (x \; and \sim a \; and \; b) \; xor \; (x \; and \; a \; and \; b)) \\ or \; (x \; and \; a \; and \; b \; is \; undecidable) \; or \; (x \; and \; b \; and \; a \; is \; undecidable) \end{array}\right)$$
$$= \left(\begin{array}{c} ((x \; and \; a) \; and \sim b) \; xor \; (\sim a \; and \; (x \; and \; b)) \; xor \; ((x \; and \; a) \; and \; (x \; and \; b)) \\ or \; ((x \; and \; a) \; and \; b \; is \; undecidable) \; or \; ((x \; and \; b) \; and \; a \; is \; undecidable) \end{array}\right)$$

And we have for any c:

$$x \Rightarrow (c \rightarrow x) = (c \leftrightarrow (x \; and \; c))$$

So:

$$(\boldsymbol{x} \; \boldsymbol{and} \; (\boldsymbol{a} \; \boldsymbol{or} \; \boldsymbol{b})) \leftrightarrow$$
$$\left(\begin{array}{c} (((x \; and \; a) \; and \sim (x \; and \; b)) \; xor \; (\sim (x \; and \; a) \; and \; (x \; and \; b)) \; xor \; ((x \; and \; a) \; and \; (x \; and \; b))) \\ or \; ((x \; and \; a) \; and \; (x \; and \; b) \; is \; undecidable) \; or \; ((x \; and \; b) \; and \; (x \; and \; a) \; is \; undecidable) \end{array}\right)$$
$$= ((\boldsymbol{x} \; \boldsymbol{and} \; \boldsymbol{a}) \; \boldsymbol{or} \; (\boldsymbol{x} \; \boldsymbol{and} \; \boldsymbol{b}))$$

So in any situation:

$$(x \; and \; (a \; or \; b)) \leftrightarrow ((x \; and \; a) \; or \; (x \; and \; b))$$

So:

$$(\boldsymbol{x} \; \boldsymbol{and} \; (\boldsymbol{a} \; \boldsymbol{or} \; \boldsymbol{b})) = ((\boldsymbol{x} \; \boldsymbol{and} \; \boldsymbol{a}) \; \boldsymbol{or} \; (\boldsymbol{x} \; \boldsymbol{and} \; \boldsymbol{b}))$$

and from this we have:

$$(\boldsymbol{x}\ \boldsymbol{or}\ (\boldsymbol{a}\ \boldsymbol{and}\ \boldsymbol{b})) = \sim(\sim x\ and\ (\sim a\ or\ \sim b)) = \sim((\sim x\ and\ \sim a)\ or\ (\sim x\ and\ \sim b))$$
$$= ((\boldsymbol{x}\ \boldsymbol{or}\ \boldsymbol{a})\ \boldsymbol{and}\ (\boldsymbol{x}\ \boldsymbol{or}\ \boldsymbol{b}))$$

Q.E.D.

For any concrete sets A, B and X we have:

$$(\boldsymbol{X} \cap (\boldsymbol{A} \cup \boldsymbol{B})) = \left\{x: \left((x \in X)\ and\ ((x \in A)\ or\ (x \in B))\right)\right\}$$
$$= \left\{x: \left(((x \in X)\ and\ (x \in A))\ or\ ((x \in X)\ and\ (x \in B))\right)\right\}$$
$$= ((\boldsymbol{X} \cap \boldsymbol{A}) \cup (\boldsymbol{X} \cap \boldsymbol{B}))$$
$$(\boldsymbol{X} \cup (\boldsymbol{A} \cap \boldsymbol{B})) = \sim(\sim X \cap (\sim A \cup \sim B)) = \sim((\sim X \cap \sim A) \cup (\sim X \cap \sim B))$$
$$= ((\boldsymbol{X} \cup \boldsymbol{A}) \cap (\boldsymbol{X} \cup \boldsymbol{B}))$$

Q.E.D.

But it is not Boolean algebra, because "*x or ~x*" is not a tautology and only "*x or ~x or x is undecidable*" is a tautology, since undecidable statements exist.

5.8 Ordered pair

For any plain thing X: $|X|$ is the cardinality of X, which is the number of all atomic parts of X. So for any set X: $|X|$ is the cardinality of set X.

Two plain things have the same cardinality then and only then, when you can pair (connect) each atom of the first thing with exactly one atom (that is not paired with other atom of the first thing) of the second thing and vice versa.

The cardinalities of finite sets are defined as:

$$0 = |S_0|,\ \textit{where}\ S_0 = \varnothing$$
$$1 = |S_1|,\ \textit{where}\ S_1 = (S_0 \cup \{|S_0|\}) = \{0\}$$
$$2 = |S_2|,\ \textit{where}\ S_2 = (S_1 \cup \{|S_1|\}) = \{0, 1\}$$
$$3 = |S_3|,\ \textit{where}\ S_3 = (S_2 \cup \{|S_2|\}) = \{0, 1, 2\}$$

and so on.

Now we can define ordered pair as:

$$(a, b) = \left\{\{1, \{a\}\}, \{2, \{b\}\}\right\}$$

5.9 Function definition

First of all we have to define function:

$$(f\ is\ a\ function\ with\ domain\ X\ and\ codomain\ Y) =$$

$$(f{:}X \to Y) =$$

$$\left(for\ any\ (x) \in X\ there\ is\ exactly\ one\ y \in Y\ such,\ that\ f(x) = y\right) =$$

$$\left(for\ any\ (x) \in X\ there\ is\ exactly\ one\ y \in Y\ such,\ that\ ([\![x]\!], y) \in f\right)$$

where $[\![z]\!]$ is a tuple even for z being a single element list, while (z) is equal to z for a single element list z, and both $[\![x]\!]$ and y can be a generalized tuples (so x can be a list), that are not limited to finite or even countable tuples. For more details about a generalized tuple see Section 5.56. Firstly, a single argument function is defined to further define a tuple and in consequence to define a multi-argument function. And we have:

$$(f^{\circ}g)(x) = f\left(g(x)\right)$$

For $(x) \in X$ and $y \in Y$, function f is invertible then and only then, when:

$$\left(f(x) = y\right) = \left(f^{-1}(y) = (x)\right)$$

If there is such a function for sets X and Y, then and only then:

$$|X| = |Y|$$

$$(f^{\circ}f^{-1})(x) = f^{0}(x) = (x)$$

And for any natural number n:

$$\left(\left(\overbrace{f^{\circ} \ldots {}^{\circ}f}^{n}\right)(x) = f^{n}(x) = y\right) = \left(\left(\overbrace{f^{-1\circ} \ldots {}^{\circ}f^{-1}}^{n}\right) = f^{-n}(y) = (x)\right)$$

So two sets have the same cardinality then and only then, when there is invertible function from one set to the other (function the image of which for one set is the other set).

In this book we will use anonymous functions in the following form:

$$(body) \leftarrow (arguments)$$

For example, for:

$$f(x) = \sin x$$

we have:

$$f = (\sin a) \leftarrow (a)$$

$$f(x) = (\sin a) \leftarrow (a)(x)$$

Since we have a function defined, we can define logic.

5.10 Logical values

Nowadays logic is incorrect and messy. So let me overhaul and clean up this the most fundamental in science field of study. I will explain the laws of correct thinking, that intelligent beings have been using since the very beginning, but until now could not explain, why and how they work.

$$(\mathit{the\ set\ of\ logical\ values}) = \left\{ \begin{array}{c} \mathit{true}\ (= \mathit{not\ false}), \\ \mathit{false}\ (= \mathit{not\ true} = \mathit{untrue}) \end{array} \right\}$$

true is value of truth.

$\bar{a}$ is either logical value of statement a or information, that statement a is undecidable, so we cannot decide, what is its logical value.

Since there are statements, that are undecidable, we have also constant *undecidable* as a possible result of the logical evaluation of a statement. And it is not another logical value, but only information, that a statement is undecidable. This is not just a logical value, that is neither true nor false, because the statements, that it is not true (untrue, false) and that it is not false (true) are first of all contradictory and also undecidable, so their conjunction is also undecidable. If you do not understand that, start from Section 2.1.5.

All logical rules presented in this book, are guaranteed to be true also for statements, that are undecidable, unless they are verified negatively to be incorrect in new logic.

To shorten notation:

$$\overline{\mathit{isdec}} = \left\{ \begin{array}{c} ((\mathit{true}), \mathit{true}), \\ ((\mathit{false}), \mathit{true}), \\ ((\mathit{undecidable}), \mathit{false}) \end{array} \right\}$$

$$(a\ \mathit{is\ decidable}) = \mathit{isdec}(a)$$

$$(a\ \mathit{is\ undecidable}) = {\sim}\mathit{isdec}(a)$$

Symbol *true* as an argument of logical operator in affirmative sentences is the abbreviation for “some true statement”, so in a negative sentence it is the abbreviation for “any true statement”.

Symbol *false* as an argument of logical operator in affirmative sentences is the abbreviation for "some false statement", so in a negative sentence it is the abbreviation for "any false statement".

Symbol *undecidable* as an argument of logical operator in affirmative sentences is the abbreviation for "some undecidable statement", so in a negative sentence it is the abbreviation for "any undecidable statement".

For this reason, if you want to state, that statement a is, for example, true, you can write:

$$a = true$$

where:

$$(a = true) = \bigvee_{x \in (True\,Decidable\,Statement)^{\Uparrow}} (a = x)$$

or:

$$\overline{a} = \overline{true}$$

or:

$$a \equiv true$$

But, for example, if b is false, then this is not true, that:

$$(a = b) = (a = false)$$

But this is true, that:

$$(a = b) \Rightarrow (a = false) = \bigvee_{x \in (False\,Decidable\,Statement)^{\Uparrow}} (a = x) \Rightarrow (a \leftrightarrow false) = (a \equiv false)$$

Similarly, for example, if b is false, then this is not true, that:

$$(a \Rightarrow b) = (a \Rightarrow false)$$

But this is true, that:

$$(a \Rightarrow b) \Rightarrow (a \Rightarrow false) = \bigvee_{x \in (False\,Decidable\,Statement)^{\Uparrow}} (a \Rightarrow x) \Rightarrow (a \rightarrow false)$$

$$= (a \equiv false)$$

Truth is the conjunction of all true statements. Truth is consistent and complete. Every statement, that is a part of truth, is a fact.

This is very important to understand, that we really consider whole truth as the conjunction of all statements of the theory of everything, because everything can be covered by this logic.

For that reason this logic is not based on a set of standalone axioms (all axioms are used only to define basic notions, e.g.: implication, negation), from which you can derive all provable sentences, but it concerns all statements, the number of which is far greater (e.g. the set of all statements of sentences in the form "*x* is a real number" for every real number *x* is uncountable) than the number of all possible sentences in any language.

Instead of a set of axioms this logic defines a statement. For example, how a statement can be constructed using other statements. For more details see Section 5.19.

$$\overline{negation} = \overline{\sim} = \left\{ \begin{matrix} ((false), true), \\ ((true), false), \\ ((undecidable), undecidable) \end{matrix} \right\}$$

$$\overline{\sim a} = \overline{\sim}\,\overline{a}$$

false is the logical value of the negation of truth.

The negation is a unary operator, that gives the opposite statement, that has the opposite meaning to given statement. So $\overline{negation}$ is the function, that gives the opposite value to given logical value.

Since by the definition of opposite things the opposite to the opposite of a thing is this thing itself, for any statement x:

$$\sim\sim x = x$$

The negation of *undecidable* is *undecidable*, because:

$$\sim x = (x\ is\ false)$$

where $(x\ is\ false)$ is undecidable for undecidable x.

Since trueness is the opposite of falseness and is the negation of falseness and vice versa, we have the new law of noncontradiction:

$$\sim(a\ and \sim a\ and\ a\ is\ decidable)$$

So also:

$$a\ or \sim a\ or\ a\ is\ undecidable$$

So always either a decidable statement or its negation is true.

$$(a\ is\ decidable) \Rightarrow (a\ xor \sim a) \Rightarrow (a\ or \sim a) \equiv true$$

And statement is always either decidable or not, so:

$$true \equiv (a \text{ is decidable xor } a \text{ is undecidable}) \Rightarrow (a \text{ is decidable or } a \text{ is undecidable})$$
$$\Rightarrow (a \text{ or } \sim a \text{ or } a \text{ is undecidable})$$

and we have:

$$\bigvee_{\{a,b\} \subset Statement^{\Uparrow}} \sim\left(a \Rightarrow ((a \, and \, b) = b)\right)$$

but only:

$$a \Rightarrow ((a \, and \, b) \equiv (true \, and \, b) \equiv b)$$

and:

$$\bigvee_{\{a,b\} \subset Statement^{\Uparrow}} \sim\left(\sim a \Rightarrow ((a \, or \, b) = b)\right)$$

but only:

$$\sim a \Rightarrow ((a \, or \, b) \equiv (false \, or \, b) \equiv b)$$

5.11 Basics of new logic

We have the following new law of excluded middle:

$$a \, or \sim a \, or \, (a \, is \, undecidable)$$

because for undecidable a:

$$(a \, or \sim a) \equiv undecidable$$

$$(a \, xor \sim a) \equiv undecidable$$

and the new law of noncontradiction:

$$\sim(a \, and \sim a \, and \, (a \, is \, decidable))$$

And we have also different law:

$$isdec(a) \, or \sim isdec(a)$$

Truth is consistent by its definition, since it always correctly informs about totality, that is consistent by its nature. For more details see Section 5.26.

Truth is also complete. For the appropriate proof you should see Section 5.14.

The Gödel's incompleteness theorem is disproven in this logic in Section 15.12.

So every true statement has a proof and, of course, every provable statement is true. So in general for every statement x:

$$x = (x \, is \, true) = (x \, is \, provable)$$

Since truth is consistent:

$$(x\ \mathit{is\ false}) = (x\ \mathit{is\ not\ true}) = \sim(x\ \mathit{is\ true}) = \sim x$$

$$(x\ \mathit{is\ true}) = (x\ \mathit{is\ not\ false}) = \sim(x\ \mathit{is\ false}) = \sim\sim x = x$$

So for any statement x:

$$\sim x = (x\ \mathit{is\ false})$$

And since statement $(x\ \mathit{is\ decidable})$ is decidable for any statement x:

$$(x\ \mathit{is\ decidable}) \neq (x\ \mathit{or} \sim x) = \sim(x\ \mathit{and} \sim x)$$

$$(x\ \mathit{is\ decidable}) \neq (x\ \mathit{xor} \sim x)$$

because $(x\ \mathit{or} \sim x)$ and $(x\ \mathit{xor} \sim x)$ are undecidable for undecidable x.

And we have:

$$(x\ \mathit{is\ decidable}) \Rightarrow (x\ \mathit{xor} \sim x) \Rightarrow (x\ \mathit{or} \sim x)$$

So decidability would be wrongly defined as:

$$(x\ \mathit{is\ decidable}) = \sim\big((x\ \mathit{is\ not\ provable})\ \mathit{and}\ (\sim x\ \mathit{is\ not\ provable})\big) = \sim(\sim x\ \mathit{and}\ x)$$

because $\sim(\sim x\ \mathit{and}\ x)$ is undecidable and $(x\ \mathit{is\ decidable})$ is decidable for undecidable x.

And it should be defined as follows:

$$\sim(x\ \mathit{is\ decidable}) = (x\ \mathit{is\ undecidable})$$
$$= (\mathit{assumption}\ x\ \mathit{and\ assumption} \sim x\ \mathit{both\ lead\ to\ contradictions})$$

So statement x is decidable if and only if assumption x or assumption $\sim x$ does not lead to a contradiction. So it is enough to prove, that assumption x and assumption $\sim x$ both lead to contradictions, to prove undecidability, and vice versa.

For a proof see Section 5.13.

5.12 Theory and a contradiction

A **theory** is a set of statements about some topic (some set of objects). A theory is a **true theory** if and only if all statements, that belong to it, are true.

The **fullness** (@*T*) of theory *T* is the conjunction of all statements, that belong to the theory. So the fullness of a theory is a statement. And any statement, that belongs to a theory, the same as any statement, that can be inferred from it, is a part of the fullness of the theory. A theory is true if and only if its fullness is true. The **potential** ((@*T*)@) of theory *T* is a theory, to which only all parts of the fullness of *T* belong.

A theory is **full** if and only if every part of its fullness belongs to it. A theory is **consistent** if its fullness is consistent. A statement is inconsistent if and only if it implicates some statement and its negation, so such a statement implicates its negation, so such a statement is equal to a self-contrary statement. An inconsistent statement as an assumption yields inconsistent (logically impossible) reality. This is logically impossible for it to be true, so such statement could not ever be true. The potential of a theory is a full theory. A theory is *Q*-complete if and only if all elements of set *Q* of statements can be inferred from it. A theory is **complete** if and only if all true statements can be inferred from it, so its fullness includes truth. A theory is **exhaustive** if and only if all the knowledge, that can be inferred from the definitions of its objects, is included in it. An exhaustive theory is full, but a full theory is not necessarily exhaustive.

Truth (*truth*) is the fullness of the true, consistent, full, exhaustive and complete **theory of everything** (*truth@*), so it is the fullness of the greatest true theory. The fullness of every true theory is a part of truth.

The **theoretical complement** of theory T is equal to $truth@ - T$. The **local theoretical complement** of theory T is equal to $(@T)@ - T$.

A **subtheory** of a theory is its part that is a theory. A **supertheory** of a theory is a theory that includes it. Every theory is a subtheory of the theory of everything and the theory of everything is a supertheory of every theory. Every subtheory of a true theory is a true theory and every supertheory of a false theory is false theory.

The logical values of contextual implication (that can be derived from the definition of contextual implication; for more details see Section 5.38) allow us to derive the following tautological equalities for any statement x:

$$(\sim x \; or \; x \; is \; undecidable) = \big((x \rightarrow false) \; or \; (x \rightarrow undecidable)\big) = (x \rightarrow \sim x)$$

Here is the proof:

For:

$$a = (\sim x \; or \; x \; is \; undecidable)$$

$$b = \big((x \rightarrow false) \; or \; (x \rightarrow undecidable)\big)$$

$$c = (x \rightarrow \sim x)$$

we have:

x	*a*	*b*	*c*
true	*false*	*false*	*false*
false	*true*	*true*	*true*
undecidable	*true*	*true*	*true*

So in any situation (for any x, for which the appropriate declarences are meaningful):

$$a \leftrightarrow b \leftrightarrow c$$

So in general:

$$a = b = c$$

Q.E.D.

Since *truth* cannot implicate any part of the core (the exterior of a statement is the greatest true part of it, while the core is the rest of it; for its definition see Section 5.21) of a statement, which is *untruth* for a false statement and the undecidable $\blacksquare_\varsigma$ for an undecidable statement, we have for any statement x:

$$(x \rightarrow \mathit{false}) = ((x \text{ and } \mathit{truth}) \Rightarrow \mathit{false}) = (x \Rightarrow \mathit{false})$$

$$(x \rightarrow \mathit{undecidable}) = ((x \text{ and } \mathit{truth}) \Rightarrow \mathit{undecidable}) = (x \Rightarrow \mathit{undecidable})$$

A statement is a true decidable statement if and only if the assumption, that it is true, does not lead to a contradiction.

First of all, for any statement x the assumption, that x is true, leads to a contradiction if and only if it contextually implicates some false or undecidable statement, because a true statement cannot implicate it and a false statement and an undecidable statement must contextually implicate some respectively false and undecidable statement.

Here is the proof:

We need only the following two true statements (that can be derived from the definition of contextual implication; for more details see Section 5.38) to derive conclusion that a statement leads to a contradiction if and only if it is not a decidable true statement:

1. $\sim(\mathit{true} \rightarrow \mathit{false})$
2. $\sim(\mathit{true} \rightarrow \mathit{undecidable})$

So:

$$(\mathit{assumption}\ x\ \mathit{leads\ to\ a\ contradiction}) \Leftarrow (x \rightarrow \mathit{false})$$

$$(\mathit{assumption}\ x\ \mathit{leads\ to\ a\ contradiction}) \Leftarrow (x \rightarrow \mathit{undecidable})$$

So:

$$(\mathit{assumption}\ x\ \mathit{leads\ to\ a\ contradiction}) \Leftarrow ((x \rightarrow \mathit{false})\ \mathit{or}\ (x \rightarrow \mathit{undecidable}))$$
$$= (\sim x\ \mathit{or}\ x\ \mathit{is\ undecidable})$$

So:

$$(\mathit{assumption}\ x\ \mathit{leads\ to\ a\ contradiction}) \Leftarrow (\sim x\ \mathit{or}\ x\ \mathit{is\ undecidable})$$

Above we simply ask what are the logical values of statements, that cannot be implicated by a true statement, because we want to know, when the assumption, that statement x is true, leads to a contradiction. In other words, we want to know when assumption x leads to a contradiction.

And, of course:

$$(x\ \mathit{and}\ x\ \mathit{is\ decidable}) \Rightarrow (\mathit{assumption}\ x\ \mathit{does\ not\ lead\ to\ a\ contradiction})$$

because, of course, the assumption, that a true decidable statement is true, does not lead to a contradiction, since it is decidable and true.

So:

$$(assumption\ x\ leads\ to\ a\ contradiction) \Rightarrow (\sim x\ or\ x\ is\ undecidable)$$

So, summing up:

$$(assumption\ x\ leads\ to\ a\ contradiction) = (\sim x\ or\ x\ is\ undecidable)$$

So:

$$(\boldsymbol{assumption\ x\ does\ not\ lead\ to\ a\ contradiction}) = (\boldsymbol{x\ is\ true\ and\ x\ is\ decidable})$$

So a statement leads to a contradiction if and only if it is false or undecidable and does not lead to a contradiction if and only if it is a decidable true statement.

If some assumption x, that can be inferred from statement y, leads to a contradiction, then assumption y leads to a contradiction. So if y does not lead to any contradiction, then any statement inferred from it also does not lead to any contradiction:

$$((x \rightarrow \sim x)\ and\ (y \rightarrow x)) \Rightarrow (y \rightarrow x \rightarrow \sim x \rightarrow \sim y) \Rightarrow (y \rightarrow \sim y)$$

And that is, how we can, for example, prove, that classical logic is false. Since anything inferred from true logic cannot make a contradiction, a contradiction in something inferred from given logic proves, that this logic is false.

5.13 Decidability

First of all, it has nothing to do with an independent statement and undecided logical value understood as neither true nor false or both true and false at the same time.

There are only two **logical values**: true and untrue (false). False is the opposite of true and vice versa. An **undecidable statement** is a statement, about which we cannot decide, what logical value it has, so we cannot assign a logical value to it. A **statement about the unpredictable future** is an undecidable statement, because we cannot decide what logical value it has, since both assumptions about its logical value contradict reality, because this unpredictable part of reality is simply unknown. For any statement x about the unpredictable future:

$$(x\ is\ decidable\ and\ x) \Rightarrow (x\ is\ predictable) \equiv false$$

$$(\sim x\ is\ decidable\ and\ \sim x) \Rightarrow (\sim x\ is\ predictable) \equiv false$$

where we assume that, of course, statements not about future are all predictable.

So:

$$true \equiv (x\ is\ unpredictable) \Rightarrow (x\ is\ undecidable\ or\ \sim x)$$

$$true \equiv (\sim x\ is\ unpredictable) \Rightarrow (\sim x\ is\ undecidable\ or\ x)$$

So:

$$(x \textit{ is undecidable or } {\sim}x) \textit{ and } ({\sim}x \textit{ is undecidable or } x)$$

So:

$$(x \textit{ is undecidable or } {\sim}x) \textit{ and } (x \textit{ is undecidable or } x)$$

So:

$$x \textit{ is undecidable or } ({\sim}x \textit{ and } x)$$

As you can see, x cannot be decidable, because a decidable statement cannot be true and false at the same time. So x is undecidable.

If we assume, that a statement has defined logical value if and only if it is either true or false, then the statement, that an undecidable statement has defined logical value is also undecidable. If we assume, that a statement has defined logical value if and only if it is decidable and either true or false, then the statement that a statement has defined logical value means the same as the statement that it is decidable.

Important note: It has also nothing to do with an undecidable problem, the way it is understood nowadays, because, as I have proved in the next section of this book, true or false decidable statement always has a proof, that it is respectively true or false and such a proof is enough to decide, what logical value it has. And such a proof can have even uncountable number of steps (it can be even inexpressible in any language), for which it can be impossible to give a computer algorithm, that by its nature has to have finite number of steps. It is this way, because if we understand a problem, then in our mind we can do any number of steps at a single moment, which we do, for example, in case of a proof by induction, while computer can do only finite number of steps in finite time. For more details see Chapter 7.

From the following true statement (for a proof of it see Section 5.12):

$$(x \textit{ is decidable and } x) = (\textit{assumption } x \textit{ does not lead to a contradiction})$$

we have:

$$(x \textit{ is undecidable or } {\sim}x) = (\textit{assumption } x \textit{ leads to a contradiction})$$

and by substitution ${\sim}x$ to x:

$$\begin{aligned}({\sim}x \textit{ is decidable and } {\sim}x) &= (x \textit{ is decidable and } {\sim}x)\\ &= (\textit{assumption } {\sim}x \textit{ does not lead to a contradiction})\end{aligned}$$

$$\begin{aligned}({\sim}x \textit{ is undecidable or } {\sim}{\sim}x) &= (x \textit{ is undecidable or } x)\\ &= (\textit{assumption } {\sim}x \textit{ leads to a contradiction})\end{aligned}$$

because:

$$(x \textit{ is decidable}) = ({\sim}x \textit{ is decidable})$$

We have only four cases of contradictions for any statement *x*:

1. Both assumptions *x* and ~*x* lead to contradictions, so (*x* is undecidable or ~*x*) and (*x* is undecidable or *x*). Since assumption *x* contradicts assumption ~*x*, they cannot be both true, so *x* must be **undecidable**.
2. Both assumptions *x* and ~*x* do not lead to a contradiction, so *x* is decidable and *x* and *x* is decidable and ~*x*. Since decidable statement cannot be true and false at the same time, it is **impossible**.
3. Assumption *x* does not lead to a contradiction and assumption ~*x* leads to a contradiction, so *x* is decidable and *x* is true and (*x* is undecidable or *x* is true), so *x* is a **decidable true** statement.
4. Assumption ~*x* does not lead to a contradiction and assumption *x* leads to a contradiction, so *x* is decidable and *x* is false and (*x* is undecidable or *x* is false), so *x* is a **decidable false** statement.

And we have obvious equalities:

$$(\textit{undecidable or true}) \equiv \textit{true}$$

$$(\textit{undecidable and false}) \equiv \textit{false}$$

$$(\textit{undecidable and true}) \equiv \textit{undecidable}$$

$$(\textit{undecidable or false}) \equiv \textit{undecidable}$$

$$(\textit{undecidable or undecidable}) \equiv \textit{undecidable}$$

$$(\textit{undecidable and undecidable}) \equiv \textit{undecidable}$$

The value *undecidable* is not a logical value. The negation of an undecidable statement is also undecidable, because $\sim(x \textit{ is decidable}) = \sim(\sim x \textit{ is decidable})$. So if the value *undecidable* was a logical value, then it would be its own opposite, which is impossible. So treat the value *undecidable* only as information that a statement is not decidable.

Remember, that undecidability of a statement cannot be defined as the fact of being neither true nor false, because:

$$(x \textit{ is neither true nor false}) = (\sim x \textit{ and } \sim\sim x) = (\sim x \textit{ and } x)$$
$$\equiv (\textit{undecidable and undecidable}) \equiv \textit{undecidable}$$

And it cannot be defined also as the fact of being true and false at the same time, because:

$$(x \textit{ is true and false at the same time}) = (x \textit{ and } \sim x)$$
$$\equiv (\textit{undecidable and undecidable}) \equiv \textit{undecidable}$$

The law of excluded middle is not fulfilled, because we have the following new law:

$$x \textit{ or } \sim x \textit{ or } x \textit{ is undecidable}$$

And from this also:

$$\sim(x \textit{ and } \sim x \textit{ and } x \textit{ is decidable})$$

So if x is undecidable then:

$$\begin{aligned}(x \textit{ or } \sim x \textit{ or } x \textit{ is undecidable}) &\equiv ((\textit{undecidable or undecidable}) \textit{ or true})\\ &\equiv (\textit{undecidable or true}) \equiv \textit{true}\end{aligned}$$

5.14 Provability

You do not need to read this section, if you read the book for the first time. Remember only, that for any statement *x*:

$$\boldsymbol{x = (x \textit{ is provable})}$$

which is proved in this section.

First of all, we have for any concrete statement x:

$$\boldsymbol{(x \textit{ is provable}) \Rightarrow x}$$

$$\boldsymbol{(x \textit{ is decidable}) = (\sim x \textit{ is decidable})}$$

Secondly, we have for any concrete statements a and b:

$$\boldsymbol{(a \Rightarrow b) \Rightarrow ((a \textit{ is decidable and } a) \Rightarrow (b \textit{ is decidable and } b))}$$

because a true decidable statement can have only a true decidable implication, since it includes neither *untruth* nor the core of an undecidable statement.

So from the following true statement (for a proof see Section 5.12):

$$(x \textit{ and } x \textit{ is decidable}) = (\textit{assumption } x \textit{ does not lead to a contradiction})$$

We have:

$$\begin{aligned}&(x \textit{ is provable and } (x \textit{ is provable}) \textit{ is decidable})\\ &\Rightarrow (x \textit{ and } x \textit{ is decidable and } (x \textit{ is provable}) \textit{ is decidable})\\ &\Rightarrow (x \textit{ and } x \textit{ is decidable}) = (\textit{assumption } x \textit{ does not lead to a contradiction})\end{aligned}$$

So:

$$\begin{aligned}&(x \textit{ is provable and } (x \textit{ is provable}) \textit{ is decidable})\\ &\Rightarrow (x \textit{ and } x \textit{ is decidable and } (x \textit{ is provable}) \textit{ is decidable})\\ &\Rightarrow (\textit{assumption } x \textit{ does not lead to a contradiction})\end{aligned}$$

And we have:

$$(\textit{assumption x does not lead to a contradiction})$$
$$\Rightarrow (\textit{x is provable and (x is provable) is decidable})$$
$$\Rightarrow (\textit{x and x is decidable and (x is provable) is decidable})$$

because if assumption x does not lead to a contradiction, then it is a true decidable statement (for a proof of it see Section 5.13). And this is a proof, that x is true, so x is provable.

By the way, a false assumption must implicate a false statement (at least itself), so a statement, that has only true conclusions, is true. So a false assumption always leads to a contradiction.

Summing up:

$$(\textbf{\textit{x is provable and (x is provable) is decidable}})$$
$$= (\textbf{\textit{assumption x does not lead to a contradiction}})$$

$$(\textit{x and x is decidable and (x is provable) is decidable})$$
$$= (\textit{assumption x does not lead to a contradiction})$$

From this also:

$$((\textit{x is provable}) \textit{ is undecidable or} \sim(\textit{x is provable}))$$
$$= (\textit{assumption x leads to a contradiction})$$

$$((\textbf{\textit{x is provable}}) \textbf{\textit{ is undecidable or x is undecidable or}} \sim \textbf{\textit{x}})$$
$$= (\textbf{\textit{assumption x leads to a contradiction}})$$

And by substituting $\sim x$ for x:

$$(\textbf{\textit{assumption}} \sim\textbf{\textit{x does not lead to a contradiction}})$$
$$= ((\sim\textbf{\textit{x is provable}}) \textbf{\textit{ is decidable and }} (\sim\textbf{\textit{x is provable}}))$$

$$(\textit{assumption} \sim x \textit{ does not lead to a contradiction})$$
$$= (\sim x \textit{ and} \sim x \textit{ is decidable and } (\sim x \textit{ is provable}) \textit{ is decidable})$$

$$(\textit{assumption} \sim x \textit{ leads to a contradiction})$$
$$= ((\sim x \textit{ is provable}) \textit{ is undecidable or} \sim(\sim x \textit{ is provable}))$$

$$(\textbf{\textit{assumption}} \sim\textbf{\textit{x leads to a contradiction}})$$
$$= ((\sim x \textit{ is provable}) \textit{ is undecidable or} \sim x \textit{ is undecidable or} \sim\sim x)$$
$$= ((\sim\textbf{\textit{x is provable}}) \textbf{\textit{ is undecidable or}} \sim\textbf{\textit{x is undecidable or x}})$$

And we have:

$$(x\ \textit{is provable}) \Rightarrow x$$

So:

$$(\boldsymbol{\sim x\ is\ provable}) \Rightarrow \boldsymbol{\sim x} \Rightarrow \sim(\boldsymbol{x\ is\ provable})$$

$$(x\ \textit{is provable}) \Rightarrow x \Rightarrow \sim(\sim x\ \textit{is provable})$$

So:

$$\begin{aligned}&(\boldsymbol{assumption \sim x\ does\ not\ lead\ to\ a\ contradiction})\\ &= ((\boldsymbol{\sim x\ is\ provable})\ \boldsymbol{is\ decidable\ and}\ (\boldsymbol{\sim x\ is\ provable}))\\ &\Rightarrow (\sim(\boldsymbol{x\ is\ provable})\ \boldsymbol{is\ decidable\ and} \sim(\boldsymbol{x\ is\ provable}))\end{aligned}$$

And:

$$\begin{aligned}&((x\ \textit{is provable})\ \textit{is undecidable or}\ (x\ \textit{is provable}))\\ &\Rightarrow (\sim(\sim x\ \textit{is provable})\ \textit{is undecidable or} \sim(\sim x\ \textit{is provable}))\\ &= (\textit{assumption} \sim x\ \textit{leads to a contradiction})\end{aligned}$$

We have only four cases of contradictions for any statement *x*:

1. Both assumptions *x* and ~*x* lead to contradictions, so *x* is undecidable. So we cannot decide what logical value it has, so we cannot decide between points 3 and 4, so we cannot decide whether *x* is provable or not, so (*x* is provable) is also **undecidable**.
2. Both assumptions *x* and ~*x* do not lead to a contradiction, so (*x* is provable) is decidable and (*x* is provable) and (~*x* is provable) is decidable and (~*x* is provable), so (*x* is provable) is decidable and (*x* is provable) and ~(*x* is provable) is decidable and ~(*x* is provable). Since a decidable statement cannot be true and false at the same time, it is **impossible**.
3. Assumption *x* does not lead to a contradiction and assumption ~*x* leads to a contradiction, so *x* is a decidable true statement. Then (*x* is provable) is decidable and (*x* is provable) and ((~*x* is provable) is undecidable or ~*x* is undecidable or *x*). And *x* is a decidable true statement, so (*x* is provable) is also a **decidable true** statement.
4. Assumption ~*x* does not lead to a contradiction and assumption *x* leads to a contradiction, so *x* is a decidable false statement. Then (~*x* is provable) is decidable and (~*x* is provable) and ((*x* is provable) is undecidable or *x* is undecidable or ~*x*), so ~(*x* is provable) is decidable and ~(*x* is provable) and ((*x* is provable) is undecidable or *x* is undecidable or ~*x*). And *x* is a decidable false statement, so (*x* is provable) is also a **decidable false** statement.

As you can see, in all possible cases *x* and $(x\ \textit{is provable})$ have the same logical value, so in every model or situation we have:

$$x = (x\ \textit{is true}) \leftrightarrow (x\ \textit{is provable})$$

For more details see Section 5.13.

So for any concrete statement x:

$$x = (\boldsymbol{x\ is\ true}) = (\boldsymbol{x\ is\ provable})$$

Q.E.D.

An **evidence** and a **proof** of statement y is any being x for which:

$$(x\ exists) = (x \in \mathcal{B}) \Rightarrow y$$

A proof of a statement, proves all statements, that can be inferred from the statement, because if $(a \Rightarrow b)$ then by the definition of implication any proof of a proves b.

Here is the proof:

For any concrete statements a and b and any thing x:

$$\big(((x \in \mathcal{B}) \Rightarrow a)\ and\ (a \Rightarrow b)\big) \Rightarrow ((x \in \mathcal{B}) \Rightarrow a \Rightarrow b) \Rightarrow ((x \in \mathcal{B}) \Rightarrow b)$$

For any concrete statements a and b:

$$\bigwedge_{(x \in \mathcal{B}) \Rightarrow a} ((x \in \mathcal{B}) \Rightarrow b) = \big(((x \in \mathcal{B}) \supset a) \Rightarrow ((x \in \mathcal{B}) \supset b)\big) = (a \supset b) = (a \Rightarrow b)$$

For more details see Sections 5.65 and 6.3.

Summing up, for any concrete statements a and b:

$$(a \Rightarrow b) = (any\ proof\ of\ a\ proves\ b)$$

Q.E.D.

5.15 Logical paradox, contrary and contradictory statements

A **logical paradox** is at least a contextual contrariety between decidable true statements, since being undecidable is not contrary to being undecidable. Since truth is consistent (which means, that truth is true; and also complete), because it states all about totality (and nothing else, because nothing else is there), that is consistent from its nature, there cannot be true decidable statements in a contrariety. So a logical paradox is something that fulfills the following condition for some set S of true decidable statements:

$$\bigwedge_{x \in S} x\ and \bigvee_{y \in S}\Big(\Big(\bigwedge_{z \in (S - \{y\})} z\Big) \rightarrow \sim y\Big)$$

which is impossible, because a true decidable statement cannot contextually implicate a false decidable statement, since a true decidable statement cannot implicate a false decidable statement. So logical paradoxes do not exist. The guise of a paradox is when we think that we have a contextual contrariety of decidable true statements and we do not see any error in our reasoning. So if you "proved" a logical paradox, then this is only the guise of a paradox and you only proved that these statements are not all decidable and true or there is not a contextual contrariety between them. So the guise of a logical paradox is always the sign of the mistake you made assuming, that such statements are decidable and true or that there is at least a contextual contrariety between them. So it should be always an impulse to correct errors in the attempts of definitions or in your reasoning.

For example, when we do not know about undecidable statements (in other words, when we assume, that all statements are decidable), then in the liar paradox we have:

$$x = (I\ am\ lying) = (x\ is\ false) = {\sim}x$$

$$ {\sim}x \rightarrow x$$

So x is true, because true statement cannot implicate false statement.

And:

$$x \rightarrow {\sim}x$$

So ${\sim}x$ is true, because true statement cannot implicate false statement.

And we have a contrariety:

$$x \rightarrow {\sim}({\sim}x)$$

So we have the guise of a paradox. Analogously we have the guises of all other paradoxes of classical logic. The mistake here is the assumption, that all statements are decidable. In classical logic we use this hidden assumption, so **it is a proof by a contradiction, that there are undecidable statements**, an example of which is the liar statement.

Q.E.D.

The following definitions of a contradiction and a contrariety are based on the square of opposition.

Two statements x and y contradict each other if and only if $(x = {\sim}y)$.

Statement x contradicts statement y if and only if statement y contradicts statement x, because:

$$(x = {\sim}y) = ({\sim}x = y)$$

Only the Liar statement x is self-contradictory, because it is defined as its own negation $(x = {\sim}x)$.

Two statements x and y contradict each other contextually if and only if $(x \leftrightarrow {\sim}y)$.

Statement x contradicts contextually statement y if and only if statement y contradicts contextually statement x, because:

$$(x \leftrightarrow {\sim}y) = ({\sim}x \leftrightarrow y)$$

In other words, two decidable statements contextually contradict each other if and only if they have contradictory (opposite) logical values. Any two undecidable statements contextually contradict each other, but still their logical situations ("logical values") are not contradictory.

Every undecidable statement x is contextually self-contradictory ($x \leftrightarrow \sim x$).

Two statements x and y are contrary to each other if and only if $(x \Rightarrow \sim y)$.

Statement x is contrary to statement y if and only if statement y is contrary to statement x, because:

$$(x \Rightarrow \sim y) = (y \Rightarrow \sim x)$$

Two statements x and y are contrary to each other contextually if and only if $(x \rightarrow \sim y)$.

Statement x is contrary to statement y contextually if and only if statement y is contextually contrary to statement x, because:

$$(x \rightarrow \sim y) = (y \rightarrow \sim x)$$

In other words, two decidable statements are contextually contrary to each other if and only if they have contrary logical values. Any two undecidable statements are contextually contrary to each other, but still their logical situations ("logical values") are not contrary.

Every false statemen x fulfills $(x \rightarrow \sim x)$, so it is contrary to itself contextually.

Every undecidable statement x fulfills $(x \leftrightarrow \sim x)$, because for any statement x:

$$(\textit{assumption } x \textit{ leads to a contradiction}) = (\sim x \textit{ or } x \textit{ is undecidable}) = (x \rightarrow \sim x)$$

So for any statement x:

$$(x \textit{ is undecidable}) = (\textit{both assumptions } x \textit{ and } \sim x \textit{ lead to a contradiction})$$
$$= \big((x \rightarrow \sim x) \textit{ and } (\sim x \rightarrow x)\big) = (x \leftrightarrow \sim x)$$

So:

$$(x \leftrightarrow \sim x) = (x \textit{ is undecidable})$$

As you can see, a statement is undecidable if and only if its negation is undecidable – just put $\sim x$ in place of x in the above equality:

$$(x \textit{ is undecidable}) = (x \leftrightarrow \sim x) = (\sim x \leftrightarrow x) = \big(\sim x \leftrightarrow \sim(\sim x)\big) = (\sim x \textit{ is undecidable})$$

A self-contrary statement x fulfills $(x \Rightarrow \sim x)$.

A self-contrary statement is equal to an inconsistent statement.

So every decidable self-contrary statement is false:

$$(x \Rightarrow \sim x) = \big(x \Rightarrow (\sim x \textit{ and } x) \equiv \textit{false}\big) \Rightarrow (x \Rightarrow \textit{false})$$

Of course, a self-contrary statement is not necessarily undecidable. There are self-contrary statements, that are decidable, e.g.:

$$a = (\textit{God is almighty})$$

If god is almighty (a), then he can create everything, so he can create a stone that he cannot lift and then he is not almighty ($\sim a$), because he cannot lift that stone. So $(a \Rightarrow \sim a)$.

Such a decidable statement is, of course, false, because a true decidable statement cannot implicate a false decidable statement.

Remember, that a paradox is always something impossible, that only seems to happen.

If statement x and its negation are contextually self-contrary then this statement is undecidable, because:

$$\big((x \rightarrow \sim x)\ and\ (\sim x \rightarrow x)\big) = (x \leftrightarrow \sim x)$$

and $(x \leftrightarrow \sim x)$ cannot be fulfilled for decidable statement x.

And undecidable statement $x\,(= (x\ is\ true))$ and its negation $\sim x\,(= (x\ is\ false))$ do not make a paradox, because being undecidable is not contrary to being undecidable. In other words, they do not make a paradox, because they cannot be both decidable true statements and only decidable true statements can make a paradox.

5.16 Liar paradox

The liar paradox is as follows:

$$x = (I\ am\ lying) = (x\ is\ false) = \sim x$$

We only need simple tautological assumption $(x \Rightarrow x)$ to get the following statement by the appropriate substitutions:

$$(x \Rightarrow \sim x)\ and\ (\sim x \Rightarrow x)$$

So both assumptions lead to a contradiction. And it is enough for the proof, that x is undecidable statement.

Which means, that it is not a paradox.

In other words, we have:

$$x = \sim x$$

Let us assume that x is decidable and true:

$$\boldsymbol{true} \equiv x = \sim x \equiv \boldsymbol{false}$$

So we have a contradiction.

Let us assume that x is decidable and false:

$$\boldsymbol{false} \equiv x = \sim x \equiv \boldsymbol{true}$$

So we have a contradiction.

So x cannot be decidable. So the statement of the Liar paradox must be undecidable.

Which means, that it is not a paradox.

Q.E.D.

You can do exactly the same in classical logic assuming the definition of decidability of this new logic.

All paradoxes can be solved the same way. In my book in Chapter 15 there are solutions to all important paradoxes of classical logic and to many other important problems.

5.17 Verified and not verified rules

I will mark not verified rules by tag [not verified] and verified negatively rules by tag [verified negatively] placed under them or the pairs of tags:

<not verified> *rules* </not verified>

<verified negatively> *rules* </verified negatively>

All proofs are verified by default, until they do not use unverified rules.

You can also very easily make any not verified or verified negatively rule to be correct in new logic, when you will convert all possible undecidable values to true or false values. To do it you have to replace all occurrences of any variable, that can be undecidable, say *x*, with one of two possible expressions:

Conversion	Replacement
To false	$((x \text{ is decidable}) \text{ and } x)$
To true	$\sim((x \text{ is decidable}) \text{ and } \sim x)$

So we can turn, for example:

$$(a \Rightarrow b) = (\sim b \Rightarrow \sim a)$$

into, for example:

$$\big(((a \text{ is decidable}) \text{ and } a) \Rightarrow ((b \text{ is decidable}) \text{ and } b)\big)$$
$$= \big(\sim((b \text{ is decidable}) \text{ and } b) \Rightarrow \sim((a \text{ is decidable}) \text{ and } a)\big)$$

or:

$$\big(\sim((a \text{ is decidable}) \text{ and } \sim a) \Rightarrow \sim((b \text{ is decidable}) \text{ and } \sim b)\big)$$
$$= \big(((b \text{ is decidable}) \text{ and } \sim b) \Rightarrow ((a \text{ is decidable}) \text{ and } \sim a)\big)$$

5.18 Sentence, statement, logical value

Now we have to distinguish between a sentence, a statement and its logical value.

A meaningful declarative sentence **(declarence)** is a statement expressed in some language. So a declarence is a sentence, that has its statement. It will be used in this meaning in the whole book.

A **declarence ("ζ")** is an expression of a statement (declaration), so it is by its definition always meaningful, because if it did not say anything meaningful, then indeed its statement (declaration) would not declare anything, which would be a contradiction. I use the name "declarence" and not the name "declaration", because a declaration is a statement for me, while a declarence is a sentence.

A **statement ("ς")** is information (meaning) of a meaningful declarative sentence. Different sentences can have the same information (meaning) – their statements can be equal, e.g.: (*Everything is fine*!) = (*Nothing is not fine*!). And similarly, a sentence in different contexts or said by different speakers can have different statements, e.g.: "This is my dog". A statement is either undecidable or decidable. A decidable statement is either true or false.

A decidable statement and its logical value do not ever change, so it is the same, when it is used by different speakers in different contexts. The statement of a sentence can be different, when the sentence is said by different speakers in different contexts.

The main theorem, on which whole new logic is based, is the following one: Every statement implicates only everything, that it states among alternatively other statements, that it states, where we assume, that a conjunction of statements states them all. So **every statement (a) implicates, what it states (b) among alternatively other statements ($allother_a$; $a = (b\ and\ allother_a)$), that it states, and states, what it implicates, among alternatively other statements, that it states,** So a statement implicates other statement if and only if it states this other statement among alternatively other statements, that it states.

For any concrete statements a and b this theorem can be written the following way:

$$(a \Rightarrow b) = \left(a = (b\ and\ allother_a) = (b\ and\ (b\ and\ allother_a)) = (b\ and\ a)\right)$$
$$= \left(a = (b\ and\ a)\right) = (a \supset b)$$

And it is **equivalent to the following theorem: A statement is the conjunction of all its necessary conditions.** This theorem is trivially true, because the conjunction of all necessary conditions of a statement is the conjunction of all conditions this statement fulfills, so this conjunction is also a sufficient condition for that statement, because this statement simply does not require any other conditions to be fulfilled. Simply, every necessary condition of a statement is a condition, that is required to be true for that statement to be true, so if there are not more required conditions than those, that are included in a given conjunction, then this conjunction is a sufficient condition for that statement. Since the main theorem is equivalent to this theorem, the main theorem is also trivially true.

We can derive the main theorem the following way:

The conjunction of the following five quite obvious conditions for any concrete statements X, Y and Z is the necessary and sufficient condition for relation $\Rightarrow$ to be implication:

$$\boldsymbol{(X = Y) = ((X \Leftarrow Y)\ and\ (Y \Leftarrow X))}$$

$$\boldsymbol{(Y \Leftarrow (X\ and\ Y))}$$

$$\boldsymbol{(Y \Leftarrow Y)}$$

$$\boldsymbol{((X\ and\ Y) \Leftarrow Z) = ((X \Leftarrow Z)\ and\ (Y \Leftarrow Z))}$$

$$\boldsymbol{(Z \Leftarrow (X\ or\ Y)) = ((Z \Leftarrow X)\ and\ (Z \Leftarrow Y))}$$

Then:

$$\boldsymbol{(Y = (X\ and\ Y))} = \Big((Y \Leftarrow (X\ and\ Y))\ and\ (Y \Rightarrow (X\ and\ Y))\Big)$$

$$= \Big((Y \Leftarrow (X\ and\ Y))\ and\ (X \Leftarrow Y)\ and\ (Y \Leftarrow Y)\Big) = \boldsymbol{(X \Leftarrow Y)}$$

where "$(Y \Leftarrow (X\ and\ Y))$" and "$(Y \Leftarrow Y)$" are tautologies; therefore we can eliminate them to get statement $(X \Leftarrow Y)$. For more details see Section 5.20.

So:

$$\boldsymbol{(Y = (X\ and\ Y)) = (X \Leftarrow Y)}$$

As you can see, the definition of implication includes the definition of inclusion (for the definition of inclusion of things see Section 5.4), so for any concrete statements X and Y:

$$\boldsymbol{(X \Rightarrow Y) = (X \subset Y) = (Y = (X\ and\ Y))}$$

Q.E.D.

First of all we can derive the second theorem from the main theorem. For more details see Section 9.2.2.

Secondly, from the second theorem we can derive the main theorem the following way:

$$(\boldsymbol{a} \Rightarrow \boldsymbol{b}) = \left(a = \bigwedge_{a \Rightarrow x} x = \left(b\ and \bigwedge_{(a \Rightarrow x)\ and\ (x \neq b)} x\right) = \left(b\ and\ b\ and \bigwedge_{(a \Rightarrow x)\ and\ (x \neq b)} x\right)\right.$$

$$\left. = \left(b\ and \bigwedge_{a \Rightarrow x} x\right) = (b\ and\ a)\right) = (\boldsymbol{a} = (\boldsymbol{b}\ \boldsymbol{and}\ \boldsymbol{a}))$$

So:

$$(a \Rightarrow b) = (a = (b\; and\; a))$$

Statement $(a \Rightarrow b)$ is true if and only if statement a among its necessary conditions has all necessary conditions of statement b.

So statements a and b implicate each other if and only if they have the same necessary conditions, so if and only if they are equal to each other.

So:

$$((a \Rightarrow b)\; and\; (b \Rightarrow a)) = (a = b)$$

So a statement is true if and only if all its necessary conditions are fulfilled. Every sentence said by some speaker in some context always expresses these conditions of its statement in that context and for that speaker and different quantifiers derived from two quantifiers *Some* (existential), *Any* (universal) are used in them to make such a conjunction of necessary condition more readable for a human. For more details see Sections 5.20, 5.35, 9.2 and 10.12.

Since a sentence is just a symbol it can be sometimes interpreted differently. Sentences, that are metaphoric or use non-strict (subjective) terms (e.g.: tall, hot, beautiful) or non-strict references (e.g.: I, me, he, his, they, these books, such a weather, the house), can be interpreted differently (e.g.: subjectively) in different contexts (e.g.: place in the text, situation, etc.), because they can depend on the context, or when said by different speakers, because each speaker can have different definitions of these non-strict terms. For more details see Chapter 8.

For these reasons sentences cannot be truth-bearers, because the same sentence can express at the same time some truth and some untruth depending on a context, in which it is used, or a speaker, that uses it. So you can only say, that given sentence in given situation (in given context and when said by given speaker) expresses some truth, but not, that it is true or false, because this would be a mistake. This is, of course, very common mistake (you can see it clearly, for example, in definitions of a sentence and a statement from English Wikipedia, 2017-10). Simply saying, any symbol is always true, since it is a symbol and exists. The words "true" and "untrue" ("false") have the special meaning only for statements, where they do not mean, that a statement is or is not real, but that it is respectively a part of truth or untruth is a part of it, so as a piece of information it is respectively correct or incorrect. Saying, that a sentence is false or true, is as incorrect as saying, that symbol "–2" is negative, because admittedly it is a symbol of a negative number (the same as a sentence is a symbol of a true or false statement), but not a negative symbol of a number (the same as a sentence is not a true or false symbol of a statement), which has not even any well-established meaning.

Summing up, a meaningful declarative sentence (declarence) is a symbol of a statement. So a declarence is a statement expressed in some language. Therefore, the same statement can be often expressed in the same language different ways by different sentences and the same declarence can often express different statements in different contexts and when said by different speakers. A declarence cannot be true or false in the meaning of respectively a true or false piece of information. Only the statement of a declarence can be a true or false piece of information.

And a sentence is a declarence if and only if it expresses some statement except only the sentence "This sentence expresses some truth", that is a declarence, because it expresses, what every declarence expresses, though it does not express any independent statement. In case of any statement *x* the declarence "*x* is true" expresses statement *x*, but the above declarence is different, because it demands the following definition of its statement:

$$x = (x \text{ is true})$$

But for any statement *x* it defines nothing, because it is a tautology, so it does not inform about the meaning of *x*. So such declarences (e.g. also: "The statement of this sentence is true") are the only cases, when a sentence is a declarence, but its statement is undefined. If I will say in this book, that every declarence has a statement, I will mean implicitly that there are only these few exceptions, because if I repeat these exceptions every time I write about it, then the text would be unreadable, and, of course, it is very important, that except these few cases every declarence has a statement.

A sentence expresses some truth then and only then, when its statement is true.

A statement is true then and only then, when it is a part of truth.

5.19 The definition of a statement

"and", "or", "xor" are words of the language for multi-ary relations, that can take statements as their arguments.

A statement is something, that fulfills the following seven conditions:

$$\bigwedge_{x \text{ is a statement}} (\sim x \text{ is a statement})$$

$$\bigwedge_{x \text{ is a statement}} (x = (x \text{ is true}) = (x \text{ is not false}))$$

$$\bigwedge_{x \text{ is a statement}} (\sim x = (x \text{ is not true}) = (x \text{ is false}))$$

$$\bigwedge_{x,y} ((x = y) \text{ is an elementary statement})$$

A conjunction ("and") of any number of statements is a statement.

A disjunction ("or") of any number of statements is a statement.

An exclusive disjunction ("xor") of any number of statements is a statement.

A statement is equal to a piece of information.

Implication is a special case of equality and contextual implication (analogy to material implication) is a special case of implication.

Elementary statements and their negations are not atomic, because:

$$(x=y)=((x\subset y)\ and\ (y\subset x))$$

$$(x=y)=\bigwedge_{a\ is\ not\ y}(x\neq a)$$

$$(x\neq y)=\sim(x=y)=\sim\bigwedge_{a\ is\ not\ y}(x\neq a)=\bigvee_{a\ is\ not\ y}(x=a)=\bigvee_{a\ is\ not\ y}((x\subset a)\ and\ (a\subset x))$$

$$(x\neq y)=\sim(x=y)=\sim\bigwedge_{a\ is\ not\ y}(x\neq a)=\bigvee_{a\ is\ not\ y}(x=a)=\bigvee_{a\ is\ not\ y}\ \bigwedge_{b\ is\ not\ a}(x\neq b)$$

Inclusion and thereby implication and their negations are not atomic, because:

$$(x\subset y)=(y=(x\ and\ y))=\Big((y\subset(x\ and\ y))\ and\ ((x\ and\ y)\subset y)\Big)$$

$$\sim(x\subset y)=(y\neq(x\ and\ y))=\bigvee_{a\ is\ not\ y}((x\ and\ y)=a)$$

$$=\bigvee_{a\ is\ not\ y}\Big((a\subset(x\ and\ y))\ and\ ((x\ and\ y)\subset a)\Big)$$

$$\sim(x\subset y)=(y\neq(x\ and\ y))=\bigvee_{a\ is\ not\ y}((x\ and\ y)=a)=\bigvee_{a\ is\ not\ y}\ \bigwedge_{b\ is\ not\ a}((x\ and\ y)\neq b)$$

For these reasons statements in general are not plain things.

For any concrete statements *X*, *Y* and *Z* and . . .:

$$(X\ xor\ Y)=(\sim(X\ and\ Y)\ and\ \sim(\sim X\ and\ \sim Y))$$

$$(X\ xor\ Y\ xor\ Z)=\left(\begin{array}{c}\sim(X\ and\ Y\ and\ Z)\\ and\ \sim(\sim X\ and\ Y\ and\ Z)\\ and\ \sim(X\ and\ \sim Y\ and\ Z)\\ and\ \sim(X\ and\ Y\ and\ \sim Z)\\ and\ \sim(\sim X\ and\ \sim Y\ and\ \sim Z)\end{array}\right)$$

and so on.

$$(X\ or\ Y)=\left(\begin{array}{c}(X\ and\ \sim Y)\ xor\ (\sim X\ and\ Y)\ xor\ (X\ and\ Y)\\ or\ (X\ is\ undecidable\ and\ Y)\ or\ (X\ and\ Y\ is\ undecidable)\end{array}\right)$$

The negation (~) of a concrete statement is something, that fulfills the following two conditions for any concrete statements *X* and *Y*:

$$\sim\sim X=X$$

$$\sim(X\,and\,Y) = \begin{pmatrix} (X\,and \sim Y)\,xor\,(\sim X\,and\,Y)\,xor\,(\sim X\,and \sim Y) \\ or\,(X\,is\,undecidable\,and \sim Y)\,or\,(\sim X\,and\,Y\,is\,undecidable) \end{pmatrix}$$
$$= \begin{pmatrix} (\sim\sim X\,and \sim Y)\,xor\,(\sim X\,and \sim\sim Y)\,xor\,(\sim X\,and \sim Y) \\ or\,(X\,is\,undecidable\,and \sim Y)\,or\,(\sim X\,and\,Y\,is\,undecidable) \end{pmatrix}$$
$$= (\sim X\,or \sim Y)$$

Then for any concrete statements X and Y:

$$(X - Y) = \sim(\sim X\,and\,Y) = (X\,or \sim Y)$$

and for any concrete statement X:

$$(X\,is\,a\,statement) = \bigwedge_{Y \subset X, X \neq Y} (Y\,is\,a\,statement)$$

5.20 Tautologies

A tautology (**"τ"**) is an expression of the tautological statement, which is equal to the empty statement.

The statement of a tautology is equal to the tautological statement.

Every part of the tautological statement is the tautological statement and a conjunction of the tautological statements is the tautological statement.

Pure tautologies and their partial and complete instantiations are tautologies. A pure tautology is a declarence, that has no constants that are not removable and its statement is true regardless of the values of the variables and regardless of the situation (the context and the speaker or alternatively the model) in which the declarence is said. A pure tautology has true statement regardless of the logical values of all statement variables used in it, so it has true statement also for all undecidable statements. The following examples are tautologies: $(a\,or \sim a\,or\,a\,is\,undecidable)$, "it is raining or it is not raining or the sentence "it is raining" has undecidable statement," while "it is raining or it is not raining" is not a tautology, because it is not an instantiation of a pure tautology. A tautology does not implicate anything about the logical values of the statement variables used in it, because it is true regardless of their values. It also does not implicate anything about the information it contains, because it is true regardless of this information. And, in general, a tautology does not implicate anything except any other tautology.

For example, the declarence "This is what it is" is a tautology, because it would be a complete instantiation of the sentence "$a = a$", which is a pure tautology. The declarence "$x \in \Omega$" is not a tautology, because it would be a partial instantiation of the declarence "$x \in X$", which is not a pure tautology.

A declarence, that in any situation, in which it is meaningful, has a true statement, is a tautology and, of course, a tautology in any situation, in which it is meaningful, has a true statement.

For that reason the following declarence is a tautology if and only if the symbol "+" is reserved for operations, that are commutative, which should be true:

$$x+y=y+x$$

Since the statement of any tautology is always true, it does not ever influence the logical value of any statement, the part of which it is. As a consequence of that any tautology has the empty statement, that is always true, because to be true it does not have to fulfill any condition, that could be false in an alternative reality. (An alternative reality is equal to a logically possible reality, which means that it is consistent.) The empty statement implicates only the empty statement and is implicated by any statement. This way we extract precisely the meaning of a sentence excluding not important parts, which we do not have to care about, because they are worthless as necessary conditions. We simply do not have to demand, that they are fulfilled, because we know, that they are always fulfilled, because they cannot be false. Due to that the statements of tautologies inform only, that all statements of tautologies are true.

Here is the proof:

For any model m and any two tautologies q and p we have:

$$\begin{aligned} true &\equiv (\varnothing = \varnothing) \\ &= \left(\{n: (n \in M)\ and \sim[n]^q\} = \{n: (n \in M)\ and \sim[n]^p\}\right) \\ &\Rightarrow \left(\left(m \in \{n: (n \in M)\ and \sim[n]^q\}\right) = \left(m \in \{n: (n \in M)\ and \sim[n]^p\}\right)\right) \\ &= \left(\sim[m]^q = \sim[m]^p\right) = \left([m]^q = [m]^p\right) \end{aligned}$$

where M is the set of all models and $[a]^x$ is the statement of sentence x in model a.

Similarly, for any logically possible situation s and any two tautologies q and p we have:

$$\begin{aligned} true &\equiv (\varnothing = \varnothing) \\ &= \left(\{t: (t \in S)\ and \sim t^q\} = \{t: (t \in S)\ and \sim t^p\}\right) \\ &\Rightarrow \left(\left(s \in \{t: (t \in S)\ and \sim t^q\}\right) = \left(s \in \{t: (t \in S)\ and \sim t^p\}\right)\right) \\ &= \left(\sim s^q = \sim s^p\right) = \left(s^q = s^p\right) \end{aligned}$$

where S is the set of all logically possible situations and a^x is the statement of sentence x in logically possible situation a.

So the statements of any two tautologies are equal and the empty statement is a tautological statement, so the statement of any tautology is equal to the empty statement. It is so, because the statement of every tautology by the definition of a tautology must be **unconditionally true** just like the empty statement.

And we have:

$$(X \Rightarrow Y) \Rightarrow (X \rightarrow Y)$$

$$(X = Y) \Rightarrow (X \leftrightarrow Y)$$

For any statement a and for any tautology t:

$$\boldsymbol{a} = (\boldsymbol{a\ and}\ {}^\wedge\mathbf{t}) = (\boldsymbol{a\ or} \sim {}^\wedge\mathbf{t}) = \sim(\sim\boldsymbol{a\ and}\ {}^\wedge\mathbf{t})$$

$$\sim\boldsymbol{a} = (\sim\boldsymbol{a\ and}\ {}^\wedge\mathbf{t}) = (\sim\boldsymbol{a\ or} \sim {}^\wedge\boldsymbol{t}) = \sim(\boldsymbol{a\ and}\ {}^\wedge\mathbf{t})$$

$${}^\wedge\mathrm{t} = \varnothing_\varsigma = \sim\Lambda_\varsigma = \bigwedge_{x \in \varnothing} (x \in \varnothing) \equiv true$$

$$\sim {}^\wedge\mathrm{t} = \sim\varnothing_\varsigma = \Lambda_\varsigma = \sim \bigwedge_{x \in \varnothing} (x \in \varnothing) = \bigvee_{x \in \varnothing} \sim(x \in \varnothing) \equiv false$$

Just as the empty set is a being and not just nothing, because it has its cardinality, the empty statement is a being and not just nothing, because it has its logical value.

$$\psi_\varnothing = \varnothing_\psi = \varnothing = (The\ Empty\ Set)$$

$$\varsigma_\varnothing = \varnothing_\varsigma = (The\ Empty\ Statement)$$

For any statement a:

$$a = (a\ and\ \varnothing_\varsigma)$$

So also:

$$(a\ or \sim\varnothing_\varsigma) = \sim(\sim a\ and\ \varnothing_\varsigma) = \sim\sim a = a$$

and:

$$(a\ and\ \Lambda_\varsigma) = \Lambda_\varsigma$$

$$(a\ or\ \varnothing_\varsigma) = \sim(\sim a\ and\ \Lambda_\varsigma) = \sim\Lambda_\varsigma = \varnothing_\varsigma$$

So every statement implicates the empty statement. Therefore, the negation of the empty statement implicates the negations of all statements, so it implicates all statements, so the negation of the empty statement is the conjunction of all statements:

$$(x \in Statement^{\Uparrow}) = (\sim x \in Statement^{\Uparrow})$$

$$\sim {}^\wedge\mathrm{t} = \sim\varnothing_\varsigma = \Lambda_\varsigma = \bigwedge_{x \in Statement^{\Uparrow}} x = \bigwedge_{\sim x \in Statement^{\Uparrow}} x = \bigwedge_{x \in Statement^{\Uparrow}} \sim x \equiv false$$

$${}^\wedge\mathrm{t} = \varnothing_\varsigma = \sim\Lambda_\varsigma = \sim \bigwedge_{x \in Statement^{\Uparrow}} x = \bigvee_{x \in Statement^{\Uparrow}} \sim x = \bigvee_{\sim x \in Statement^{\Uparrow}} \sim x = \bigvee_{x \in Statement^{\Uparrow}} x$$

$$= (at\ least\ one\ false\ statement\ exists)$$

$$= (at\ least\ one\ true\ statement\ exists)$$

$$= (at\ least\ one\ decidable\ statement\ exists) \equiv true$$

$$\hat{}t = \varnothing_{\varsigma} = \bigvee_{x \in Statement^{\Uparrow}} x = \bigvee_{x \in Statement^{\Uparrow}} (x \; or \sim x \; or \; x \; is \; undecidable)$$

$$= \bigvee_{x \in Statement^{\Uparrow}} \varnothing_{\varsigma} = \varnothing_{\varsigma}$$

$$\sim\hat{}t = \Lambda_{\varsigma} = \bigwedge_{x \in Statement^{\Uparrow}} x = \bigwedge_{x \in Statement^{\Uparrow}} (\sim x \; and \; x \; and \; x \; is \; decidable)$$

$$= \bigwedge_{x \in Statement^{\Uparrow}} \Lambda_{\varsigma} = \Lambda_{\varsigma}$$

So this is logically impossible, that the statement of a tautology would be false, the same as this is logically impossible, that the negation of the statement of a tautology would be true, because then all statements would be false and true at the same time, so we would have logically impossible reality. And remember, that if only this is logically impossible, that a decidable statement is false or true, then it is respectively necessarily true or necessarily false, so it is respectively the statement of a tautology or the negation of the statement of a tautology, because it is respectively true in every situation or false in every situation.

5.21 Truth and untruth

$$truth = @\{x: x \subset truth\} = @\{x: \sim x \supset \sim truth\} = @\{\sim x: x \supset untruth\}$$

$$untruth = \sim truth$$

In other words, every part of truth is a true statement, so truth itself is true, and untruth is a part of every untrue (false) statement, so untruth is untrue (false) and every statement, that includes untruth, is false.

So truth is the conjunction of all true decidable statements and untruth is the disjunction of all false decidable statements:

$$truth = \bigwedge_{x \subset truth} x$$

$$untruth = \bigvee_{x \supset untruth} x$$

So every true decidable statement as a conjunction can be only a conjunction of true decidable statements, while every false decidable statement as a disjunction can be only a disjunction of false decidable statements:

$$x = (x \subset truth) = \left(x = \bigwedge_{a \subset x \subset truth} a\right) = (truth = (x \; and \; truth)) = (x = (truth \; or \; x))$$

$$\sim x = (x \supset untruth) = \left(x = \bigvee_{a \supset x \supset untruth} a\right) = (untruth = (x \; or \; untruth))$$

$$= (x = (untruth \; and \; x))$$

You do not need read the rest of this section, if you are reading the book for the first time.

And we have, that $(truth\ or \sim truth\ or\ truth\ is\ undecidable)$ is an instantiation of tautology "*x or ~x or x is undecidable*" and *truth* includes *truth is decidable*, so:

$$\begin{aligned}(truth\ or \sim truth\ or\ truth\ is\ undecidable)\\ &= \sim(untruth\ and\ truth\ and\ truth\ is\ decidable)\\ &= \sim(truth\ and\ untruth) = \varnothing_\varsigma\end{aligned}$$

So:

$$\bigwedge_{x\ is\ a\ statement} (x \Rightarrow \sim(truth\ and\ untruth))$$

So also:

$$\begin{aligned} true \equiv & \bigwedge_{x\ is\ a\ statement} (\sim\sim(truth\ and\ untruth) \Rightarrow \sim x)\\ = & \bigwedge_{\sim x\ is\ a\ statement} ((truth\ and\ untruth) \Rightarrow x)\\ = & \bigwedge_{x\ is\ a\ statement} ((truth\ and\ untruth) \Rightarrow x)\end{aligned}$$

In other words, since *untruth* is a part of every false decidable statement, the rest of a false decidable statement must be a part of *truth*. So every false decidable statement is a conjunction of *untruth* and some true decidable statement. So every true decidable statement is a disjunction of *truth* and some false decidable statement.

And we have for any true statement *t*:

$$true \equiv (truth \Rightarrow t) = (t = (t\ or\ truth))$$

So:

$$\begin{aligned}(truth\ or \sim(truth - t)) &= (truth\ or \sim(truth\ or \sim t)) = (truth\ or\ (untruth\ and\ t))\\ &= ((truth\ or\ t)\ and\ (truth\ or\ untruth)) = ((truth\ or\ t)\ and\ \varnothing_\varsigma)\\ &= (truth\ or\ t) = t\end{aligned}$$

So:

$$\boldsymbol{t = (truth\ or \sim(truth\ or \sim t)) = (truth - (truth - t))}$$

So:

$$\sim t = \sim(truth\ or \sim(truth - t)) = \sim(truth\ or \sim(truth\ or \sim t))$$

So for any false statement *f*:

$$f = \sim(\mathit{truth}\ \mathit{or} \sim(\mathit{truth}\ \mathit{or}\ f)) = (\mathit{untruth}\ \mathit{and}\ (\mathit{truth}\ \mathit{or}\ f))$$
$$= (\mathit{untruth}\ \mathit{and}\ (\mathit{truth} - \sim f)) = (\mathit{untruth}\ \mathit{and} \sim(\mathit{untruth}\ \mathit{and} \sim f))$$

So:

$$\boldsymbol{f} = (\mathbf{untruth\ and}\ (\mathbf{truth} - \sim \boldsymbol{f})) = (\mathbf{untruth\ and} \sim(\mathbf{untruth\ and} \sim \boldsymbol{f}))$$

Q.E.D.

And, of course, we have:

$$\sim(a - b) = \sim(a -_{and} b) = \sim(a\ \mathit{or} \sim b) = (b\ \mathit{and} \sim a) = (b -_{or} a)$$

So:

$$\boldsymbol{f} = (\mathit{untruth}\ \mathit{and}\ (\mathit{truth} - \sim f)) = (\mathit{untruth}\ \mathit{and}\ (\mathit{truth}\ \mathit{or}\ f))$$
$$= (\mathit{untruth}\ \mathit{and} \sim(\mathit{untruth}\ \mathit{and} \sim f))$$
$$= \sim((\mathbf{untruth\ and} \sim \boldsymbol{f}) - \mathbf{untruth})$$

So:

$$\sim f = ((\mathit{untruth}\ \mathit{and} \sim f) - \mathit{untruth})$$

So for any true statement *t*:

$$\boldsymbol{t} = ((\mathbf{untruth\ and}\ \boldsymbol{t}) - \mathbf{untruth})$$

And we have:

$$(\mathit{truth}\ \mathit{and}\ t) = \mathit{truth}$$
$$(\mathit{truth}\ \mathit{or}\ t) = t$$
$$(\mathit{untruth}\ \mathit{and}\ f) = f$$
$$(\mathit{untruth}\ \mathit{or}\ f) = \mathit{untruth}$$

and:

$$(\mathit{truth}\ \mathit{or}\ f) = \sim(\mathit{untruth} -_{or} f)$$
$$(\mathit{truth}\ \mathit{and}\ f) = \sim(\mathit{untruth} - f) = (\mathit{truth}\ \mathit{and} \sim f\ \mathit{and}\ f\ \mathit{and}\ f\ \mathit{is\ decidable})$$
$$= (\mathit{truth}\ \mathit{and}\ \Lambda_\varsigma) = \Lambda_\varsigma$$
$$(\mathit{untruth}\ \mathit{or}\ t) = \sim(\mathit{truth} -_{or} t) = (\mathit{untruth}\ \mathit{or} \sim t\ \mathit{or}\ t\ \mathit{or}\ t\ \mathit{is\ undecidable})$$
$$= (\mathit{untruth}\ \mathit{or}\ \varnothing_\varsigma) = \varnothing_\varsigma$$
$$(\mathit{untruth}\ \mathit{and}\ t) = \sim(\mathit{truth} - t)$$

And we have:

$$truth \Rightarrow (truth\ is\ decidable)$$

So:

$$(truth\ and\ untruth) = (truth\ and\ untruth\ and\ truth\ is\ decidable) = \Lambda_{\varsigma}$$

A **trivial statement** is a statement, that any statement implicates. The empty statement is the only trivial statement, because for any trivial statement q:

$$\mathbf{true} \equiv \left(\bigwedge_{p\ is\ a\ statement} (p \Rightarrow q) \right) \Rightarrow \left(\varnothing_{\varsigma} = \bigvee_{p\ is\ a\ statement} p \Rightarrow q \right) = (\varnothing_{\varsigma} \Rightarrow q)$$

$$= \left(\varnothing_{\varsigma} = (q\ and\ \varnothing_{\varsigma}) = q \right) = (\mathbf{q} = \varnothing_{\varsigma})$$

The **exterior** $x_{\#}$ of a statement x is the greatest true part of it, while the **core** $x_{\blacksquare}$ of x is the rest of it.

So the core $x_{\blacksquare}$ of statement x is the conjunction of all its nonempty parts, that do not implicate any nonempty true statement. The core of a true statement is the empty statement. The core of a decidable false statement is, as is proved earlier in this section, equal to *untruth*. The core of an undecidable statement is also an undecidable statement, because if it was decidable, then the source statement would be decidable too. The exterior $x_{\#}$ of statement x is the conjunction of all its parts, that are true. The exterior of a decidable true statement is equal to it.

For any statement x:

$$x_{\blacksquare} = \bigwedge_{(x \Rightarrow a)\ and\ \sim \mathrm{V}_{p\ is\ a\ true\ statement}(a \Rightarrow p \neq \varnothing_{\varsigma})} a$$

$$x_{\#} = \bigwedge_{(x \Rightarrow a)\ and\ a} a$$

So:

$$x = (x_{\blacksquare}\ and\ x_{\#})$$

$$truth_{\blacksquare} = \varnothing_{\varsigma}$$

$$untruth_{\blacksquare} = untruth$$

$$truth_{\#} = truth$$

$$untruth_{\#} = \varnothing_{\varsigma}$$

For any false statement f:

$$f_{\#} = (truth - \sim f)$$

For any undecidable statement x:

$$(x \to \sim x) = ((truth\ and\ x) \Rightarrow (truth\ and \sim x))$$
$$= ((truth\ and\ x_{\#}\ and\ x_{\blacksquare}) \Rightarrow truth\ and\ (\sim x)_{\#}\ and\ (\sim x)_{\blacksquare})$$
$$= \left((truth\ and\ x_{\blacksquare}) \Rightarrow (truth\ and\ (\sim x)_{\blacksquare})\right) = (x_{\blacksquare} \Rightarrow (\sim x)_{\blacksquare})$$

where *truth* cannot contain any part of $x_{\blacksquare}$ and any part of $(\sim x)_{\blacksquare}$

So:

$$true \equiv (x \leftrightarrow \sim x) = ((x \to \sim x)\ and\ (\sim x \to x)) = \left((x_{\blacksquare} \Rightarrow (\sim x)_{\blacksquare})\ and\ ((\sim x)_{\blacksquare} \Rightarrow x_{\blacksquare})\right)$$
$$= (x_{\blacksquare} = (\sim x)_{\blacksquare})$$

For any undecidable statement x:

$$(truth\ and\ untruth) = \Lambda_{\varsigma} \Rightarrow x_{\blacksquare}$$

But *truth* cannot implicate any nonempty part of $x_{\blacksquare}$, so *untruth* must implicate whole $x_{\blacksquare}$:

$$untruth \Rightarrow x_{\blacksquare}$$

So *untruth* includes the conjunction of all cores of undecidable statements:

$$untruth \Rightarrow \bigwedge_{u\ is\ an\ undecidable\ statement} u_{\blacksquare} = \blacksquare_{\varsigma}$$

So every false statement f includes it too:

$$f \Rightarrow untruth \Rightarrow \blacksquare_{\varsigma}$$

Let us take the core of any undecidable statement x and make false statement y of it by the conjunction with some false statement f:

$$y = (x_{\blacksquare}\ and\ f)$$

As a false statement it must include $\blacksquare_{\varsigma}$, but we also know, that it includes only $x_{\blacksquare}$ and $\blacksquare_{\varsigma}$ is not smaller than $x_{\blacksquare}$, so $\blacksquare_{\varsigma} = x_{\blacksquare}$ for any undecidable statement x. So all cores of undecidable statements are the same and equal to $\blacksquare_{\varsigma}$.

Q.E.D.

And we define *undecidables* as follows:

$$undecidables = \bigwedge_{x \text{ is an undecidable statement}} x = \bigwedge_{x \text{ is an undecidable statement}} (x_\blacksquare \text{ and } x_\#)$$
$$= \left(\bigwedge_{x \text{ is an undecidable statement}} x_\blacksquare \text{ and } \bigwedge_{x \text{ is an undecidable statement}} x_\# \right)$$
$$= (\blacksquare_\varsigma \text{ and } truth)$$

And we define *falses* as follows:

$$falses = \sim trues = \bigwedge_{x \text{ is a false statement}} x = \bigwedge_{x \text{ is a false statement}} (x_\blacksquare \text{ and } x_\#)$$
$$= \left(\bigwedge_{x \text{ is a false statement}} x_\blacksquare \text{ and } \bigwedge_{x \text{ is a false statement}} x_\# \right)$$
$$= (untruth \text{ and } truth) = \Lambda_\varsigma = (untruth \text{ and } \blacksquare_\varsigma \text{ and } truth)$$
$$= (untruth \text{ and } undecidables)$$

$$trues = \sim falses = \sim\Lambda_\varsigma = \varnothing_\varsigma = \sim \bigwedge_{x \text{ is a false statement}} x = \bigvee_{x \text{ is a false statement}} \sim x$$
$$= \bigvee_{x \text{ is a true statement}} x$$

The rest of this section is for advanced readers.

Any nonempty part of $\blacksquare_\varsigma$ cannot be true (because it is the core of a statement) and cannot be false (because otherwise $\blacksquare_\varsigma$ would be false, which is not true), so must be undecidable. But if there was an undecidable part of $\blacksquare_\varsigma$ different from $\blacksquare_\varsigma$, then it would have smaller core, which is impossible, because all undecidable statements have the same core. So $\blacksquare_\varsigma$ is indivisible.

The only false implication of *untruth* is *untruth*, because for any decidable false statement f, that is a conclusion of *untruth*, we have:

$$true \equiv (untruth \Rightarrow f) \Rightarrow (untruth = (untruth \text{ and } f) = f) \Rightarrow (f = untruth)$$

And *untruth* does not have any nonempty true implication, since it is a core of any false statement (the only false statement, that implicates *truth*, is Λ_ς, the negation of which is empty).

So the only other implication of *untruth* is $\blacksquare_\varsigma$, but still $\blacksquare_\varsigma \neq untruth$, because $\sim(\blacksquare_\varsigma \Rightarrow untruth)$, since $\blacksquare_\varsigma$ is not a false decidable statement.

So:

$$(\sim\blacksquare_\varsigma \Rightarrow truth)$$

which means, that $(\sim\blacksquare_\varsigma)_\blacksquare = \blacksquare_\varsigma \neq \sim\blacksquare_\varsigma$.

So we have:

$$untruth = (untruth\ and\ \blacksquare_\varsigma) = (untruth -_{or} \sim\blacksquare_\varsigma)$$

So this is the only way *untruth* can be expressed as a conjunction of different nonempty statements. So also $(truth\ or \sim\blacksquare_\varsigma)$ is the only way *truth* can be expressed as a disjunction of different nonempty statements.

And:

$$truth = (truth\ or \sim\blacksquare_\varsigma) = (truth - \blacksquare_\varsigma)$$

Now let us prove, that the liar statement implicates some nonempty true statement, so it is not equal to the core of an undecidable statement.

Assume x is the liar statement, so:

$$x = \sim x$$

And x must implicate some nonempty true statement, because otherwise $\blacksquare_\varsigma = x = \sim x = \sim\blacksquare_\varsigma$, which is impossible, because then $untruth \Rightarrow \blacksquare_\varsigma = \sim\blacksquare_\varsigma \Rightarrow truth$, so then also every false statement would implicate every true statement, which is not true.

Q.E.D.

Here is another proof:

Assume x is the liar statement, so:

$$x = \sim x$$

And for some true statement t we have:

$$x = (\blacksquare_\varsigma\ and\ t) = \sim x = (\sim\blacksquare_\varsigma\ or \sim t)$$

And we have:

$$(x\ is\ decidable) \Rightarrow (x\ or \sim x)$$

So:

$$\sim(x\ or \sim x) = (x\ and \sim x) \Rightarrow \sim(x\ is\ decidable) = (x\ is\ undecidable)$$

Then we have the following nonempty true implication of x, that is implicated as a true statement only by the liar statement:

$$x = (x\ is\ false) = (x\ and\ x) = (x\ and \sim x) \Rightarrow (x\ is\ undecidable) \equiv true$$

And $\sim((x\ is\ undecidable) \Rightarrow x)$.

So $t \Rightarrow (x\ is\ undecidable) \Rightarrow (x\ is\ a\ statement)$, where "*x is undecidable*" and "*x is a statement*" are not tautologies.

So t is not empty and is divisible.

Q.E.D.

In other words, $\blacksquare_\varsigma \neq \sim\blacksquare_\varsigma$, because $(\blacksquare_\varsigma = \sim\blacksquare_\varsigma)$ implicates $(\blacksquare_\varsigma = x)$, since then $\blacksquare_\varsigma$ and x would have the same definition, which as proved above is not true.

Q.E.D.

This can be proved in a much simpler way as follows:

We have that:

$$\sim\blacksquare_\varsigma \Rightarrow truth \Rightarrow (\sim\blacksquare_\varsigma)_\#$$

So:

$$\sim\blacksquare_\varsigma = (\sim\blacksquare_\varsigma\ \boldsymbol{and\ truth}) = ((\sim\blacksquare_\varsigma)_\blacksquare\ and\ (\sim\blacksquare_\varsigma)_\#\ and\ truth) = (\blacksquare_\varsigma\ \boldsymbol{and\ truth}) \neq \blacksquare_\varsigma$$

because $\sim(\blacksquare_\varsigma \Rightarrow truth)$, by the definition of a core.

Q.E.D.

So we have:

$$\sim\blacksquare_\varsigma = (\sim\blacksquare_\varsigma\ and\ truth) = (\blacksquare_\varsigma\ and\ truth)$$

So we can verify that:

$$\blacksquare_\varsigma = \sim\sim\blacksquare_\varsigma = \sim(\sim\blacksquare_\varsigma\ and\ truth) = (\blacksquare_\varsigma\ or\ untruth)$$

which is true, because $(untruth \Rightarrow \blacksquare_\varsigma)$.

And we have:

$$\blacksquare_\varsigma = \sim\sim\blacksquare_\varsigma = \sim(\blacksquare_\varsigma\ and\ truth) = (\sim\blacksquare_\varsigma\ or\ untruth)$$

So also:

$$\begin{aligned}\blacksquare_\varsigma = (\sim\blacksquare_\varsigma\ \boldsymbol{or\ untruth}) &= ((\blacksquare_\varsigma\ and\ truth)\ or\ untruth)\\ &= ((\blacksquare_\varsigma\ or\ untruth)\ and\ (truth\ or\ untruth))\\ &= ((\blacksquare_\varsigma\ or\ untruth)\ and\ \varnothing_\varsigma) = (\blacksquare_\varsigma\ \boldsymbol{or\ untruth})\end{aligned}$$

So:

$$\blacksquare_\varsigma = (\sim\blacksquare_\varsigma\ or\ untruth) = (\blacksquare_\varsigma\ or\ untruth)$$

And we have:

$$\sim\blacksquare_\varsigma = (\blacksquare_\varsigma\ and\ truth) = undecidables$$

So:

$$\sim undecidables = \blacksquare_\varsigma = (\sim\blacksquare_\varsigma\ or\ untruth) = (\blacksquare_\varsigma\ or\ untruth)$$

$$\sim falses = trues = \sim\Lambda_\varsigma = \varnothing_\varsigma = \sim(untruth\ and\ undecidables) = (\blacksquare_\varsigma\ or\ truth)$$

$$(\sim undecidables\ and\ truth) = (\blacksquare_\varsigma\ and\ truth) = undecidables$$

$$\begin{aligned}(trues\ and\ untruth) &= ((\blacksquare_\varsigma\ or\ truth)\ and\ untruth)\\ &= ((\blacksquare_\varsigma\ and\ untruth)\ or\ (truth\ and\ untruth))\\ &= ((\blacksquare_\varsigma\ and\ untruth)\ or\ \Lambda_\varsigma) = (\blacksquare_\varsigma\ and\ untruth) = untruth\end{aligned}$$

$$\begin{aligned}(undecidables\ or\ truth) &= ((\blacksquare_\varsigma\ and\ truth)\ or\ truth) = ((\blacksquare_\varsigma\ or\ truth)\ and\ truth)\\ &= (trues\ and\ truth)\end{aligned}$$

Summing up, we have:

$$undecidables = (\blacksquare_\varsigma\ and\ truth)$$

$$\sim undecidables = \blacksquare_\varsigma$$

$$(\blacksquare_\varsigma\ or\ truth) = \varnothing_\varsigma$$

And for any undecidable statement u we have:

$$true \equiv (undecidables \Rightarrow u) = (u = (undecidables\ or\ u))$$

So:

$$\begin{aligned}&(undecidables\ or \sim(undecidables\ or \sim u))\\ &= (undecidables\ or\ (\sim undecidables\ and\ u))\\ &= (undecidables\ or\ (\blacksquare_\varsigma\ and\ u))\\ &= ((undecidables\ or\ u)\ and\ (undecidables\ or\ \blacksquare_\varsigma))\\ &= (u\ and\ ((\blacksquare_\varsigma\ and\ truth)\ or\ \blacksquare_\varsigma)) = (u\ and\ (\blacksquare_\varsigma\ and\ (\blacksquare_\varsigma\ or\ truth)))\\ &= (u\ and\ (\blacksquare_\varsigma\ and\ \varnothing_\varsigma)) = (u\ and\ \blacksquare_\varsigma) = u\end{aligned}$$

So:

$$\begin{aligned}\boldsymbol{u} &= (\textbf{undecidables} - (\textbf{undecidables} - \boldsymbol{u}))\\ &= (\textbf{undecidables or} \sim(\textbf{undecidables or} \sim\boldsymbol{u}))\end{aligned}$$

We can also derive it the simplest way:

$$\begin{aligned}\boldsymbol{u} &= (undecidables\ or\ u) = (undecidables\ or \sim\sim u)\\ &= (undecidables\ or \sim(undecidables\ or \sim u))\\ &= (\textbf{undecidables} - (\textbf{undecidables} - \boldsymbol{u}))\end{aligned}$$

And we have also:

$$\sim u = (undecidables\ or \sim (undecidables\ or\ u))$$

So:

$$\boldsymbol{u = \sim (undecidables\ or \sim (undecidables\ or\ u))}$$

5.22 Relation between mind and logic

All statements and all other external things are outside any mind and in a mind only their more or less precise reflections exist, where the senses of living beings are simply measuring tools with a specific measurement error. If a statement was something, that exists in someone's mind, then this statement could not exist in someone else's mind, because these minds are separated (do not have intersection) beings. Truth and untruth are directly accessible to every mind and are the same for every mind, so statements as subthings (parts) of truth and superthings of untruth exist outside mind with their logical values. And every mind has some reflections of statements, that are elementary elements of someone's knowledge (consciousness) and thoughts and can exist alone in the mind, but can be also expressed in some language, in which case they exist before they are expressed, and can be created in a mind as an interpretation of someone's else expression of thoughts. Thoughts are operations of a mind on its reflections of statements, that discover truth and untruth.

Every time you focus your intellect on some part of totality, you consider some truth. So truth is, what hypothetical all-knowing, pure and perfect mind would know about all parts of totality. The existence of truth is equivalent to the potential of totality to be understood by minds of intelligent beings. So it is a necessary condition for the existence of intelligence and a sufficient condition for the possibility of the existence of intelligent beings. The potential of totality to be understood comes from its nonrandomness – from the fact, that everything has its cause and reason.

For example, to correctly answer a question one has to understand its meaning, so he has to have in his mind the reflection of the statement implied by the question, because in many cases the correct answer is not contained in any way in the sentence of the question alone. So, for example, AI, that can correctly answer questions (this section was written before the invention of LLMs – Large Language Models), for example, about any image, has to operate on some reflections of statements, that are elementary elements of thoughts, so without a doubt it should be called an artificial reason. The existence of such an AI itself is an experimental proof, that AI can really understand something, because the reflection of a statement is the reflection of a meaning and when you know the meaning of something, then, of course, you understand it. So *Chinese room argument* is wrong, because, as has been just shown, understanding proved to be possible by the application of very sophisticated and mathematical rules on symbols, that are realized by artificial neural networks. By the way, that is, why the knowledge of mathematics, and especially logic, can improve thinking. But do not get me wrong – the man in the Chinese room does not understand Chinese just as a processor in a computer machine would not understand it, but the whole room understands it.

5.23 Meaningful and nonmeaningful sentences, decidable and undecidable statements

A basic declarative sentence is a syntactically (grammatically) correct sentence, that consists of the predicate and at least the implied subject.

So a basic declarative sentence always expresses some relation between some things.

A basic declarence is a semantically correct (meaningful, sensible) basic declarative sentence, so it is such a basic declarative sentence, that the relation used in it is defined (as either undecidable or either true or false) for the arguments passed to this relation in this sentence. In other words, a basic declarative sentence is a basic declarence if and only if the arguments passed to the relation used in it are from the domain of the relation treated as a logical function.

A basic declarence is always meaningful (has a statement), because any relation is defined for any arguments from its domain.

A declarence, that uses logical operators on the statements of basic declarences, is not basic and is also meaningful. So a declarence, that uses quantifiers *Some* and *Any*, is not basic, because the quantifier *Some* is a disjunction of the statements of declarences and the quantifier *Any* is a conjunction of the statements of declarences.

And you have to remember, that some relations are opposite to each other only for some domain, e.g.:

$$(\textit{The girl is undressed}) = \sim(\textit{The girl is dressed})$$

But:

$$\begin{aligned} \textit{false} \equiv (\textit{Number one is undressed}) &\neq \sim(\textit{Number one is dressed}) \\ &= (\textit{Number one is not dressed}) \equiv \textit{true} \end{aligned}$$

because number one is not undressed, because it does not have a body, since it is not a physical object in the first place, and it is not dressed for the same reason. That is, why relations $\textit{isDressed}(X)$ and $\textit{isUndressed}(X)$ are opposite to each other only for being X, that has a body, e.g. the girl.

All relations of a natural language are defined for all things, so, for example, the sentence "The Sun is smiling" $(= \textit{isSmiling}(\textit{the Sun}))$ is meaningful, but its statement is false, since the Sun cannot smile. So in natural language all syntactically correct sentences are also semantically correct and thereby meaningful.

A sentence, that is a declarence, in given context and said by given speaker always has a statement, that is either decidable or undecidable.

A decidable statement is either true or false, but it does not mean that a statement, that is either true or false, is decidable. So for any statement x:

$$(x\ \textit{is decidable}) \Rightarrow (x\ \textit{xor} \sim x) \Rightarrow (x\ \textit{or} \sim x)$$

But:

$$\sim((x\ \textit{xor} \sim x) \Rightarrow (x\ \textit{is decidable}))$$

$$\sim((x\ \textit{or} \sim x) \Rightarrow (x\ \textit{is decidable}))$$

because, in case of undecidable x, true statement $(x\ \textit{is undecidable})$ cannot implicate undecidable statements $\sim(x\ \textit{or} \sim x)$ and $\sim(x\ \textit{xor} \sim x)$.

We can decide, what logical value (either true or false) a statement has if and only if it is decidable. Otherwise it is in the logical situation, in which we cannot decide, what logical value it has.

A decidable statement **("ν")** is not a statement about the unpredictable future and does not depend on the logical value of an undecidable statement. So a decidable statement is not an element of a cycle of dependencies or an element of an infinite chain of dependencies, because if the logical value of its statement depends on the logical values of some statements, then they are all decidable, so the directed graph of dependencies has only finite paths (so does not have any cycles), at the ends of which are decidable statements, that do not depend on any other statements. So we can decide, what is the logical value of such a statement. That is, why it is decidable.

The above definition of a decidable statement proves, what are undecidable statements and why they are undecidable, so using it we can always very easily recognize, whether given sentence has a decidable statement.

Remember, that a **recurrent definition** is not incorrect for the reason that it can be called a circular definition. For example, the following definitions are circular, but are correct (and are widely used in programing languages), because they define, what they intended to define, not contradictorily and completely:

A **node** is something, that can be connected to other <u>nodes</u>.

A **list** is a node, that can be connected to a <u>list</u>.

A **graph** is a node, that can be connected to many <u>graphs</u>.

A **tree** is a node, that can be connected to many <u>trees</u>, that are not yet connected to the tree.

And so on.

An **undecidable statement ("μ")** that is not a statement about the unpredictable future has a recurrent definition, in which it is defined as a statement, the logical value of which depends on the logical value of an <u>undecidable statement</u>, where any statement in a cycle of dependencies and in the infinite chain of dependencies is undecidable, because both structures have no end. This is a proper and successful recurrent definition (if you think that this is not a proper definition, then ask yourself whether it defines an undecidable statement completely and not contradictorily, as a single rooted graph of dependencies of logical values of statements the root of which is an element of a cycle of undecidable statements or an element of an infinite chain of undecidable statements). So an undecidable statement depends on a statement, that is an element of a cycle of dependencies (like, for example, in the Liar paradox and the Card paradox; for more details see Sections 15.2 and 15.3), or is an element of an infinite chain of dependencies (like, for example, in the Yablo's paradox; for more details see Section 15.25), that, of course, does not have a decidable statement at the end, because it does not have the end in the first place. That is, why such a statement is undecidable. In other words, to answer the question about the logical value of such a statement, you have to first answer the question about the logical value of the same or another statement, which you cannot ever accomplish, because there is always another such statement, the value of the current statement depends on, and so on to infinity. So, as a result, you cannot answer the question about the logical value of an undecidable statement. In yet another words, in an infinite chain of dependencies and in a cycle we cannot answer the question about the logical value of any statement, because for any statement there is always another statement, the question about logical value of which we must answer before we will be able to answer the previous question and so on without end to infinity, so it is impossible to answer the question about the logical values of these statements. Additionally, in case of a cycle to answer the question about the logical value of a statement, you must first answer the same question and so on to infinity, which is also impossible, because then you would have to answer the question before you can answer the same question.

For a proof, that those are not incorrect (nonmeaningful) sentences see Section 5.25.

Statement $f(a)$ depends on the logical value of statement a if and only if $f(unknown) = unknown$. For more details see Section 5.30.

A conjunction of statements is undecidable if and only if except undecidable statements there are only true statements in this conjunction. A disjunction of statements is undecidable if and only if except undecidable statements there are only false statements in this disjunction.

The statement of a self-referential declarence, the logical value of which depends on its logical value, is also undecidable by this definition, because to answer the question, what is the logical value of such a statement, you first need to answer the same question, which is not possible, so you cannot answer this question at all. The simplest example of such a statement is the statement of the sentence from Liar paradox, that can be defined as $x = (x\ is\ false)$ or equivalently as $x = (x\ is\ not\ true)$. For more details see Section 5.29.

But remember, that a self-referential declarence, the question about the logical value of the statement of which can be reduced to an implication, that its logical value is true, if its logical value is true, or exactly the same, that its logical value is false, if its logical value is false, does not have a statement at all, because both these conditions are always true for any statement. So, for example, equality $x = (x\ is\ true)$, the same as equivalent equality $x = (x\ is\ not\ false)$, is true for every statement x, so does not define any independent statement.

We cannot decide, what is the logical value of an undecidable statement, because every statement about the logical value of the undecidable statement is also undecidable by the definition of an undecidable statement. You cannot even say, that an undecidable statement x is neither true nor false $((\sim a\ and\ \sim\sim a) = (\sim a\ and\ a) \equiv undecidable)$ or both true and false at the same time $((a\ and\ \sim a) \equiv undecidable)$, because the statements of such declarences are undecidable.

A part of the future is unpredictable if and only if someone's free will is among all its causes. I am not deciding whether free will exists. I included it in the theory in case it turns out that it exists. If free will does not exist, then there are no potential beings, so some of them are future beings and the rest of them are impossible beings.

We cannot define, what is the logical value of the statement of a declarence about the unpredictable future, because such a future is not determined, so it is yet undecided, what will happen; thus such a statement can yet be both either true or false, until this future will be predictable or it will eventually happen.

For:

$$Y = (set\ of\ all\ statements,\ on\ the\ logical\ value\ of\ which\ logical\ value\ of\ x\ depends)$$

we have:

$$(x\ is\ decidable) = \bigwedge_{y \in Y} (y\ is\ decidable)$$

$$(x\ is\ undecidable) = \sim(x\ is\ decidable) = \sim \bigwedge_{y \in Y} (y\ is\ decidable) = \bigvee_{y \in Y} \sim(y\ is\ decidable)$$

$$= \bigvee_{y \in Y} (y\ is\ undecidable)$$

Of course, a nonmeaningful sentence, e.g.: $Cat \in Dog$, does not have any meaning, so does not have any statement.

A nonmeaningful sentence does not have any meaning, because it does not have any correct interpretation, that would define its meaning.

A syntactically incorrect sentence is not meaningful.

A syntactically correct sentence is not meaningful if and only if it is a basic nonmeaningful sentence or its logical value depends on a basic nonmeaningful sentence.

A syntactically correct not meaningful sentence does not have a statement, so as an expression it will return an undefined statement, where the undefined statement was meant to be the statement of a declarative sentence, but is not a statement, since the attempt to define it failed. So if x is an undefined statement:

$$\sim x \textit{ is undefined}$$

$$(x \Rightarrow y) \textit{ is undefined}$$

$$(y \Rightarrow x) \textit{ is undefined}$$

$$(x = y) \textit{ is undefined}$$

$$(x \rightarrow y) \textit{ is undefined}$$

$$(y \rightarrow x) \textit{ is undefined}$$

$$(x \textit{ xor } y) \textit{ is undefined}$$

$$((x \textit{ or } y) \textit{ is defined}) = (y \textit{ and } y \textit{ is decidable})$$

$$((x \textit{ and } y) \textit{ is defined}) = (\sim y \textit{ and } y \textit{ is decidable})$$

The above possibilities of the use of undefined statement are important, because some relations may be not defined for some arguments and then they become undefined statements, that are used in formulas, that are still meaningful and about which we want to know, what their logical values are (whether they are either undecidable or either true or false). Such formulas are still meaningful, because their not meaningful parts are unimportant, since the logical values of the statements of these formulas do not depend on the logical values of the statements of these unimportant parts.

If the assumption, that a statement is true, does not lead to a contradiction and the assumption, that it is false, leads to a contradiction or the assumption, that it is false, does not lead to a contradiction and the assumption, that it is true, leads to a contradiction, then the statement is decidable. Here is an example of such a statement:

$$x = (x \textit{ is decidable})$$

We have, that:

$$\sim((x \textit{ is decidable}) \Rightarrow (x \textit{ is not true}))$$

So:

$$\sim(x \Rightarrow (x \textit{ is not true}))$$

So the assumption, that x is true, does not lead to a contradiction.

On the other hand assume that $\sim x$. Then:

$$\textit{false} \equiv x = (x \textit{ is decidable}) \equiv \textit{true}$$

which is impossible, so x is not false.

Another such example is $x = (x \textit{ is undecidable})$.

The simplest example of an undecidable statement is the statement of the sentence "The statement of this sentence is false" $\left(\text{“}\hat{x}\text{”}\right)$:

$$x = (x\ is\ false)$$

for which we have:

$$x = (x\ is\ false) = \sim x$$

So $x = \sim x$.

We can also demonstrate, that the negation of x has the same definition:

$$p = \sim x = \sim(x\ is\ false) = (x\ is\ not\ false) = (\sim x\ is\ false) = (p\ is\ false)$$

The next example can be as follows: "The statement of the next sentence is true" $\left(\text{“}\hat{x}\text{”}\right)$, "The statement of the previous sentence is false" $\left(\text{“}\hat{y}\text{”}\right)$:

$$x = (y\ is\ true) = \sim y$$

$$y = (x\ is\ false) = \sim x$$

So:

$$x = \big((x\ is\ false)\ is\ true\big) = (x\ is\ false) = \sim x$$

$$y = \big((y\ is\ true)\ is\ false\big) = (y\ is\ false) = \sim y$$

So both statements are equal to the liar statement.

5.24 Proof, that zero-valued and double-valued statements are not the solution to the problem

In the first place, a statement cannot be neither true nor false, because these assumptions are contradictory. For the same reason a statement cannot be true and false at the same time. But let us forget that for a moment.

Here is a proof, that the statement of the Liar's sentence "The statement of this sentence is false" is neither zero-valued (neither true nor false) nor double-valued (true and false at the same time):

$$x = (x\ is\ false)$$

Assume, that x is zero valued. Then:

$$zero \equiv x = (x\ is\ false) \equiv false$$

But the same statement cannot be not false and false at the same time.

Q.E.D.

Assume, that x is double valued. Then:

$$double \equiv x = (x\ is\ false) \equiv true$$

But the same statement cannot be false and not false at the same time.

Q.E.D.

Some people may believe, that a statement, that other statement is not true or is not false, does not mean, that it is respectively false or true, because it can be sometimes undecidable.

So assume, that we have undecidable statement x and that the statement, that it is not true is true, then:

$$y = (x\ is\ not\ true) \equiv true$$

$$\sim y = (x\ is\ true) = x \equiv undecidable$$

But the negation of a true statement is always false, so cannot be undecidable, so we have a contradiction. In other words, y states, that x is false, and is undecidable.

Q.E.D.

Now assume, that we have undecidable statement x and that the statement, that it is not false is true, then:

$$y = (x\ is\ not\ false) \equiv true$$

$$\sim y = (x\ is\ false) = \sim x \equiv undecidable$$

But the negation of a true statement is always false, so cannot be undecidable, so we have a contradiction. In other words, y states, that x is true, and is undecidable.

Q.E.D.

5.25 Proof, that those are not incorrect sentences

A declarence is a correct declarence if and only if it has a statement. In other words, every statement expressed in a language is a correct declarence, because then it expresses some information, that is contained in that statement.

The simplest proof is as follows:

Assume, that the sentence "This sentence expresses a false statement" does not express any statement. Then automatically this sentence expresses a false statement, because if this sentence did not express any statement, then thereby it would not express a false statement. So we have a contradiction. So the sentence has its statement.

Q.E.D.
Let us assume, that all sentences, that supposedly have undecidable statement, are incorrect sentences.

Then the sentence "The statement of this sentence is false" ($= X$), which is the simplest example of a sentence, that supposedly has undecidable statement, is incorrect declarence, so it does not have a statement. And "The statement of the sentence X is false" is also an incorrect declarence, but since we know, that the sentence X does not have a statement, we can evaluate the logical value of the statement of the second sentence to be false, because the statement of X cannot be false, because it does not have a statement. So the second sentence has a statement, so it is a correct declarence. So we have a contradiction. So the assumption, that all declarences, that supposedly have undecidable statements, are incorrect declarences, is false.

Q.E.D.
In other words, all declarences have their statements, so they are correct.

5.26 Consistency of totality and reality

Totality is the conjunction of all things, definition of each of which by its definition fulfills condition, that is not contrary to any other definition of a thing, because any attempt to define something to succeed cannot be self-contrary or contrary to any other definition of a thing and otherwise such a thing is defined only as a nobeing, that had this attempt of definition. That is, why totality (and thereby also reality, that is included in it) is consistent.

More information about a contradiction and a contrariety you can find in Section 5.15 and Section 5.12.

Of course, there are statements, that are contrary, but a statement is not its attribute, because its attribute is in such a case only the appropriate conversion of the statement, that this statement states, what it states. In other words, a statement (that can be converted to an attribute), that some statement states something, is not equal to, what this statement states. So such statements do not make contrary attributes, even if these statements, that are "mentioned" by them, are contrary.

5.27 Counterfactual noncontradiction

More information about a contradiction and a contrariety can be found in Section 5.15 and Section 5.12.

In general remember, that:

$$(a \, and \, x) \Rightarrow \sim(a \Rightarrow \sim x)$$

$$a \Rightarrow (x \rightarrow \sim(a \Rightarrow \sim x))$$

$$\boldsymbol{a \Rightarrow ((x \Rightarrow \sim a) \rightarrow \sim x)}$$

which is **the law of counterfactual noncontradiction**, that says, that for counterfactual assumption a every statement, that implicates $\sim a$, has to be assumed (sometimes also counterfactually) to be false.

And by putting $\sim y$ in place of x we have the same different way:

$$\boldsymbol{a \Rightarrow ((a \Rightarrow y) \rightarrow y)}$$

So all implications of counterfactual assumption must also be assumed (sometimes also counterfactually) to be true.

Remember, that if you counterfactually assume something, then you treat it as some truth of a new consistent counterfactual reality, so this is not counterfactual in this counterfactual reality and all implications of such an assumption must also be parts of truth in this reality, because truth is consistent by its definition in a consistent reality. A consistent reality is the same as a logically possible reality, which is the same as an alternative reality. And a reality is consistent if and only if its truth is consistent. In a logically possible reality all tautologies must have the empty statement, which is true by its definition.

This is rather an open question, whether every alternative reality is alternatively possible. For more details see Section 5.1.4.

If a counterfactual assumption is a self-contrary statement, which means, that it is logically impossible, that this assumption is true, then it yields inconsistent reality.

In case of a not removable contradiction in a correct indirect proof an assumption proves to be false or undecidable because of its false or undecidable consequence, and thereby it is the negation of a true statement or is an undecidable statement.

For example, if a square circle existed, then beings could have contrary attributes, because:

$$\begin{aligned}&(\textit{a square circle exists})\\ &\quad\Rightarrow (\textit{a square circle, that has contrary attributes, is a being})\\ &\quad\Rightarrow (\textit{there is a being, that has contrary attributes})\\ &\quad\Rightarrow (\textit{beings can have contrary attributes})\end{aligned}$$

which is contrary to truth, so it is a not removable contradiction. It means, that such a reality simply is not our reality, because both realities differ from each other due to the contradiction. Simply, some statement, that is true in such a reality is false in our reality and that is the difference.

Remember also, that we have:

$$(a \Rightarrow b) = (b\ is\ true\ in\ all\ logically\ possible\ situations\ in\ which\ a\ would\ be\ true)$$
$$= (b\ is\ true\ in\ all\ consistent\ realities\ in\ which\ a\ is\ true)$$

Since a counterfactual assumption always is contrary to truth of some reality, it always yields a reality, that is different from previous reality, so it always makes a not removable contradiction. For this reason an indirect proof is always backed by a not removable contradiction and every such contradiction in an indirect proof always proves some indirect proof inside a main proof or proves a main indirect proof itself, while a removable contradiction proves only, that at least two assumptions, that are not intentionally counterfactual to each other, are contrary to each other, so to keep some of them you must only resign from these, that are contrary to them. So you can always remove such a contradiction, therefore it does not prove anything.

So not removable contradictions have important role in indirect proofs, because using them we can simply eliminate some dead ends inside a proof and by drawing conclusions from it we can narrow down the possible ways to the final solution.

This is very important to not mistake removable contradiction for not removable one, because removable contradictions are always only the sign of an error in a proof and not a proof of anything, so to prove something you must first eliminate them all.

In any consistent reality any false decidable statement, that is not self-contrary, can be assumed to be true in new consistent reality. A decidable counterfactual assumption x yields logically impossible (inconsistent) reality if and only if it is logically impossible, that x is true, so if and only if x is self-contrary statement, where the negation of the statement of a tautology is self-contrary. So a counterfactual assumption can be self-contrary and then it yields inconsistent reality.

5.28 Solving the system of statements about the unpredictable future

Any statements about the unpredictable future can be treated as undecided yet statements, because they are undecidable only temporarily, so they will not be undecidable forever. They are not counterfactual assumptions, because we will not assume, that they are true or false, but we only want to get to know, what setting of their logical values is possible in the future, so is not self-contrary. They are very useful, when you want to draw all conclusions about the possible (not self-contrary) future.

When you want to use set U of undecided statements, that have yet not assigned logical values, then you can set to them respectively only values v_s for $s \in U$, for which:

$$\sim\left(\bigwedge_{s \in U} (s \equiv v_s) \Rightarrow \sim \bigwedge_{s \in U} (s \equiv v_s)\right)$$

which means, that $\sim x$ (to eliminate all contradictions in given version of the future) for every such statement x, that:

$$\left(x \Rightarrow \sim \bigwedge_{s \in U} (s \equiv v_s)\right)$$

For example, for undecided statement p different statement q for $q = \sim p$ is also undecided, but they cannot have the same value at the same time, because they are contradictory to each other.

In other words, it is the problem of solving a system of statements about the unpredictable future to get to know, what assumptions about their logical values do not create self-contrariety of such a system.

More information about a contradiction and a contrariety you can find in Section 5.15 and Section 5.12.

5.29 About Liar paradox

For s = ^"The statement of this sentence is false":

$$s = (s\ is\ false)$$

For proof, that s is undecidable see Section 5.16.

This sentence has a meaning, because we came to the conclusion, that its statement is undecidable, by logical implication of its meaning. And this sentence has the statement, that is equal to the statement of the different sentence "The statement of this sentence is not true". And this sentence has information, which informs simply, that the statement of some sentence has given logical value. All this proves, that a statement can be undecidable.

Another example of undecidable statements can be as follows: ^"The statement of the next sentence is true", ^"The statement of the previous sentence is false".

So the **Liar paradox** (^"I am lying", ^"The statement of this sentence is false", ^"This sentence expresses some untruth") is an example of an undecidable statement, because we cannot say anything about its logical value. Similarly, for example, the statement of the Gödel sentence is undecidable, so the Gödel's first incompleteness theorem cannot be proved. For more details see Section 15.12.

Remember also, that the definition $s = (s\ is\ true)$ does not define any concrete statement, because it is true for all statements.

The Liar paradox can have different nonobvious forms. For example, let us assume, that someone has just played a game with himself and is stating now, that he has won that game:

$$x = (I\ have\ won\ with\ myself) = (I\ have\ lost\ with\ myself)$$
$$= (I\ have\ not\ won\ with\ myself) = \sim x = (x\ is\ false)$$

So $x = (x \text{ is } false)$.

5.30 Unknown logical value

If you do not know logical values of some statements, then you can use iteratively the following formula:

$$F(unknown) = \left(either\ F(undecidable)\ or\ \left(either\ F(true)\ or\ F(false)\right)\right)$$

where the result should be treated as an exclusive disjunction and not as a logical expression to evaluate. And:

$$(either\ false\ or\ false) = false$$

$$(either\ true\ or\ true) = true$$

$$(either\ undecidable\ or\ undecidable) = undecidable$$

$$(either\ true\ or\ false) = (either\ false\ or\ true) = unknown$$

$$(either\ undecidable\ or\ true) = (either\ true\ or\ undecidable) = unknown$$

$$(either\ undecidable\ or\ false) = (either\ false\ or\ undecidable) = unknown$$

$$(either\ unknown\ or\ x) = (either\ x\ or\ unknown) = unknown$$

For example, for two arguments:

$$F(unknown, unknown) = \begin{pmatrix} either\ F(undecidable,\ unknown) \\ or\ (either\ F(true,\ unknown)\ or\ F(false,\ unknown)) \end{pmatrix}$$

and so on.

5.31 Realization and materialization – an evidence, that a statement is true and the definition of truth

The result of the attempt of a realization of a statement, that we will call shortly a **realization** of a statement, is an impossible or potential or existent evidence, that this statement is true. So the existence of a realization of a statement is a direct proof of this statement. The result of the failed attempt of a realization, that we will call shortly a **failed realization** (or **potential or impossible realization**) of a statement, is always a nobeing. The result of the successful attempt of a realization, that we will call shortly a **successful realization** (or **actual realization**) of a statement, is always a being, so it was at least potential being before its existence. A successful realization of a statement proves the statement. A failed realization of a statement does not prove the statement.

An impossible thing can be an evidence of a false statement, e.g.: the existence of a square circle is a proof of the false statement (*A square circle is a circle*), because if a square circle existed, then it would be a circle according to its definition. And, of course, if all realizations of a statement are impossible, then it is a false statement.

A realization of a piece of information (a statement) is equal to a part of totality referenced by that statement.

A successful realization of true information (a true statement) is equal to a part of reality referenced by that statement.

So a successful realization of a true statement is always a **logical proof** of it or something, that takes place (happens) in reality – some **state of affairs**.

A realization, that is an impossible or potential or existent material being (a part of material totality), can be called a **materialization**.

So we have the following **definition of truth**: a statement is true if and only if it has at least one successful realization.

So, for example, the statement of the following declarence could be a true information and thereby could reference a part of reality:

"This apple will be riper tomorrow [in the predictable future], than it was yesterday [in the past]"

where at the current moment the relation of being riper would exist and be fulfilled for "the same" apple from the past and from the prediction of the future. And this relation, that would happen (take place) in reality at the current moment, would be here a realization of this true statement.

Since the assumption, that a false statement is true, is counterfactual, the existence of any realization of a false statement must also be counterfactual. Remember, that the fact, that a realization belongs to the set of all realizations of a statement, does not implicate, that the realization exists.

Undecidable statements and their negations have not any realizations.

A realization of a statement is separate, irreducible, distinct and different from all other notions in logic.

For example, when the statement of the sentence "John is running" is true, then the meaning of this sentence "references" that part of reality, where John is running, and running John is a successful materialization of this statement.

For the statement $(1+1=2)$ any set of two different elements a and b is its realization, because the existence of such a set implicates the existence of all its parts and thanks to this we have from the natural number definition: $2=|\{a,b\}|=|\{a\}\cup\{b\}|=|\{a\}|+|\{b\}-\{a\}|=|\{a\}|+|\{b\}|=1+1$. For more details see Section 16.2.

Truth is the conjunction of all true statements. The existence of at least some of all realizations of a true statement is not counterfactual. Untruth is the disjunction of all negations of true statements. The existence of any realization of a false statement is thereby counterfactual. Truth is true information about whole reality. Untruth is false information.

For any statement s the set of all its realizations is defined as follows:

$$[s] = \left\{ r \in \Omega : ((r \in \mathcal{B}) \Rightarrow s) \; and \left(\left((t \subset r) \; and \; ((t \in \mathcal{B}) \Rightarrow s) \right) \Rightarrow (t = r) \right) \right\}$$

This definition of the set of all realizations of a statement guarantees, that realizations are not surplus.

And we have:

$$a = \bigvee_{x \in [a]} (x \in \mathcal{B})$$

Here is a proof:

For any statement a:

$$a = (a \; is \; provable) = \bigvee_{p \; is \; a \; proof} ((p \in \mathcal{B}) \Rightarrow a) \Rightarrow \bigvee_{p \in [a]} (p \in \mathcal{B})$$

and by the definition of a realization:

$$a \Leftarrow \bigvee_{x \in [a]} (x \in \mathcal{B})$$

So, summing up:

$$a = \bigvee_{x \in [a]} (x \in \mathcal{B})$$

Q.E.D.

Remember, that for any concrete things x and Y:

$$(x \subset Y) = (Y = (x \; and \; Y))$$

And we have the following two operators:

$$@_K X = (the \; conjunction \; of \; all \; elements \; of \; set \; X, \; that \; are \; representatives \; of \; kind \; K)$$

$$X @_K = (the \; set \; of \; all \; parts \; of \; X, \; that \; are \; representatives \; of \; kind \; K)$$

$$Totality@ = \Omega$$

$$@\Omega = Totality$$

$$Reality@ = \mathcal{B}$$

$$@\mathcal{B} = Reality$$

$$\Phi@ = \varnothing$$

$$@\varnothing = \Phi$$

For any concrete sets X, Y and any K:

$$(X \subset Y) \Rightarrow (@X \subset @Y) \Rightarrow (@_K X \subset @_K Y)$$

For any things X, Y and any K:

$$(X \subset Y) = (X@ \subset Y@) = (X \in Y@) \Rightarrow (X@_K \subset Y@_K)$$

And for any thing X and any K:

$$\int_X f(x)\ by\ x\ of\ kind\ K = \int_{X@_K} f(x)\ by\ x$$

There is a substantial difference, for example, between the set of all points of a disk and a disk itself. A disk itself is a being built from its points, so points of the disk taken together as one whole without specified elements are equal to the disk. So the following definition of a disk d, that has center at point p and radius r, is incorrect:

$$d = \left\{ (x,y)\colon (x - p.x)^2 + (y - p.y)^2 \leq r \right\}$$

because if it was correct definition of disk d, then we could define it also as a different set:

$$d = \left\{ circle\colon \left(circle = \left\{ (x,y)\colon (x - p.x)^2 + (y - p.y)^2 = R \right\} \right) and\ (R \leq r) \right\}$$

But these two concrete sets are not equal, so cannot define the same being. So each one of them cannot define disk d, because they are equivalent in this, that elements of each of them builds the same disk.

A correct definition can be as follows:

$$(disk\ d) = d = \left(conjuction\ of\ all\ elements\ of\ set \left\{ (x,y)\colon (x - p.x)^2 + (y - p.y)^2 \leq r \right\} \right)$$

$$= @\left\{ (x,y)\colon (x - p.x)^2 + (y - p.y)^2 \leq r \right\}$$

$$= @@\left\{ circle\colon \left(circle = \left\{ (x,y)\colon (x - p.x)^2 + (y - p.y)^2 = R \right\} \right) and\ (R \leq r) \right\}$$

Similarly, for example, a point is a special case of a line segment, so a line segment cannot be a set of points, because a point is not the set of a point. So a line segment must be a conjunction of points.

Q.E.D.

And we have for any decidable statements *a*, *b*:

$$([a] \subset [b]) \Rightarrow \bigwedge_{x \in [a]} \bigvee_{y \in [b]} (x \supset y) = (a \Rightarrow b) = (a \supset b) = (a@ \supset b@)$$

If statement *a* is false, then $[a]$ is the set of all things, that would be evidences of *a*, if *a* was true.

In order to prove some statement you have to demonstrate, that at least one of its realization is a being.

Some people may be confused, how, for example, a realization of a statement of some action (process) can be a part of reality, but it is by the definition of reality. Reality itself is a collective process, thus it includes all processes. To understand it better you can turn every verb into a noun, e.g.: "John was running today at 7:00 PM" into (state of subject):

$$(John\text{'}s\,Run) \subset Reality(today\;at\;7{:}00\;PM)$$

or (subject in state):

$$(Running\,John) \subset Reality(today\;at\;7{:}00\;PM)$$

Both these relations will be equivalent to:

$$John\;was\;runinng\;today\;at\;7{:}00\;PM$$

Operator $p \subset B$ means: *p* is a part of *B*.

The same applies to all other attributes, e.g.: turn "Those roses are red" into:

$$(Redness\;of\;Those\;Roses) \subset Reality(now)$$

or:

$$(Those\;Red\;Roses) \subset Reality(now)$$

For example, the statement (*A horse is an animal*) can be transformed into:

$$\begin{pmatrix} Inclusion\;of\;the\;Set\;of\;All\;Members\;of\;the\;Class\;of\;Horse \\ in\;the\;Set\;of\;All\;Members\;of\;the\;Class\;of\;Animal \end{pmatrix} \subset Reality(always)$$

which corresponds to (explained in more detail further in this book in the part about theory of things):

$$((Representation\ of\ the\ Kind\ of\ Horse) \subset (Representation\ of\ the\ Kind\ of\ Animal))$$

$$= (Horse^{\uparrow} \subset Animal^{\uparrow}) \subset Reality(always)$$

5.32 Realization - examples

^"Any/Every [existing member of the class of] horse is sometimes running" is true, because at least one horse is running from time to time. And every horse, that was running or is running now is a successful materialization of this statement.

^"Any/Every [existing member of the class of] horse does not ever run" (=~^"Some [existing member of the class of] horse is sometimes running") is false, because at least one horse is running from time to time. There is no successful realization of this statement.

^"Any/Every [existing member of the class of] horse is always running" is false, because at least one horse is sometimes not running. There is no successful realization of this statement.

^"Any/Every [existing member of the class of] horse is not always running" (=~^"Some [existing member of the class of] horse is always running") is true, because every horse is sometimes not running. Every horse, that was not running or is not running now or will be not running for sure is a successful realization of this statement.

^"Any/Every [existing member of the class of] horse is running now" is almost always false, because almost always some horses are not running. Almost always there is not any successful realization of this statement. But the conjunction of all horses running would be a successful realization of this statement.

^"Any/Every [existing member of the class of] horse is not running now" (=~^"Some [existing member of the class of] horse is running") is almost always false, because almost always some horses are running. Almost always there is no successful realization of this statement. But the conjunction of all horses standing would be a successful realization of this statement.

^"The (/given/certain) horse is running now" – such kind of a statement depends on the context, so can have the assigned probability, what is the chance, that a randomly chosen horse, about which will be that statement, is running now. For more details see Chapter 12.

^"The (/given/certain) horse is not running now" (=~^"The (/given/certain) horse is running now") – such kind of a statement depends on the context, so can have the assigned probability, what is the chance, that a randomly chosen horse, about which will be that statement, is not running now. For more details see Chapter 12.

^"Some horse runs" is true, because there is at least one horse, that runs from time to time. Every horse, that was running or is running now or will be running for sure is a successful realization of this statement.

^"Neither of horses runs" (=~^"Some horse runs") is false, because there is at least one horse, that runs from time to time. There is no successful realization of this statement.

^"Some horse is running" is almost always true, because almost always there is a horse, that is running. Every running horse is a successful realization of this statement.

^"Neither of horses is running" (=~^"Some horse is running") is almost always false, because almost always there is a horse, that is running. Almost always there is no successful realization of this statement. But the conjunction of all horses standing would be a successful realization of this statement.

5.33 More about logical values

It is worth noticing here, that it is not by accident that symbols "true" and "real" mean almost the same.

A statement is *true* if and only if this statement is a part of truth.

A statement is *false* if and only if untruth is a part of this statement.

A statement is *undecidable* if and only if this statement is neither a part of truth nor untruth is a part of it.

A decidable statement is always either true or false, so it is never true and false at the same time and it is never neither true nor false at the same time, because trueness and falseness are opposite to each other. So for any concrete statement x:

$$(x\ \textit{is not false}) = (x\ \textit{is true}) = x = (\sim x\ \textit{is false}) = (\sim x\ \textit{is not true})$$

$$(x\ \textit{is not true}) = (x\ \textit{is false}) = \sim x = (\sim x\ \textit{is true}) = (\sim x\ \textit{is not false})$$

Now we will prove that $(x\ \textit{is not true}) = (x\ \textit{is false})$ for decidable statement x.

Let us assume that x is decidable. So x is either true or false and it has to be false, since it is not true. On the other hand, if x is false, then x is not true, since it is false. The same way we can prove $(x\ \textit{is not false}) = (x\ \textit{is true})$ for decidable x.

The logical value of a statement is equal to *true* if and only if this statement is true.

The logical value of a statement is equal to *false* if and only if this statement is false.

The logical value of a statement is named *undecidable* if and only if we cannot decide what logical value it has. It is not the third logical value, but only the information, that we cannot decide what logical value a statement has. This is an abbreviation of the logical situation of a statement. For any two undecidable statements their logical situation is the same.

The simplest example of a declarence is "x", where x is the symbol of a statement (variable).

For any two statements a, b:

$a = b$ means, that these statements are equal (have the same meaning).

$\overline{a} = \overline{b}$ means, that these statements have the same logical situation (which is for simplicity often abbreviated in this book as "logical value", which is the exact meaning of the logical situation for decidable statements).

$$(a \equiv b) = (\overline{a} = \overline{b})$$

For decidable statements a and b: $(a \equiv b) = \overline{\sim(a\,xor\,b)}$, so for decidable statements also $(a \equiv b)$ is the complement of $\overline{a\,xor\,b}$.

$$a = (\overline{a} = \overline{true}) = (a \equiv true) = (a\ is\ true) = (a \subset truth)$$

$$\sim a = (\overline{a} = \overline{false}) = (a \equiv false) = (a\ is\ false) = (a \supset untruth)$$

$\hat{x}$ should be read as the phrase "*x is true*" without quotation marks. So you can treat it as insertion of the text of the sentence, that is the expression of the statement x in the current language.

$\check{x}$ should be read as the phrase "*x is false*" without quotation marks. So you can treat it as insertion of the text of the sentence, that is the expression of the negation of the statement x in the current language.

And we have:

$$\hat{x} = \widetilde{\sim x}$$

$$\check{x} = \widehat{\sim x}$$

For example:

$$a = (it\ is\ often\ cold)$$

$$b = (sun\ is\ shining)$$

$$\left(\hat{a}\ if\ \check{b}\right) = (a\ is\ true\ if\ b\ is\ false) = (it\ is\ often\ cold\ if\ sun\ is\ not\ shining)$$

When you state some statement, then you assume, that it is true:

$$a = (a\ is\ true) = \left(this\ is\ true,\ that\ \hat{a}\right)$$

$$\sim a = (a\ is\ false) = \left(this\ is\ false,\ that\ \hat{a}\right) = \left(this\ is\ true,\ that\ \check{a}\right)$$

For example:

^"*A horse is an animal*"

= (*The statement of the sentence "A horse is an animal" is true*)

= ^ "*This is true, that a horse is an animal*"

5.34 Logical functions

Negation is its inversion, so $\sim(\sim(x)) = negation(negation^{-1}(x)) = x$.

$$\sim\sim x = x$$

$$\overline{conjunction} = \overline{and} = (\overline{\wedge}) = \left\{\begin{array}{c} ((false, false), false), \\ ((false, true), false), \\ ((true, false), false), \\ ((true, true), true), \\ ((undecidable, false), false), \\ ((undecidable, true), undecidable), \\ ((false, undecidable), false), \\ ((true, undecidable), undecidable), \\ ((undecidable, undecidable), undecidable) \end{array}\right\}$$

$$\overline{and\ (a, b)} = \overline{and}(\overline{a}, \overline{b}) = \overline{and}\left(\hat{\overline{a}}, \hat{\overline{b}}\right)$$

where "$and(a, b)$" is a language structure (a sentence) and $and(a, b)$ is a statement, while $\overline{and}(\overline{a}, \overline{b})$ is a binary logical function.

$$or(a, b) = \sim and(\sim a, \sim b) = \sim(\sim a \text{ and } \sim b)$$

$$\overline{or(a, b)} = \overline{or}(\overline{a}, \overline{b}) = \overline{or}\left(\hat{\overline{a}}, \hat{\overline{b}}\right)$$

So:

$$and(a, b) = \sim or(\sim a, \sim b) = \sim(\sim a \text{ or } \sim b)$$

$$\overline{disjunction} = \overline{or} = (\overline{\vee}) = \left\{\begin{array}{c} ((false, false), false), \\ ((false, true), true), \\ ((true, false), true), \\ ((true, true), true), \\ ((undecidable, false), undecidable), \\ ((undecidable, true), true), \\ ((false, undecidable), undecidable), \\ ((true, undecidable), true), \\ ((undecidable, undecidable), undecidable) \end{array}\right\}$$

$$\overline{exclusive\ or} = \overline{xor} = \left\{ \begin{array}{c} ((false, false), false), \\ ((false, true), true), \\ ((true, false), true), \\ ((true, true), false), \\ ((undecidable, false), undecidable), \\ ((undecidable, true), undecidable), \\ ((false, undecidable), undecidable), \\ ((true, undecidable), undecidable), \\ ((undecidable, undecidable), undecidable) \end{array} \right\}$$

So:

$$xor(a, b) = or(and\,(a, \sim b), and(\sim a, b)) = ((a\ and \sim b)\ or\ (\sim a\ and\ b))$$

$$\overline{xor(a, b)} = \overline{xor}(\overline{a}, \overline{b}) = \overline{xor}\left(\hat{\overline{a}}, \hat{\overline{b}}\right)$$

$$and(a, b) = and\left(\hat{a}, \hat{b}\right) = (a\ and\ b) = (a\ is\ true\ and\ b\ is\ true)$$

$$= \left(not\ only\ \hat{a}\ but\ also\ \hat{b}\right) = \left(\hat{a}\ whereas\ \hat{b}\right) = (a \wedge b)$$

$$or(a, b) = or\left(\hat{a}, \hat{b}\right) = (a\ or\ b) = (a\ is\ true\ or\ b\ is\ true) = (a \vee b)$$

$$xor(a, b) = xor\left(\hat{a}, \hat{b}\right) = (a\ xor\ b) = \left(either\ \hat{a}\ or\ \hat{b}\right) = (either\ a\ is\ true\ or\ b\ is\ true)$$

$$and(\sim a, \sim b) = \left(neither\ \hat{a}\ nor\ \hat{b}\right)$$

If we have the following logical functions from classical logic: $\overline{CONJUNCTION}$, $\overline{DISJUNCTION}$, $\overline{EXCLUSIVE\ OR}$, then:

$$\overline{CONJUNCTION} \subset \overline{conjunction}$$

$$\overline{DISJUNCTION} \subset \overline{disjunction}$$

$$\overline{EXCLUSIVE\ OR} \subset \overline{exclusive\ or}$$

5.35 Implication operator

For any concrete statements *a* and *b*:

$$implication(a,b) = (a\ implicates\ b) = \left(\hat{b}\ if\ only\ \hat{a}\right) = \left(\hat{b}\ when\ only\ \hat{a}\right)$$
$$= (from\ only\ a\ you\ can\ draw\ conclusion\ b)$$
$$= (b\ is\ logical\ consequence\ of\ a) = (a\ entails\ b)$$
$$= (b\ can\ be\ inferred\ from\ a) = (\Rightarrow (a,b)) = (a \Rightarrow b) = (a\ includes\ b)$$
$$= (b\ is\ a\ part\ of\ a) = (a\ is\ not\ weaker\ than\ b)$$
$$= (b\ is\ not\ stronger\ than\ a) = (a\ is\ sufficient\ condition\ for\ b)$$
$$= (b\ is\ necessary\ condition\ for\ a)$$
$$= \left(if\ it\ is\ enough\ that\ \hat{b}\ then\ it\ is\ enough\ that\ \hat{a}\right)$$

For any x:

$$(a \Rightarrow b) = ((x\ is\ a\ proof\ of\ a) \Rightarrow (x\ is\ a\ proof\ of\ b)) = \begin{pmatrix} \left(x \in (Proof\ of\ a)^{\Uparrow}\right) \\ \Rightarrow \left(x \in (Proof\ of\ b)^{\Uparrow}\right) \end{pmatrix}$$
$$= \left((Proof\ of\ a)^{\Uparrow} \subset (Proof\ of\ b)^{\Uparrow}\right)$$
$$= ((x\ is\ a\ sufficient\ reason\ for\ a) \Rightarrow (x\ is\ a\ sufficient\ reason\ for\ b))$$
$$= \begin{pmatrix} \left(x \in (Sufficient\ Reason\ for\ a)^{\Uparrow}\right) \\ \Rightarrow \left(x \in (Sufficient\ Reason\ for\ b)^{\Uparrow}\right) \end{pmatrix}$$
$$= \left((Sufficient\ Reason\ for\ a)^{\Uparrow} \subset (Sufficient\ Reason\ for\ b)^{\Uparrow}\right)$$
$$= (any\ proof\ of\ a\ is\ some\ proof\ of\ b) = (a\ proof\ of\ a\ is\ a\ proof\ of\ b)$$
$$= (any\ proof\ of\ a\ proves\ b)$$
$$= (any\ sufficient\ reason\ for\ a\ is\ some\ sufficient\ reason\ for\ b)$$
$$= (a\ sufficient\ reason\ for\ a\ is\ a\ sufficient\ reason\ for\ b)$$
$$\neq ((there\ exists\ a\ proof\ of\ a) \Rightarrow (there\ exists\ a\ proof\ of\ b))$$

$$(a \Rightarrow b) = (a\ fulfils\ all\ necessary\ conditions\ that\ b\ fulfils)$$
$$= (all\ conclusions,\ that\ you\ can\ draw\ from\ b,\ you\ can\ draw\ from\ a)$$

$$(a \Rightarrow b) = (a\ contains\ logical\ conclusion\ b)$$
$$= (a\ includes\ all\ information\ that\ b\ includes) = (a\ states\ not\ less\ than\ b)$$

$$(a \Rightarrow b) \Leftarrow (a\ contains\ all\ cases\ of\ b)$$

$$((a \Rightarrow b)\ and\ (a \neq b)) \Leftarrow (a\ is\ generalization\ of\ b) = (b\ is\ special\ case\ of\ a)$$

Since every statement implicated by some statement is a part of it, a true statement cannot implicate a false statement, because if a true statement implicated a false statement, then this true statement would be false, since every statement is the conjunction of all its parts (a proof of it is further in this section). So we would have a contradiction. Since a true statement cannot implicate a false statement, a true statement cannot be false and the conjunction of a true statement and an undecidable statement is undecidable, so if a true statement implicated an undecidable statement, then it would be undecidable. So we would have a contradiction. So a true statement cannot implicate an undecidable statement and by the inversion of implication an undecidable statement cannot implicate a false statement.

Q.E.D.

Since every statement implicates all its necessary conditions and every statement is implicated by all its sufficient conditions, we have, that a implicates b then and only then, when b is necessary condition for a and a is sufficient condition for b. And b is necessary condition for a then and only then, when a is sufficient condition for b.

Let us prove, that:

$$((x\ is\ a\ proof\ of\ a) \Rightarrow (x\ is\ a\ proof\ of\ b)) \neq$$
$$((there\ exists\ a\ proof\ of\ a) \Rightarrow (there\ exists\ a\ proof\ of\ b))$$

Here is the proof:

For $b = (there\ exist\ a\ proof\ of\ a)$ we have:

$$(there\ exist\ a\ proof\ of\ a) \Rightarrow b$$

So we have a proof of b from the assumption that $(there\ exist\ a\ proof\ of\ a)$.

So:

$$((there\ exists\ a\ proof\ of\ a) \Rightarrow (there\ exists\ a\ proof\ of\ b))$$

But a proof of a does not necessarily prove that $(there\ exist\ a\ proof\ of\ a)$ $(= b)$, because not every proof proves its existence. So a proof of a does not necessarily prove b.

Q.E.D.

First of all, implication (the same as equality) between two statements is decidable in any situation, because it is a special case of equality, which is decidable in any situation.

Secondly, the logical value of the implication between two given statements does not ever change, but the logical value of the implication between the statements of two given sentences can change depending on the situation – the speaker of the sentences and the context (e.g. time, space, etc.) – in which they are used.

If a is false, then we have counterfactual conditional:

$$((a \Rightarrow b)\ and \sim a) = (if\ only\ a\ was\ true\ then\ b\ would\ be\ true)$$

$$= \left(if\ only\ it\ was\ true\ that\ \hat{a}\ then\ it\ would\ be\ true\ that\ \hat{b}\right)$$

So:

$$\sim a \Rightarrow ((a \Rightarrow b) \leftrightarrow (if\ only\ a\ was\ true\ then\ b\ would\ be\ true))$$

Since $(a \Rightarrow b)$ is the inclusion of information b in information a, for any concrete statements a and b:

$$(\boldsymbol{a} \Rightarrow \boldsymbol{b}) = (\boldsymbol{a} \supset \boldsymbol{b}) = (\boldsymbol{a@} \supset \boldsymbol{b@}) = ((\boldsymbol{a\ and\ b}) = (\boldsymbol{a} \cup \boldsymbol{b}) = \boldsymbol{a}) = ((\boldsymbol{a\ or\ b}) = \boldsymbol{b})$$

And then also:

$$(a = b) = ((a \supset b)\ and\ (b \supset a)) = ((a \Leftarrow b)\ and\ (a \Rightarrow b))$$

So:

$$(\boldsymbol{a} = \boldsymbol{b}) = ((\boldsymbol{a} \Leftarrow \boldsymbol{b})\ \boldsymbol{and}\ (\boldsymbol{a} \Rightarrow \boldsymbol{b}))$$

Relation $(y \supset x)$ means $(x\ is\ a\ part\ of\ y)$. More details about this relation can be found in Sections 9.1 and 9.2.

Above definition of implication is correct, because statement a implicates statement b if and only if it states, what statement b states, and alternatively something else. So:

$$a = (b\ and\ a)$$

So by the definition of inclusion, implication $(a \Rightarrow b)$ is inclusion $(a \supset b)$. It means, that piece of information a implicates piece of information b if and only if piece of information a includes piece of information b, which is also a correct definition of true implication.

Then we have:

$$\boldsymbol{a} = \bigwedge_{\boldsymbol{a} \supset \boldsymbol{b}} \boldsymbol{b} = \bigvee_{\boldsymbol{b} \supset \boldsymbol{a}} \boldsymbol{b}$$

which is very easy to be proved, since for every b such, that $(a \Rightarrow b)$ $(= (a \supset b))$, we have recurrence $a = (b\ and\ a)$.

Here is the proof:

$$\left(\bigwedge_{a\supset b}(a=(a\,and\,b))\right)\Rightarrow\left(\boldsymbol{a}=\left(a\,and\,\widehat{@}\left(a\widehat{@}\right)\right)=\widehat{@}\left(a\widehat{@}\right)=\widehat{@}\left(\bigcup_{a\supset b}\{b\}\right)=\bigwedge_{a\supset b}\boldsymbol{b}\right)$$

Important: For statements, that have uncountable number of implications, that **proof has uncountable number of steps**, that are simplified to single step, that you see above. As you can see, it does not matter, that we cannot write down a conjunction of the uncountable number of arguments as an expression in the form "$x_1\,and\,x_2\,and\ldots\,and\,x_n$", because we can do this uncountable number of steps at once in our mind, just like in case of any proof by induction. Remember, that this is all about statements and not sentences, where a statement can be a conjunction or a disjunction of any number of statements. We simply are not limited by the form of a sentence anymore, so we can do much more.

So it is enough for a statement to have one false part to be false, so a true statement has not any false parts.

So every statement is equal to the conjunction of all statements (all its necessary conditions) implicated by it. That is, why every true statement implicates only true statements, because logical value of a conjunction of statements is true, if and only if all these statements (all information included in them) are true too. So every true statement cannot implicate contrary statements. And also a statement is true if and only if all statements implicated by this statement are true.

Every true statement has at least one true statement, that implicates it, and never implicates any false statement. And every false statement does not have any true statement, that implicates it, and always implicates at least one false statement.

In other words, it is impossible to prove any untruth without making a mistake.

And for every a for $(a\Rightarrow b)(=(a\supset b))$ we have recurrence $b=(a\,or\,b)$.

So:

$$\left(\bigwedge_{b\supset a}(a=(a\,or\,b))\right)\Rightarrow\left(\boldsymbol{a}=\left(a\,or\,\breve{@}\left(a\breve{@}\right)\right)=\breve{@}\left(a\breve{@}\right)=\breve{@}\left(\bigcup_{b\supset a}\{b\}\right)=\bigvee_{\boldsymbol{b}\supset\boldsymbol{a}}\boldsymbol{b}\right)$$

So it is enough for a statement to have one true superthing to be true, so a false statement does not have any true superthing.

So we have also:

$$a=\bigwedge_{a\Rightarrow b}b=\bigvee_{b\Rightarrow a}b=\sim\bigvee_{a\Rightarrow b}\sim b$$

So:

$$\sim a=\bigvee_{a\Rightarrow b}\sim b=\bigvee_{\sim b\Rightarrow\sim a}\sim b$$

$$a=\bigwedge_{a\Rightarrow b}b=\bigwedge_{\sim b\Rightarrow\sim a}b$$

And:

$$(a \Rightarrow b) = ((a \, is \, true) \Rightarrow (b \, is \, true))$$

$$(a \Rightarrow (a \, is \, true) \Rightarrow a)$$

For any concrete statements a and b (for a proof see Section 5.4):

$$(a \Rightarrow b) \Rightarrow (\sim b \Rightarrow \sim a)$$

In other words, if b is necessary condition of a, then, of course, $\sim b$ implicates $\sim a$.

Then:

$$(a \Rightarrow b) \Rightarrow (\sim \boldsymbol{b} \Rightarrow \sim \boldsymbol{a}) \Rightarrow (\sim\sim a \Rightarrow \sim\sim b) = (\boldsymbol{a} \Rightarrow \boldsymbol{b})$$

So:

$$(\boldsymbol{a} \Rightarrow \boldsymbol{b}) = (\sim \boldsymbol{b} \Rightarrow \sim \boldsymbol{a})$$

And we have, that:

$$(a \, and \, (a \Rightarrow b)) \Rightarrow b$$

Here is the proof:

$$(a \, and \, (a \Rightarrow b)) = \left(a \, and \, (a = (b \, and \, a))\right) = \left(b \, and \, a \, and \, (a = (b \, and \, a))\right) \Rightarrow b$$

Q.E.D.

And we have, that:

$$(\sim b \, and \, (a \Rightarrow b)) \Rightarrow \sim a$$

Here is the proof:

$$(\sim b \, and \, (a \Rightarrow b)) = (\sim b \, and \, (\sim b \Rightarrow \sim a)) = \left(\sim b \, and \, (\sim b = (\sim a \, and \sim b))\right)$$

$$= \left(\sim a \, and \sim b \, and \, (\sim b = (\sim a \, and \sim b))\right) \Rightarrow \sim a$$

Q.E.D.

And we have, that:

$$(a\,and\,(a \Rightarrow b)) \Rightarrow (a\,and\,b) \Rightarrow \sim(a \Rightarrow \sim b)$$

So:

$$\boldsymbol{a} \Rightarrow ((\boldsymbol{a} \Rightarrow \boldsymbol{b}) \rightarrow \sim(\boldsymbol{a} \Rightarrow \sim\boldsymbol{b}))$$

In other words, this is another proof, that every true statement cannot have contrary implications.

And:

$$(\boldsymbol{a} \Rightarrow \boldsymbol{b}) = ((\boldsymbol{a}\,\boldsymbol{or}\,\boldsymbol{b}) = \boldsymbol{b})$$

Here is a proof of it:

$$(a \Rightarrow b) = (\sim b \Rightarrow \sim a) = (\sim b = (\sim a\,and \sim b) = \sim(a\,or\,b)) = ((a\,or\,b) = b)$$

And we can derive from this assumption the equality $((a \Rightarrow b) = (\sim b \Rightarrow \sim a))$:

$$(a \Rightarrow b) = (a = (b\,and\,a)) = (\sim a = (\sim a\,or \sim b)) = (\sim b \Rightarrow \sim a)$$

And we have:

$$\boldsymbol{a} = \bigvee_{\boldsymbol{r} \in [\boldsymbol{a}]} (\boldsymbol{r} \in \mathcal{B}) = \sim \bigwedge_{\boldsymbol{r} \in [\boldsymbol{a}]} \sim(\boldsymbol{r} \in \mathcal{B}) = \bigwedge_{\boldsymbol{r} \in [\sim\boldsymbol{a}]} \sim(\boldsymbol{r} \in \mathcal{B})$$

Summing up, we have, for example, the following rules:

1. $(a\,and \sim b) \Rightarrow \sim(a \Rightarrow b)$
2. $(a \Rightarrow b) = (\sim b \Rightarrow \sim a)$
3. $(a\,and\,(a \Rightarrow b)) \Rightarrow b$
4. $(\sim b\ and\ (a \Rightarrow b)) = (\sim b\ and\ (\sim b \Rightarrow \sim a)) \Rightarrow \sim a$

We have, for example, also the following additional rules:

1. $((a \Rightarrow b)\,and\,(b \Rightarrow c)) \Rightarrow (a \Rightarrow c)$
   For example: The statement of the sentence "I will not stay in the house, if it is a sunny day" is true for the assumption "I will play in the garden, if the sun shines", because the statement of the sentence "it is a sunny day" implicates the statement of the sentence "the sun shines" and the statement of the sentence "I will play in the garden" implicates the statement of the sentence "I will not stay in the house". You have to use the rule twice to get this result. For a proof of this rule see Section 5.37.
2. $((a \Rightarrow (x\,and\,y))\,and\,(x \rightarrow \sim y)) \Rightarrow \sim(a\,and\,a\,is\,decidable)$
   Contextually contrary statements can never be inferred from a decidable true statement.

5.36 Intuitionistic logic unnecessary and wrong

Intuitionistic logic not only is not necessary at all, but it is also completely wrong. For example, it is not true, that:

$$\bigvee_{a \in Statement^{\Uparrow}} \sim(\sim\sim a \Rightarrow a)$$

because:

$$\sim\sim a = \left(\Lambda_\varsigma - (\Lambda_\varsigma - a)\right) = \left(\Lambda_\varsigma \; or \sim(\Lambda_\varsigma \; or \sim a)\right) = \left(\Lambda_\varsigma \; or \; (\varnothing_\varsigma \; and \, a)\right)$$
$$= (\Lambda_\varsigma \; or \; a) = a$$

5.37 Useful rules for implication

First of all for any concrete statements *a*, *b* and *c*:

$$(a \Rightarrow b \Rightarrow c) = ((a \Rightarrow b) \; and \; (b \Rightarrow c))$$

and in general for any binary relations *R*, *S*:

$$(a \, R \, b \, S \, c) = ((a \, R \, b) \; and \; (b \, S \, c))$$

and we have, that $\Rightarrow (a\,,\, b)$ is transitive relation:

$$((a \Rightarrow b) \; and \; (b \Rightarrow c)) \Rightarrow (a \Rightarrow c)$$

Here is the proof:

$$((a \Rightarrow b) \; and \; (b \Rightarrow c)) = \left((a = (b \; and \; a)) \; and \; (b = (c \; and \; b))\right)$$
$$\Rightarrow \left(a = (c \; and \; b \; and \; a) = (c \; and \; a)\right) = (a \Rightarrow c)$$

and left-distributive over “and” (axiom):

$$((a \Rightarrow b) \; and \; (a \Rightarrow c)) = (a \Rightarrow (b \; and \; c))$$

Here is the proof:

$$((a \Rightarrow b) \; and \; (a \Rightarrow c))$$
$$= \left((a = (b \; and \; a)) \; and \; (a = (c \; and \; a))\right) \Rightarrow \left(a = (b \; and \; c \; and \; a) = ((b \; and \; c) \; and \; a)\right)$$
$$= (a \Rightarrow (b \; and \; c))$$

And left-distributive over “or” in one direction:

$$((a \Rightarrow b)\ or\ (a \Rightarrow c)) \Rightarrow (a \Rightarrow (b\ or\ c))$$

Here is the proof:

$$\begin{aligned}&((a \Rightarrow b)\ or\ (a \Rightarrow c))\\ &\quad = \big((a = (b\ and\ a))\ or\ (a = (c\ and\ a))\big)\\ &\quad \Rightarrow \big(a = ((b\ and\ a)\ or\ (c\ and\ a)) = ((b\ or\ c)\ and\ a)\big) = (a \Rightarrow (b\ or\ c))\end{aligned}$$

So as a consequence also:

$$((a \Rightarrow c)\ and\ (b \Rightarrow c)) = ((a\ or\ b) \Rightarrow c)$$

$$((a \Rightarrow c)\ or\ (b \Rightarrow c)) \Rightarrow ((a\ and\ b) \Rightarrow c)$$

So also:

$$\sim(a \Rightarrow (b\ and\ c)) = (\sim(a \Rightarrow b)\ or\ \sim(a \Rightarrow c))$$

$$\sim((a\ or\ b) \Rightarrow c) = (\sim(a \Rightarrow c)\ or\ \sim(b \Rightarrow c))$$

$$\sim(a \Rightarrow (b\ or\ c)) \Rightarrow (\sim(a \Rightarrow b)\ and\ \sim(a \Rightarrow c))$$

$$\sim((a\ and\ b) \Rightarrow c) \Rightarrow (\sim(a \Rightarrow c)\ and\ \sim(b \Rightarrow c))$$

Of course, universal and existential quantifiers are respectively generalized conjunction and generalized disjunction, so we have for any concrete statement S and any concrete set X of concrete statements:

$$\bigwedge_{x \in X} (S \Rightarrow x) = \Big(S \Rightarrow \bigwedge_{x \in X} x\Big)$$

$$\bigwedge_{x \in X} (x \Rightarrow S) = \Big(\bigvee_{x \in X} x \Rightarrow S\Big)$$

$$\bigvee_{x \in X} (x \Rightarrow S) \Rightarrow \Big(\bigwedge_{x \in X} x \Rightarrow S\Big)$$

$$\bigvee_{x \in X} (S \Rightarrow x) \Rightarrow \Big(S \Rightarrow \bigvee_{x \in X} x\Big)$$

$$\Big(S \Rightarrow \bigwedge_{x \in X} x\Big) = \bigwedge_{x \in X} (S \Rightarrow x) \Rightarrow \bigvee_{x \in X} (S \Rightarrow x) \Rightarrow \Big(S \Rightarrow \bigvee_{x \in X} x\Big)$$

$$\Big(\bigvee_{x \in X} x \Rightarrow S\Big) = \bigwedge_{x \in X} (x \Rightarrow S) \Rightarrow \bigvee_{x \in X} (x \Rightarrow S) \Rightarrow \Big(\bigwedge_{x \in X} x \Rightarrow S\Big)$$

And for any concrete sets A and B of concrete statements:

$$\left(\bigvee_A \Rightarrow \bigvee_B\right) = \left(\bigwedge_{a\in A}\left(a \Rightarrow \bigvee_B\right)\right) \Leftarrow \left(\bigwedge_{a\in A}\bigvee_{b\in B}(a \Rightarrow b)\right)$$

$$\left(\bigwedge_A \Rightarrow \bigwedge_B\right) \Leftarrow \left(\bigvee_{a\in A}\left(a \Rightarrow \bigwedge_B\right)\right) = \left(\bigvee_{a\in A}\bigwedge_{b\in B}(a \Rightarrow b)\right)$$

$$\left(\bigvee_A \Rightarrow \bigwedge_B\right) = \left(\bigwedge_{a\in A}\left(a \Rightarrow \bigwedge_B\right)\right) = \left(\bigwedge_{a\in A}\bigwedge_{b\in B}(a \Rightarrow b)\right)$$

$$\left(\bigwedge_A \Rightarrow \bigvee_B\right) \Leftarrow \left(\bigvee_{a\in A}\left(a \Rightarrow \bigvee_B\right)\right) \Leftarrow \left(\bigvee_{a\in A}\bigvee_{b\in B}(a \Rightarrow b)\right)$$

We also have for any concrete statements a, b, x and y:

$$((a \Rightarrow x)\ and\ (b \Rightarrow y)) \Rightarrow ((a\ and\ b) \Rightarrow (x\ and\ y))$$

$$((a \Rightarrow x)\ and\ (b \Rightarrow y)) \Rightarrow ((a\ or\ b) \Rightarrow (x\ or\ y))$$

and for any concrete statements a_i for $i \in I$, where I is a set of consecutive natural numbers:

$$\left(\bigwedge_{i\in I} a_i\right) \Rightarrow \left(\bigvee_{i\in I} a_i\right)$$

So:

$$\left(\bigwedge_{i\in I}(a_i \Rightarrow b_i)\right) \Rightarrow \left(\left(\bigwedge_{i\in I} a_i\right) \Rightarrow \left(\bigwedge_{i\in I} b_i\right) \Rightarrow \left(\bigvee_{i\in I} b_i\right)\right)$$

$$\left(\bigwedge_{i\in I}(a_i \Rightarrow b_i)\right) \Rightarrow \left(\left(\bigvee_{i\in I} a_i\right) \Rightarrow \left(\bigvee_{i\in I} b_i\right)\right)$$

And we have, that:

If $F(a) \Rightarrow G(a)$ is a tautology and $F(a)$ is a tautology, then $G(a)$ is a tautology.

And if $F(a) \Rightarrow G(a)$ is a tautology and if $F(a)$ is satisfiable (satisfied for some a), so it is not the negation of a tautology, then $G(a)$ is also satisfiable (at least for this a), so it is not the negation of a tautology.

But it is not true, that:

$$((a\ and\ b) \Rightarrow x) = (a \Rightarrow (b \Rightarrow x)) = (b \Rightarrow (a \Rightarrow x))$$

$$(x \Rightarrow (a\ or\ b)) = (\sim(x \Rightarrow a) \Rightarrow b) = (\sim(x \Rightarrow b) \Rightarrow a)$$

Here is a naïve proof of the opposite:

Assume:

$$((a\ and\ b) \Rightarrow x)$$

Then:

$$a \Rightarrow (b \Rightarrow (a\ and\ b)) \Rightarrow (b \Rightarrow (a\ and\ b) \Rightarrow x) \Rightarrow (b \Rightarrow x)$$

So:

$$a \Rightarrow (b \Rightarrow x)$$

The obvious mistake lies in the assumption, that $a \Rightarrow (b \Rightarrow (a\ and\ b))$, which is not true, because if it was true, then all true statements would be equal, because then:

$$a \Rightarrow (b \Rightarrow (a\ and\ b))$$

$$b \Rightarrow (a \Rightarrow (a\ and\ b))$$

So for two true statements a and b we would have:

$$b \Rightarrow (a\ and\ b)$$

$$a \Rightarrow (a\ and\ b)$$

So:

$$(b \Rightarrow a)\ and\ (b \Rightarrow b)$$

$$(a \Rightarrow a)\ and\ (a \Rightarrow b)$$

So:

$$(b \Rightarrow a)\ and\ (a \Rightarrow b)$$

So:

$$a = b$$

which is false for all pairs of different true statements a and b, so we have a contradiction to our assumption.

So:

$$\bigvee_{\{a,b\} \subset Statement^{\Uparrow}} \sim\left(\boldsymbol{a} \Rightarrow ((\boldsymbol{a\ and\ b}) = \boldsymbol{b})\right)$$

$a \Rightarrow ((a\ and\ b) \leftrightarrow b) = (\overline{a\ and\ b = \overline{b}}) = ((a\ and\ b) \equiv b)$ is only true.

So if a is true, then $(a\ and\ b)$ and b have equal logical values, but they are not the same statements.

So:

$$\bigvee_{\{a,b,x\}\subset Statement^{\Uparrow}} \sim\left(\left((\boldsymbol{a}\ \boldsymbol{and}\ \boldsymbol{b})\Rightarrow \boldsymbol{x}\right)=\left(\boldsymbol{a}\Rightarrow(\boldsymbol{b}\Rightarrow \boldsymbol{x})\right)\right)$$

One of the exceptions is for $a=(b\Rightarrow x)$:

$$\left(\left((b\Rightarrow x)\ and\ b\right)\Rightarrow x\right)=\left((b\Rightarrow x)\Rightarrow(b\Rightarrow x)\right)=\varnothing_{\zeta}$$

Q.E.D.
So we have only:

$$\begin{aligned}(a\Rightarrow(b\Rightarrow x))\Rightarrow((a\ and\ b)\Rightarrow x)\Rightarrow(a\Rightarrow(b\rightarrow x))\Rightarrow(a\rightarrow(b\rightarrow x))\\ =((a\ and\ b)\rightarrow x)\end{aligned}$$

$$\begin{aligned}(\sim(x\Rightarrow a)\Rightarrow b)\Rightarrow(x\Rightarrow(a\ or\ b))\Rightarrow(\sim(x\rightarrow a)\Rightarrow b)\Rightarrow(\sim(x\rightarrow a)\rightarrow b)\\ =(x\rightarrow(a\ or\ b))\end{aligned}$$

Below are proofs of:

$$(a\Rightarrow(b\Rightarrow x))\Rightarrow((a\ and\ b)\Rightarrow x)$$

$$(\sim(x\Rightarrow a)\Rightarrow b)\Rightarrow(x\Rightarrow(a\ or\ b))$$

Assume:

$$(a\Rightarrow(b\Rightarrow x))$$

Then:

$$(a\ and\ b)\Rightarrow((b\Rightarrow x)\ and\ b)\Rightarrow x$$

So:

$$\left((\boldsymbol{a}\Rightarrow(\boldsymbol{b}\Rightarrow \boldsymbol{x}))\Rightarrow((\boldsymbol{a}\ \boldsymbol{and}\ \boldsymbol{b})\Rightarrow \boldsymbol{x})\right)$$

Q.E.D.
Assume:

$$(\sim(x\Rightarrow a)\Rightarrow b)$$

Then:

$$(\sim(x\Rightarrow a)\Rightarrow b)=(\sim b\Rightarrow(\sim a\Rightarrow\sim x))\Rightarrow((\sim a\ and\ \sim b)\Rightarrow\sim x)=(x\Rightarrow(a\ or\ b))$$

So:

$$(\sim(x \Rightarrow a) \Rightarrow b) \Rightarrow (x \Rightarrow (a \text{ or } b))$$

Q.E.D.

But fortunately there is a quite general exception from that rule, for which:

$$(a \Rightarrow (b \Rightarrow x)) = ((a \text{ and } b) \Rightarrow x)$$

For more details see Section 5.65.

5.38 Contextual implication – facilitators for a reasoning in proofs

5.38.1 Derivation of "material implication" from implication

Consider the following statement for any concrete statements a and b:

$$(a \rightarrow b) = ((a \text{ and } truth) \Rightarrow b)$$

where $untruth = \sim truth$ and $truth$ (TR) is the conjunction of all true statements.

This is implication in the context of whole truth. That is, why it will be called contextual implication.

Remember, that if you assume in some reality some statement to be true or false, then that statement (and all its implications) or its negation (and all implications of its negation), respectively, becomes a part of truth of a new possibly counterfactual reality. Simply, if you assume in some reality some statement x, then you assume $x \subset truth$. If, what you assumed in some reality as true or false, is in this reality respectively false or true, then this assumption is counterfactual in this reality and you start to operate in a new counterfactual reality. And that is, why then you must get a contradiction with the previous reality, that proves this assumption to be wrong in the previous reality, so the conjunction of all assumptions of this previous reality (its truth) does not implicate this assumption. For more details see Section 5.27.

First of all, we have the following definition of two-way contextual implication for any concrete statements a and b:

$$(\boldsymbol{a} \leftrightarrow \boldsymbol{b}) = ((\boldsymbol{a} \rightarrow \boldsymbol{b}) \text{ } \boldsymbol{and} \text{ } (\boldsymbol{b} \rightarrow \boldsymbol{a}))$$

and we have:

$$((\boldsymbol{a} \text{ } \boldsymbol{and} \text{ } \boldsymbol{b}) \rightarrow \boldsymbol{a}) = \left((a \text{ and } (b \text{ and } truth)) \Rightarrow a\right) = \varnothing_{\varsigma}$$

$$(\boldsymbol{c} \rightarrow (\boldsymbol{a}\ \textbf{\textit{and}}\ \boldsymbol{b})) = ((c\ and\ truth) \Rightarrow (a\ and\ b))$$
$$= \Big(((c\ and\ truth) \Rightarrow a)\ and\ ((c\ and\ truth) \Rightarrow b)\Big)$$
$$= ((\boldsymbol{c} \rightarrow \boldsymbol{a})\ \textbf{\textit{and}}\ (\boldsymbol{c} \rightarrow \boldsymbol{b}))$$

So we have:

$$(\boldsymbol{a} \leftrightarrow (\boldsymbol{b}\ \textbf{\textit{and}}\ \boldsymbol{a})) = \Big((a \rightarrow (b\ and\ a))\ and\ ((b\ and\ a) \rightarrow a)\Big) = ((a \rightarrow b)\ and\ (a \rightarrow a))$$
$$= (\boldsymbol{a} \rightarrow \boldsymbol{b})$$

and we have:

$$(a \rightarrow b) = ((a\ and\ TR) \Rightarrow b) = ((a\ and\ TR) = (b\ and\ a\ and\ TR))$$
$$= \Big((a\ and\ TR) = ((b\ and\ a\ and\ TR)\ or\ {\sim}\varnothing_{\varsigma})\Big)$$
$$= \Big((a\ and\ TR) = ((b\ and\ a\ and\ TR)\ or\ {\sim}(TR\ or\ {\sim}a\ or\ {\sim}TR))\Big)$$
$$= \Big((a\ and\ TR) = ((b\ and\ a\ and\ TR)\ or\ ({\sim}TR\ and\ a\ and\ TR))\Big)$$
$$= \Big((a\ and\ TR) = ((b\ or\ {\sim}TR)\ and\ a\ and\ TR)\Big)$$
$$= ((a\ and\ TR) \Rightarrow (b\ or\ {\sim}TR))$$

where “*TR or ~TR or ~a*” is, of course, a tautology, because “~(~ *TR and TR and TR is decidable*)” is a tautology and (*TR is decidable*) is a part of *TR*. And, of course, the statement of a tautology is equal to the empty statement.

So:

$$(a \rightarrow b) = ((a\ and\ TR) \Rightarrow (b\ or\ {\sim}TR)) = ({\sim}(b\ or\ {\sim}TR) \Rightarrow {\sim}(a\ and\ TR))$$
$$= (({\sim}b\ and\ TR) \Rightarrow ({\sim}a\ or\ {\sim}TR)) = ({\sim}b \rightarrow {\sim}a)$$

So contextual contrariety is a symmetric relation, because:

$$(a \rightarrow {\sim}b) = (b \rightarrow {\sim}a) = (a\ is\ contextually\ contrary\ to\ b)$$
$$= (b\ is\ contextually\ contrary\ to\ a)$$
$$= (a\ and\ b\ are\ contextually\ contrary\ to\ each\ other)$$
$$= (a\ and\ b\ are\ in\ a\ contextual\ contrariety)$$

and we have:

$$(a \rightarrow b) = (b \leftarrow a) = \big((a \, and \, truth) \Rightarrow b\big) = \big((a \, and \, truth) \Rightarrow (b \, or \, {\sim}truth)\big)$$
$$= \big((a \, and \, truth) \Rightarrow (b \, or \, untruth)\big)$$

and:

$$(a \Rightarrow b) = \big(a = (b \, and \, a)\big)$$
$$\Rightarrow \Big((a \rightarrow b) = \big((a \, and \, truth) \Rightarrow b\big) = \big((a \, and \, b \, and \, truth) \Rightarrow b\big)$$
$$\equiv true\Big)$$

So:

$$(a \Rightarrow b) \Rightarrow (a \rightarrow b)$$

and we have:

$$(\Phi \rightarrow b) = (\rightarrow b) = (b \, is \, true)$$
$$(a = b) \Rightarrow (a \leftrightarrow b) = \big((a \rightarrow b) \, and \, (a \leftarrow b)\big) \Rightarrow (a \rightleftharpoons b)$$

For the logical values of arguments a and b respectively equal to:
- true and true, $(a \rightarrow b)$ is true, because true b is a part of *truth* (a proof of it is given below),
- false and true, $(a \rightarrow b)$ is true, because true b is a part of *truth* (a proof of it is given below),
- undecidable and true, $(a \rightarrow b)$ is true, because true b is a part of *truth*,

$$true \equiv (truth \Rightarrow b)$$
$$= \big(truth = (b \, and \, truth)\big)$$
$$\Rightarrow \big((a \, and \, truth) = (a \, and \, b \, and \, truth)\big)$$
$$= \big((a \, and \, truth) \Rightarrow b\big) = (a \rightarrow b)$$

- false and false, $(a \rightarrow b)$ is true, because $(a \rightarrow b)$ is true for true and true arguments,
- true and false, $(a \rightarrow b)$ is false, because $(a \, and \, truth)$ becomes *truth*, which is true, and b is false, so it cannot be implicated by a true statement, since all implications of a true statement are true,

$$\begin{aligned} true &\equiv ((truth \Rightarrow a)\ and\ (b \Rightarrow untruth)) \\ &= \Big((truth = (a\ and\ truth))\ and\ (b \Rightarrow untruth)\Big) \\ &\Rightarrow \Big(((a\ and\ truth) \Rightarrow b\big) \Rightarrow (truth = (a\ and\ truth) \Rightarrow b \Rightarrow untruth) \\ &\Rightarrow (truth \Rightarrow untruth) \equiv false\Big) \Rightarrow \sim\big((a\ and\ truth) \Rightarrow b\big) = \sim(a \rightarrow b) \end{aligned}$$

– true and undecidable, $(a \rightarrow b)$ is false, because $(a\ and\ truth)$ becomes *truth*, which is true, and b is undecidable, so it cannot be implicated by a true statement, since all implications of a true statement are true:

$$true \equiv (truth \Rightarrow a) = \big(truth = (a\ and\ truth)\big)$$

Assume $(a \rightarrow b)$ is true. Then:

$$true \equiv (a \rightarrow b) = \big((a\ and\ truth) \Rightarrow b\big) = (truth \Rightarrow b) = b$$

But b is undecidable, so we have a contradiction. So $\sim(a \rightarrow b)$, since contextual implication is always decidable.

Q.E.D.

– undecidable and false, $(a \rightarrow b)$ is false, because $(a \rightarrow b)$ is false for true and undecidable arguments,
– false and undecidable, $(a \rightarrow b)$ is true, because $(a \rightarrow b)$ is true for undecidable and true arguments,
– undecidable and undecidable, $(a \rightarrow b) = (a_{\blacksquare} \Rightarrow b_{\blacksquare}) = (\blacksquare_{\varsigma} \Rightarrow \blacksquare_{\varsigma}) \equiv true$, because there is only one and the same core $\blacksquare_{\varsigma}$ for all undecidable statements, so $\blacksquare_{\varsigma} = a_{\blacksquare} = b_{\blacksquare}$ and thereby $(a \rightarrow b)$ is true and a proof of it is trivial.

We have:

$$truth \Rightarrow a_{\#}$$
$$truth \Rightarrow b_{\#}$$

So:

$$truth = (a_{\#}\ and\ truth)$$
$$truth = (b_{\#}\ and\ truth)$$

So:

$$(a \rightarrow b) = ((a_{\blacksquare}\ and\ a_{\#}\ and\ truth) \Rightarrow (b_{\blacksquare}\ and\ b_{\#}\ and\ truth))$$
$$= ((a_{\blacksquare}\ and\ truth) \Rightarrow (b_{\blacksquare}\ and\ truth))$$
$$= ((a_{\blacksquare}\ and\ truth) = (b_{\blacksquare}\ and\ a_{\blacksquare}\ and\ truth)) = ((a_{\blacksquare}\ and\ truth) \Rightarrow b_{\blacksquare})$$

where *truth* does not implicate any nonempty part of $b_{\blacksquare}$, so:

$$(\boldsymbol{a} \rightarrow \boldsymbol{b}) = ((a_{\blacksquare}\ and\ truth) \Rightarrow b_{\blacksquare}) = (\boldsymbol{a}_{\blacksquare} \Rightarrow \boldsymbol{b}_{\blacksquare})$$

So:

$$(\boldsymbol{a} \rightarrow \boldsymbol{b}) = (a_{\blacksquare} \Rightarrow b_{\blacksquare}) = (\blacksquare_{\varsigma} \Rightarrow \blacksquare_{\varsigma}) \equiv \boldsymbol{true}$$

Q.E.D.

So:

$$(a \Rightarrow b) \Rightarrow (a \rightarrow b) = (a \rightharpoondown b)$$

So for contextual implication you can use all rules, that apply to possibility of implication. For more details see Section 5.40.

For contextual implication you can also use almost all rules, that applies to implication, because for $T(x) = (x\ and\ truth)$:

$$(\boldsymbol{a} \rightarrow \boldsymbol{b}) = ((a\ and\ truth) \Rightarrow b) = ((a\ and\ truth) = (b\ and\ a\ and\ truth))$$
$$= ((a\ and\ truth) = ((b\ and\ truth)\ and\ a\ and\ truth))$$
$$= ((a\ and\ truth) \Rightarrow (b\ and\ truth)) = (\boldsymbol{T}(\boldsymbol{a}) \Rightarrow \boldsymbol{T}(\boldsymbol{b}))$$

This is true, because for any concrete statements a, b and c:

$$(a\ and\ b) \supset a$$

$$((a\ and\ b) \supset c) = (((a\ and\ b) \supset c)\ and\ ((a\ and\ b) \supset a)) = ((a\ and\ b) \supset (a\ and\ c))$$

And we have:

$$(\sim\boldsymbol{a} \rightarrow \sim\boldsymbol{b}) = (b \rightarrow a) = (T(b) \Rightarrow T(a)) = (\sim\boldsymbol{T}(\boldsymbol{a}) \Rightarrow \sim\boldsymbol{T}(\boldsymbol{b}))$$

$$(\boldsymbol{a} \rightarrow \sim\boldsymbol{b}) = ((a\ and\ truth) \Rightarrow (\sim b\ or\ \sim truth)) = ((a\ and\ truth) \Rightarrow \sim(b\ and\ truth))$$
$$= (\boldsymbol{T}(\boldsymbol{a}) \Rightarrow \sim\boldsymbol{T}(\boldsymbol{b}))$$

The only exception is for:

$$(\sim \boldsymbol{a} \to \boldsymbol{b}) = (\boldsymbol{T}(\sim \boldsymbol{a}) \Rightarrow \boldsymbol{T}(\boldsymbol{b})) = (\sim \boldsymbol{b} \to \boldsymbol{a}) = ((\sim \boldsymbol{b}\ \boldsymbol{and}\ \boldsymbol{truth}) \Rightarrow (\boldsymbol{a}\ \boldsymbol{and}\ \boldsymbol{truth}))$$
$$= (\sim(a\ and\ truth) \Rightarrow \sim(\sim b\ and\ truth)) = (\sim \boldsymbol{T}(\boldsymbol{a}) \Rightarrow \sim \boldsymbol{T}(\sim \boldsymbol{b}))$$

and we have:

$$\boldsymbol{T}(\boldsymbol{a}\ \boldsymbol{and}\ \boldsymbol{b}) = ((a\ and\ b)\ and\ truth) = ((a\ and\ truth)\ and\ (b\ and\ truth)) = (\boldsymbol{T}(\boldsymbol{a})\ \boldsymbol{and}\ \boldsymbol{T}(\boldsymbol{b}))$$

$$\boldsymbol{T}(\boldsymbol{a}\ \boldsymbol{or}\ \boldsymbol{b}) = ((a\ or\ b)\ and\ truth) = ((a\ and\ truth)\ or\ (b\ and\ truth)) = (\boldsymbol{T}(\boldsymbol{a})\ \boldsymbol{or}\ \boldsymbol{T}(\boldsymbol{b}))$$

$$\boldsymbol{a} \leftrightarrow (a\ and\ truth) = \boldsymbol{T}(\boldsymbol{a})$$

So all the rules from Section 9.3 except the rule number 29, which would be unverified, are true for contextual implication in place of inclusion and implication.

So, for example:

$$((a \to b)\ and\ (b \to c)) \Rightarrow (a \to c)$$

because:

$$((a \to b)\ and\ (b \to c)) = ((T(a) \Rightarrow T(b))\ and\ (T(b) \Rightarrow T(c))) \Rightarrow (T(a) \Rightarrow T(c))$$
$$= (a \to c)$$

5.38.2 More about statements

Any statement exists in every situation, since it can arise in every situation as the result of interpretation of a sentence, that expresses the statement in this situation in some language. So just as any number any statement is abstract being, the class of which has past, present, future and potential members, that shares the same identity, and when we will assume, that we refer to a concrete statement, then just as for numbers we will have on mind only the current one existing member. Of course, we have past, present, future and potential decidable statements due to their existence in physical reality, so they respectively **were, are, will be or potentially are** true or false.

Since a decidable statement does not depend on a situation, it exists the same way as a number in every situation and its logical value do not ever change, because only the process of interpretation and not the result of interpretation depends on a situation. In other words, interpretation is making a piece of information, that is understandable in every situation (situation-independent, absolute), of understandable often only in some situations (situation-dependent, relative) sentence, so that a mind can express back the same piece of information in different situation for other mind to interpret it again. The logical value of a decidable statement does not ever change, because a decidable statement states always something about the past, present time or predictable future, that does not ever change. Remember, that the statement of a sentence has always defined absolute time (dates and times), even if the sentence has relative time (e.g.: “now”, “today”, “tomorrow”, “next week”, etc.). Remember also, that a declarence is not strict, if it has no strict time quantifier, and then its interpretation depends on a speaker. There are non-strict time quantifiers (e.g.: often, rarely, usually, etc.) and strict time quantifiers (e.g.: always, all the time, never, ever, sometimes, etc.). For more details see Chapter 8.

Truth consists always of present true decidable statements, that, of course, are not only statements about present time. Decidable statements do not ever change their logical value. Only undecidable statements about the unpredictable future will become decidable, when this future will become predictable or will happen.

The time of existence of a statement is the time it potentially arises from interpretation and it is completely different thing from the time the action it contains happens. Of course, these times can be equal, but they have completely different meanings.

To better understand, how and why the logical value of a statement does not change, remind yourself, that every statement of a sentence is a piece of information, that incorporates all details, that its logical value depends on and that can be default in the sentence and then they come from the situation, in which the statement is expressed.

For example, the statement of a strict simple present continuous sentence with default time and space quantifier “here and now” will be different for the same speaker in every situation, because for the same speaker any two different situations have different time or place.

So here the most important will be the fact, that if function $(s, c)^x$ for some constant x is one-to-one correspondence, then the result, which in this case is a statement, of such a bijective function strictly depends on the arguments s, c given to this function. In other words, a bijective function is invertible, so we can decode all arguments from a result of such a function. So since this function in this case is interpretation of sentence x in situation (s, c) and its inversion is expression of a statement $(s, c)^x$ in situation (s, c), these arguments will be encoded in a result, which is a statement, by the rules of interpretation for the process of expression (about which we can assume for simplicity, that it is unambiguous) to decode them.

It happens the following way for any declarence x, any speaker s and any context c:

$$(s,c)^x = \begin{pmatrix} \textit{the statement of } x \textit{ is true,} \\ \textit{if only } x \textit{ is said by speaker } s \textit{ in context } c \end{pmatrix}$$

$$= ((x \textit{ is said by speaker } s \textit{ in context } c) \Rightarrow (\textit{such statement of } x \textit{ is true}))$$

$$= \Big((\textit{the current speaker is } s \textit{ and the current context is } c)$$

$$\Rightarrow (\textit{the statement of } x \textit{ is true}) = (\_,\_)^x\Big)$$

Here is the proof:

First of all we have the following rule for any relation $F(x)$ and any tuples (a) and (b) that fits into tuple (x):

$$\boldsymbol{F}(\boldsymbol{b}) = \bigwedge_{(c)=(b)} F(c) = \left(((\boldsymbol{a}) = (\boldsymbol{b})) \Rightarrow \boldsymbol{F}(\boldsymbol{a})\right) = \bigvee_{(c)=(b)} F(c) = \sim\left(((\boldsymbol{a}) = (\boldsymbol{b})) \Rightarrow \sim\boldsymbol{F}(\boldsymbol{a})\right)$$

For more details see Section 5.65.

So for any sentence x, any speaker s and any context c:

$$(s,c)^x = \left(((p,q) = (s,c)) \Rightarrow (p,q)^x\right) = \sim\left(((p,q) = (s,c)) \Rightarrow \sim(p,q)^x\right)$$

for:

$$p = (\textit{the speaker by which x is said})$$

$$q = (\textit{the context in which x is said})$$

We get:

$$(s,c)^x = ((\textit{x is said by speaker s in context c}) \Rightarrow (p,q)^x)$$
$$= \sim((\textit{x is said by speaker s in context c}) \Rightarrow \sim(p,q)^x)$$

Q.E.D.

Of course, two different statements can have the same identity. Moreover, two statements of the same sentence in different situations can be the same statement.

We can verify, that for any sentence x, the interpretation of which does not depend on the situation, in which it is interpreted, and any speaker s and any context c:

$$(s,c)^x = \bigwedge_{\substack{p \in Speaker^{\uparrow}, \\ q \in Context^{\uparrow}}} (p,q)^x = \bigwedge_{\substack{p \in Speaker^{\uparrow}, \\ q \in Context^{\uparrow}}} \left(((s,c) = (p,q)) \Rightarrow (s,c)^x\right)$$

$$= \left(\bigvee_{\substack{p \in Speaker^{\uparrow}, \\ q \in Context^{\uparrow}}} ((s,c) = (p,q)) \Rightarrow (s,c)^x \right) = (any, any)^x$$

where *any* means the same as “does not matter which one”,

which means, that a sentence is situation independent, so it has the same statement in every situation, if and only if its statement has the same logical value in all situations, which is true only for the empty statement of tautologies and for its negation.

Remember also the simplest fact:

$$(a = b) \Rightarrow (a \equiv b)$$

So we also have:

$$\sim(a \equiv b) \Rightarrow (a \neq b)$$

which means, that different logical values of any two statements always mean, that these statements are different.

5.38.3 Contextual implication as a tautology and implication as a quantifier

Here is a proof that quantifiers can be replaced by the appropriate use of implication:

For any tuple $(y) \in D_{A,B}$ and any relations $A(x)$ and $B(x)$, each of which as the restriction of a quantifier makes tuple (x) a tuple of bound variables and is decidable for any tuple (x) of arguments, for which these relations are defined (where $D_{A,B}$ is the common domain of $A(x)$ and $B(x)$):

$$\begin{aligned}\bigwedge_{A(x)} \boldsymbol{B}(\boldsymbol{x}) &= \bigwedge_{(x) \in \{(e): A(e)\}} ((x) \in \{(e): B(e)\})\\ &= \bigwedge_{(x) \in D_{A,B}} \Big(((x) \in \{(e): A(e)\}) \rightarrow ((x) \in \{(e): B(e)\}) \Big)\\ &= (\{(\boldsymbol{e}): \boldsymbol{A}(\boldsymbol{e})\} \subset \{(\boldsymbol{e}): \boldsymbol{B}(\boldsymbol{e})\}) =\\ &= \Big(((y) \in \{(e): A(e)\}) \Rightarrow ((y) \in \{(e): B(e)\}) \Big) = \big(\boldsymbol{A}(\boldsymbol{y}) \Rightarrow \boldsymbol{B}(\boldsymbol{y})\big)\end{aligned}$$

For more details and a proof of the above steps see Section 6.3.

So:

$$\bigwedge_{A(x)} \boldsymbol{B}(\boldsymbol{x}) = (\{(\boldsymbol{e}): \boldsymbol{A}(\boldsymbol{e})\} \subset \{(\boldsymbol{e}): \boldsymbol{B}(\boldsymbol{e})\}) = \big(\boldsymbol{A}(\boldsymbol{y}) \Rightarrow \boldsymbol{B}(\boldsymbol{y})\big)$$

So:

$$\bigwedge_{A(x)} \boldsymbol{B}(\boldsymbol{x}) = \big(\boldsymbol{A}(\boldsymbol{y}) \Rightarrow \boldsymbol{B}(\boldsymbol{y})\big) = \big(\sim\boldsymbol{B}(\boldsymbol{y}) \Rightarrow \sim\boldsymbol{A}(\boldsymbol{y})\big) = \bigwedge_{\sim B(x)} \sim\boldsymbol{A}(\boldsymbol{x})$$

$$\bigvee_{A(x)} \boldsymbol{B}(\boldsymbol{x}) = \sim \bigwedge_{A(x)} \sim\boldsymbol{B}(\boldsymbol{x}) = \sim\big(\boldsymbol{A}(\boldsymbol{y}) \Rightarrow \sim\boldsymbol{B}(\boldsymbol{y})\big) = \sim\big(\boldsymbol{B}(\boldsymbol{y}) \Rightarrow \sim\boldsymbol{A}(\boldsymbol{y})\big) = \bigvee_{B(x)} \boldsymbol{A}(\boldsymbol{x})$$

So universal and existential quantifiers can be replaced by the appropriate use of implications.

Q.E.D.

So for any declarative sentences q and p and any set S of all logically possible situations, for which both sentences q and p are meaningful (I assume, that only any logically possible situation x can be an argument of operation x^y for any declarence y), and any b that belongs to S we have:

$$\bigwedge_{a^p} a^q = \bigwedge_{a \in S} (a^p \rightarrow a^q) = \left(b^p \Rightarrow b^q\right)$$

So:

$$\bigwedge_{\boldsymbol{a^p}} \boldsymbol{a^q} = \bigwedge_{\boldsymbol{a \in S}} (\boldsymbol{a^p} \rightarrow \boldsymbol{a^q}) = \bigwedge_{\boldsymbol{a \in S}} \left(\boldsymbol{a^p} \Rightarrow \boldsymbol{a^q}\right)$$

And:

$$\bigwedge_{\boldsymbol{a \in S}} (\boldsymbol{a^p} \leftrightarrow \boldsymbol{a^q}) = \bigwedge_{a \in S} ((a^p \rightarrow a^q)\ and\ (a^q \rightarrow a^p)) = \left(\bigwedge_{a \in S} (a^p \rightarrow a^q)\ and\ \bigwedge_{a \in S} (a^q \rightarrow a^p)\right)$$

$$= \left(\bigwedge_{a \in S} \left(a^p \Rightarrow a^q\right)\ and\ \bigwedge_{a \in S} \left(a^q \Rightarrow a^p\right)\right)$$

$$= \bigwedge_{a \in S} \left(\left(a^p \Rightarrow a^q\right)\ and\ \left(a^q \Rightarrow a^p\right)\right) = \bigwedge_{\boldsymbol{a \in S}} (\boldsymbol{a^p} = \boldsymbol{a^q})$$

So:

$$\bigwedge_{\boldsymbol{a \in S}} (\boldsymbol{a^p} \leftrightarrow \boldsymbol{a^q}) = \bigwedge_{\boldsymbol{a \in S}} (\boldsymbol{a^p} = \boldsymbol{a^q})$$

Q.E.D.

A part of this can be proved also the following way:

For any declarative sentences q and p and any set S of all logically possible situations, for which both sentences q and p are meaningful (I assume, that only any logically possible situation x can be an argument of operation x^y for any declarence y), and any b that belongs to S we have:

$$\bigwedge_{\boldsymbol{a^p}} \boldsymbol{a^q} = \bigwedge_{\boldsymbol{a \in S}} (\boldsymbol{a^p} \rightarrow \boldsymbol{a^q}) = (\{a\colon (a \in S)\ and\ a^p\} \subset \{a\colon (a \in S)\ and\ a^q\})$$

$$= \left(\{a\colon (a \in S)\ and\ a^q\} = (\{a\colon (a \in S)\ and\ a^p\} \cup \{a\colon (a \in S)\ and\ a^q\})\right)$$

$$\Rightarrow \Bigg((b \in \{a\colon (a \in S)\ and\ a^q\})$$

$$= \left(\left(b \in (\{a\colon (a \in S)\ and\ a^p\} \cup \{a\colon (a \in S)\ and\ a^q\})\right)\right)\Bigg)$$

$$= \Bigg((b \in \{a\colon (a \in S)\ and\ a^q\})$$

$$= \left((b \in \{a\colon (a \in S)\ and\ a^p\})\ or\ (b \in \{a\colon (a \in S)\ and\ a^q\})\right)\Bigg)$$

$$= (b^q = (b^p\ or\ b^q)) = \left(\boldsymbol{b^p} \Rightarrow \boldsymbol{b^q}\right)$$

So:

$$\bigwedge_{a^p} a^q = \bigwedge_{a \in S} (a^p \to a^q) \Rightarrow \left(b^p \Rightarrow b^q\right) \Rightarrow (b^p \to b^q)$$

So:

$$\bigwedge_{a^p} a^q = \bigwedge_{a \in S} (a^p \to a^q) \Rightarrow \bigwedge_{a \in S} \left(a^p \Rightarrow a^q\right) \Rightarrow \bigwedge_{a \in S} (a^p \to a^q)$$

So:

$$\bigwedge_{\boldsymbol{a^p}} \boldsymbol{a^q} = \bigwedge_{\boldsymbol{a \in S}} (\boldsymbol{a^p} \to \boldsymbol{a^q}) = \bigwedge_{\boldsymbol{a \in S}} \left(\boldsymbol{a^p} \Rightarrow \boldsymbol{a^q}\right)$$

And:

$$\bigwedge_{\boldsymbol{a \in S}} (\boldsymbol{a^p} \leftrightarrow \boldsymbol{a^q}) = \bigwedge_{\boldsymbol{a \in S}} (\boldsymbol{a^p} = \boldsymbol{a^q})$$

So, first of all, a declarence, that contains implication of two statements, is a tautology if and only if a declarence, that contains the appropriate contextual implication, is a tautology.

Secondly, a declarence, that contains equality of two statements, is a tautology if and only if a declarence, that contains the appropriate contextual equality, is a tautology.

Q.E.D.

So we have also for any tuple $(y) \in D_{A,B}$ and any relations $A(x)$ and $B(x)$, each of which as the restriction of a quantifier makes tuple (x) a tuple of bound variables and is decidable for any tuple (x) of arguments, for which these relations are defined (where $D_{A,B}$ is the common domain of $A(x)$ and $B(x)$):

$$\bigwedge_{\boldsymbol{A(x)}} \boldsymbol{B(x)} = \bigwedge_{(x) \in D_{A,B}} (A(x) \to B(x)) \Rightarrow \left((A(y) \to B(y)) = \varnothing_{\varsigma}\right)$$
$$= \left(\left(A(y) \Rightarrow B(y)\right) = \varnothing_{\varsigma}\right) \Rightarrow \left(\boldsymbol{A(y)} \Rightarrow \boldsymbol{B(y)}\right)$$

So:

$$\bigwedge_{\boldsymbol{A(x)}} \boldsymbol{B(x)} \Rightarrow \left(\boldsymbol{A(y)} \Rightarrow \boldsymbol{B(y)}\right)$$

Q.E.D.

5.38.4 Contextual implication as a tautology – continuation

For any y and any relations $A(x)$ and $B(x)$, each of which as the restriction of a quantifier makes x a bound variable and is decidable for any x, we have:

$$\begin{aligned}\bigwedge_{A(x)} B(x) = \bigwedge_{x}(A(x) \rightarrow B(x)) &= (\{x{:}A(x)\} \subset \{x{:}B(x)\})\\ &= \left(\{x{:}B(x)\} = (\{x{:}A(x)\} \cup \{x{:}B(x)\})\right)\\ &\Rightarrow \left((y \in \{x{:}B(x)\}) = \left(y \in (\{x{:}A(x)\} \cup \{x{:}B(x)\})\right)\right)\\ &= \left((y \in \{x{:}B(x)\}) = \left((y \in \{x{:}A(x)\}) \text{ } or \text{ } (y \in \{x{:}B(x)\})\right)\right)\\ &= \left(B(y) = (A(y) \text{ } or \text{ } B(y))\right)\\ &= (A(y) \Rightarrow B(y))\end{aligned}$$

So:

$$\bigwedge_{x}(A(x) \rightarrow B(x)) \Rightarrow (A(y) \Rightarrow B(y)) \Rightarrow (A(y) \rightarrow B(y))$$

So:

$$\bigwedge_{x}(A(x) \rightarrow B(x)) \Rightarrow \bigwedge_{x}(A(x) \Rightarrow B(x)) \Rightarrow \bigwedge_{x}(A(x) \rightarrow B(x))$$

So:

$$\bigwedge_{x}(\boldsymbol{A}(x) \rightarrow \boldsymbol{B}(x)) = \bigwedge_{x}(\boldsymbol{A}(x) \Rightarrow \boldsymbol{B}(x))$$

5.38.5 Why "material implication" is not self-sufficient implication

This is the right place, where "material implication" can take its place, and not as self-sufficient (independent) implication between two arguments, that are statements, but as a **contextual implication**, that says, that a premise only together with a context implicates some statement. The word "imply" is close to the word "contain" (www.merriam-webster.com, the third definition: "to contain potentially"), so it cannot be used, when we assume, that the whole truth (context) together with our premise contains/implicates some statement, because then our premise alone does not have to contain/implicate this statement.

Remember, that for any statement a weaker than truth $((a \rightarrow b) = (a \Rightarrow b))$ is false.

As you can see, "material implication" is derived from new implication, which is true logical consequence, so logical consequence cannot belong to meta-language distinct from language, that describes object language, to which such "material implication" belongs. For more details see Chapter 14 and Section 15.13. What is more, classical logic, that cannot describe everything, including itself, is simply handicapped and does not deserve to be called true logic, especially when there is new logic, that is described in this book and can describe everything.

5.38.6 Factual and nonfactual conditionals

And only if for any concrete statements a, b and c we assume, that the following rules are true:

$$\left(\boldsymbol{if}\ \hat{a}\ \boldsymbol{and}\ \hat{b}\ \boldsymbol{then}\ \hat{c}\right) \Rightarrow \left(\boldsymbol{if}\ \hat{a}\ \boldsymbol{then}\ \left(\boldsymbol{if}\ \hat{b}\ \boldsymbol{then}\ \hat{c}\right)\right)$$

or equivalently:

$$\left(\boldsymbol{if}\ \hat{a}\ \boldsymbol{then}\ \hat{b}\ \boldsymbol{or}\ \hat{c}\right) \Rightarrow \left(\boldsymbol{if}\ \hat{a}\ \boldsymbol{then}\ \left(\boldsymbol{if}\ \check{b}\ \boldsymbol{then}\ \hat{c}\right)\right)$$

and some obvious rules, that sentence ***if* . . . *then*** demands to be true for the sentence to be still some kind of implication:

$$\left(if\ \hat{a}\ then\ \hat{b}\right) = \left(if\ \check{b}\ then\ \check{a}\right)$$

$$a = \left(if\ \widehat{truth}\ then\ \hat{a}\right) \Rightarrow \left(if\ \widehat{truth}\ and\ \hat{b}\ then\ \hat{a}\right)$$

$$(\sim b\ and\ a) \Rightarrow \sim\left(if\ \hat{a}\ then\ \hat{b}\right)$$

So also:

$$\begin{aligned}\sim a &= \left(if\ \hat{a}\ then\ \widehat{untruth}\right) = \left(if\ \widehat{truth}\ then\ \check{a}\right) \Rightarrow \left(if\ \widehat{truth}\ and\ \check{b}\ then\ \check{a}\right)\\ &= \left(if\ \hat{a}\ then\ \widehat{untruth}\ or\ \hat{b}\right)\end{aligned}$$

So:

$$\sim a = \left(if\ \hat{a}\ then\ \widehat{untruth}\right) \Rightarrow \left(if\ \hat{a}\ then\ \widehat{untruth}\ or\ \hat{b}\right)$$

where all rules except the first two are obvious.

Then:

$$(\boldsymbol{a} \rightarrow \boldsymbol{b}) = \left(\boldsymbol{if}\ \hat{a}\ \boldsymbol{then}\ \hat{b}\right)$$

But it still is not self-sufficient implication.

Here is a proof, that is simplified for simplicity to decidable statements:

1′

Assume:

$$(a \to b) = \left(if\ \hat{a}\ then\ \hat{b}\right)$$

Then:

$$\left(if\ \hat{a}\ and\ \hat{b}\ then\ \hat{c}\right) \Rightarrow \left(if\ \hat{a}\ then\ \left(if\ \hat{b}\ then\ \hat{c}\right)\right)$$

will become:

$$L = ((a\ and\ b) \to c) \Rightarrow (a \to (b \to c)) = R$$

which is a simple tautology, that can be easily proved:

As you can easily see, L the same as R is false only if c is false and a and b is true.

Q.E.D.

And:

$$\left(if\ \hat{a}\ and\ \hat{b}\ then\ \hat{c}\right) \Rightarrow \left(if\ \hat{a}\ then\ \left(if\ \check{b}\ then\ \hat{c}\right)\right)$$

will become:

$$L = (a \to (b\ or\ c)) \Rightarrow (a \to (\sim b \to c)) = R$$

which is a simple tautology, that can be easily proved:

As you can easily see, L the same as R is false only if c and b are false and a is true.

Q.E.D.

2′

Now assume:

$$\left(if\ \hat{a}\ then\ \hat{b}\ or\ \hat{c}\right) \Rightarrow \left(if\ \hat{a}\ then\ \left(if\ \hat{b}\ then\ \hat{c}\right)\right)$$

and remember, that we have the following four obvious truths for any concrete statements a, b and c:

$$\left(if\ \hat{a}\ then\ \hat{b}\right) = \left(if\ \check{b}\ then\ \check{a}\right)$$

$$a = \left(if\ \widehat{truth}\ then\ \hat{a}\right) \Rightarrow \left(if\ \widehat{truth}\ and\ \hat{b}\ then\ \hat{a}\right)$$

$$\sim a = \left(if\ \hat{a}\ then\ \widehat{untruth}\right) \Rightarrow \left(if\ \hat{a}\ then\ \widehat{untruth}\ or\ \hat{b}\right)$$

$$(\sim b\ and\ a) \Rightarrow \sim\left(if\ \hat{a}\ then\ \hat{b}\right)$$

So for any concrete statements a and b:

$$b = \left(if\ \widehat{truth}\ then\ \hat{b}\right) \Rightarrow \left(if\ \widehat{truth}\ and\ \hat{a}\ then\ \hat{b}\right) \Rightarrow \left(if\ \widehat{truth}\ then\ \left(if\ \hat{a}\ then\ \hat{b}\right)\right)$$

$$= \left(if\ \hat{a}\ then\ \hat{b}\right) = \left(if\ \check{b}\ then\ \check{a}\right)$$

So:

$$b \Rightarrow \left(if\ \hat{a}\ then\ \hat{b}\right) = \left(if\ \check{b}\ then\ \check{a}\right)$$

So also for any concrete statements a and b:

$$\sim b \Rightarrow \left(if\ \hat{b}\ then\ \hat{a}\right) = \left(if\ \check{a}\ then\ \check{b}\right)$$

We would get exactly the same from the equivalent rule:

$$\left(if\ \hat{a}\ then\ \hat{b}\ or\ \hat{c}\right) \Rightarrow \left(if\ \hat{a}\ then\ \left(if\ \check{b}\ then\ \hat{c}\right)\right)$$

Then:

$$\sim b = \left(if\ \hat{b}\ then\ \widehat{untruth}\right) \Rightarrow \left(if\ \hat{b}\ then\ \widehat{untruth}\ or\ \hat{a}\right) \Rightarrow \left(if\ \hat{b}\ then\ \left(if\ \widecheck{untruth}\ then\ \hat{a}\right)\right)$$

$$= \left(if\ \hat{b}\ then\ \left(if\ \widehat{truth}\ then\ \hat{a}\right)\right) = \left(if\ \hat{b}\ then\ \hat{a}\right) = \left(if\ \check{a}\ then\ \check{b}\right)$$

Summing up, for any concrete statements a and b:

$$b \Rightarrow \left(if\ \hat{a}\ then\ \hat{b}\right)$$

$$\sim a \Rightarrow \left(if\ \hat{a}\ then\ \hat{b}\right)$$

$$(b\ or \sim a) \Rightarrow \left(if\ \hat{a}\ then\ \hat{b}\right)$$

and we know that:

$$(\sim b\ and\ a) \Rightarrow \sim\left(if\ \hat{a}\ then\ \hat{b}\right)$$

So:

$$(b \; or \sim a) \Leftarrow \left(if\ \hat{a}\ then\ \hat{b}\right)$$

Summing up:

$$(a \to b) = (b \; or \sim a) = \left(if\ \hat{a}\ then\ \hat{b}\right)$$

Q.E.D.

Then for any concrete statements *a* and *b*:

$$(a \to b) = (if\ a\ is\ fulfilled\ then\ b\ is\ fulfilled) = \left(if\ you\ assume\ that\ \hat{a}\ then\ \hat{b}\right)$$
$$= \left(if\ \hat{a}\ then\ \hat{b}\right) = \left(\hat{b}\ if\ \hat{a}\right) = \left(if\ \hat{a}\ then\ it\ means\ that\ \hat{b}\right)$$

where $\left(if\ \hat{a}\ then\ \hat{b}\right)$ concerns only factual conditionals, because nonfactual conditionals have a more complex structure, e.g.:

(**if** it rains, **then** I will stay at home) = (**if** it <u>will</u> rain, **then** I <u>will</u> stay at home) = **((it <u>will</u> rain) → (I <u>will</u> stay at home))** ≠ ((it rains) → (I will stay at home))

(**If** Earth was in Andromeda galaxy, **then** there would not be Milky Way in the sky) = ((**If** Earth <u>is</u> in Andromeda galaxy, **then** there <u>is</u> not Milky Way in the sky) and (Earth <u>is</u> in Andromeda galaxy) is false) ⇒ **((Earth <u>is</u> in Andromeda galaxy) → (there <u>is</u> not Milky Way in the sky))** ≠ ((Earth was in Andromeda galaxy) → (there would not be Milky Way in the sky))

(**If** she had been studying before the exam, **then** she would have passed it) = ((**if** she <u>has</u> been studying before the exam, **then** she <u>has</u> passed it) and (she <u>has</u> been studying before the exam) is false) ⇒ **((she <u>has</u> been studying before the exam) → (she <u>has</u> passed it))** ≠ ((she had been studying before the exam) → (she would have passed it)).

5.38.7 The final solution to all "problems with implication"

The following rule:

$$\left(if\ \hat{a}\ and\ \hat{b}\ then\ \hat{c}\right) \Rightarrow \left(if\ \hat{a}\ then\ \left(if\ \hat{b}\ then\ \hat{c}\right)\right)$$

comes from the simple fact, that we often say in a sentence, that something can be inferred from something, not stating all conditions, that would be necessary, if we did not know anything about the situation in which the sentence is said. For example, we will admit, that the following inference is true:

If It is raining **then** the earth is wet

which would not be true, for example, in cosmic space far away from a planet. So we understand both when saying and when interpreting, that we have some situation, that we consider, when we are using sentence ***if* . . . *then***. In other words, in this example we assume, that it is true, that we are on Earth, so there is the ground under our feet. So we should say:

if we are on Earth **and** it is raining **then** the earth is wet.

to make it understandable for some aliens from cosmic space, because it would be then interpreted the same way in any context.

And that is correct true implication and correct inference, so we can write it as:

$$(\textit{We are on Earth and it is raining}) \Rightarrow (\textit{The earth is wet})$$

And since using clause *if x* of a sentence we assume, that *x* is true in the clause *then y*, we can make this implication understandable for aliens from cosmic space also in the following way:

if we are on Earth **then if** it is raining **then** the earth is wet.

From this we have our general rule:

$$\left(\textit{if } \hat{a} \textit{ and } \hat{b} \textit{ then } \hat{c}\right) \Rightarrow \left(\textit{if } \hat{a} \textit{ then } \left(\textit{if } \hat{b} \textit{ then } \hat{c}\right)\right)$$

and that is all. This is the final explanation of all problems with implication.

And that is, why if contextual implication (called in classical logic "material implication") of two statements is true in all situations (in all models or in all contexts and for all speakers), then implication (called in classical logic "logical consequence") is true, because then it is situation-independent, which means that it states in the premise of implication all conditions the inference of the conclusion of this implication depends on.

Q.E.D.

5.38.8 Operator "so"

For any concrete statements *a* and *b*:

If *a* is true and $a \rightarrow b$, then *a* in the context of truth is a sufficient reason (contextual sufficient reason) for *b*. In other words, $a \rightarrow b$ in the context of truth is a proof (contextual proof) of *b*, because then we can always explain, that $\hat{b}$ *since* $\hat{a}$.

We will assume, that in formal language for any concrete statements *a* and *b*:

$$(a \looparrowright b) = (b \looparrowleft a) = (a \textit{ and } b \textit{ and } (a \rightarrow b)) = \left(\hat{a} \textit{ which means that } \hat{b}\right) = \left(\hat{a} \textit{ that is } \hat{b}\right)$$

$$= \left(\hat{a} \textit{ so } \hat{b}\right) = \left(\hat{a} \textit{ thus } \hat{b}\right) = \left(\hat{a} \textit{ hence } \hat{b}\right) = \left(\hat{b} \textit{ since } \hat{a}\right) = \left(\textit{since } \hat{a}, \hat{b}\right)$$

$$= \left(\hat{b} \textit{ inasmuch as } \hat{a}\right) = \left(\hat{b} \textit{ forasmuch as } \hat{a}\right)$$

$$(a \textit{ and } b \textit{ and } (a \rightarrow b)) \leftrightarrow (a \textit{ and } (a \rightarrow b))$$

For example, the statement of the sentence “a triangle is a figure” ($= a$) implicates, that “a triangle has an area” ($= b$), so all the above sentences have the same true statement, e.g.: “a triangle is a figure, which means that it has an area”.

And we have for any concrete statements a, b and c:

$$(a \looparrowright b \looparrowright c) = ((a \looparrowright b)\ \mathit{and}\ (b \looparrowright c))$$

And we have, that $\looparrowright (a, b)$ is a transitive relation:

$$((a \looparrowright b)\ \mathit{and}\ (b \looparrowright c)) \Rightarrow (a \looparrowright c)$$

and left-distributive over “and”:

$$((a \looparrowright b)\ \mathit{and}\ (a \looparrowright c)) = (a \looparrowright (b\ \mathit{and}\ c))$$

and left-distributive over “or”:

$$((a \looparrowright b)\ \mathit{or}\ (a \looparrowright c)) = (a \looparrowright (b\ \mathit{or}\ c))$$

and we have the following rules:

$$\bigwedge_{\{a,x,y\} \subset \mathit{Statement}^{\Uparrow}} \sim((a \looparrowright x)\ \mathit{and}\ (a \looparrowright y)\ \mathit{and}\ (x \rightarrow \sim y))$$

and:

$$\bigwedge_{\{a,b,x\} \subset \mathit{Statement}^{\Uparrow}} \sim((a \looparrowright x)\ \mathit{and}\ (b \looparrowright x)\ \mathit{and}\ (a \rightarrow \sim b))$$

5.38.9 Operator “If . . . then . . . else”

For any concrete statements a, b and c:

$$\left(\mathit{if}\ \hat{a}\ \mathit{then}\ \hat{b}\ \mathit{else}\ \hat{c}\right) = \left(\mathit{if}\ \hat{a}\ \mathit{then}\ \hat{b},\ \mathit{otherwise}\ \hat{c}\right) = (a\ ??\ b :: c) = ((a \rightarrow b)\ \mathit{and}\ (\sim a \rightarrow c))$$

So:

$$\left(a\ ??\ ((a\ ?\ b : c) = b) :: ((a\ ?\ b : c) = c)\right)$$

Where $(a\ ?\ b : c)$ is a well-known from C++ programming language ternary operator, that means, that if a is true, then b is the result of that expression, and otherwise c is the result of that expression.

5.38.10 Two-way factual conditional

For any concrete statements a and b:

$$(a \leftrightarrow b) = ((a \rightarrow b)\ and\ (a \leftarrow b)) = \left(\hat{a}\ if\ and\ only\ if\ \hat{b}\right)$$

$$= \left(if\ \hat{a}\ then\ and\ only\ then\ \hat{b}\right) = \left(\hat{a}\ then\ and\ only\ then\ when\ \hat{b}\right)$$

$$= \left(\hat{a}\ only\ when\ \hat{b}\right) = \left(\hat{a}\ only\ if\ \hat{b}\right) = \left(\hat{b}\ as\ far\ as\ \hat{a}\right)$$

$$= \left(\hat{b}\ only\ as\ far\ as\ \hat{a}\right) = \left(\hat{b}\ as\ long\ as\ \hat{a}\right) = \left(\hat{b}\ only\ as\ long\ as\ \hat{a}\right)$$

$$= \left(\hat{b}\ under\ condition\ that\ \hat{a}\right) = \left(\hat{b}\ provided\ that\ \hat{a}\right) = \left(\breve{a}\ until\ \hat{b}\right)$$

$$= \left(\breve{a}\ uptil\ \hat{b}\right) = \left(\breve{a}\ unless\ \hat{b}\right) = \left(\breve{a}\ except\ when\ \hat{b}\right)$$

5.39 Equality operator

For any concrete statements a and b:

$$(statements\ a\ and\ b\ are\ the\ same) = (a = b) = ((a \Rightarrow b)\ and\ (b \Rightarrow a)) = \left(\hat{a}\ only\ if\ only\ \hat{b}\right)$$

$$(a = b) = ((a \Rightarrow b)\ and\ (b \Rightarrow a)) = ((\sim b \Rightarrow \sim a)\ and\ (\sim a \Rightarrow \sim b)) = (\sim a = \sim b)$$

$$(a = b) = \left(a = b = (a \cup b) = (a\ and\ b) = \left(a\ \hat{\cap}\ b\right)\right)$$

$$(a = b) \Rightarrow \left(\overline{a} = \overline{b}\right) = (a \equiv b) = (a \rightleftharpoons b)$$

$$(a = b = c) = ((a = b)\ and\ (b = c))$$

And we have, that $= (a, b)$ is transitive relation:

$$((a = b)\ and\ (b = c)) \Rightarrow (a = c)$$

And:

$$(a \Rightarrow a) = (a = a)$$

Two statements are the same if and only if both are necessary and sufficient conditions of each other.

5.40 Possibility of implication operator

For any concrete statements a and b:

$$possibility\ of\ implication(a,b) = (\rightharpoonup(a,b)) = (a \rightharpoonup b) = (b \leftharpoondown a)$$
$$= (a\ and\ b\ have\ logical\ values\ permissible\ for\ implication)$$

$$(a \rightharpoonup b) = (\sim(a\ is\ true\ and\ b\ is\ not\ true)\ or\ (a\ is\ undecidable\ and\ b\ is\ undecidable))$$
$$= (a\ is\ not\ true\ or\ b\ is\ true\ or\ (a\ is\ undecidable\ and\ b\ is\ undecidable))$$

In other words, $(a \rightharpoonup b)$ means, that the logical values of these statements are not impossible for the implication between a and b. So, this function, unlike real implication, abstracts completely from the meaning of the appropriate sentences (information they express), that makes the statement of the second one to be inferred from the statement of the first one. For more details, see the definition of operator $\rightarrow$ in Section 5.38.

So, $(a \Rightarrow b)$ implicates, that this is impossible, that $\sim(a \rightharpoonup b)$, so this is necessary, that $(a \rightharpoonup b)$.

It is sometimes called "material implication" and it is not self-sufficient implication at all. For example, in classical logic, it is only a plain disjunction with negated left argument. In new logic, it is just a disjunction, that does not consider the meaning of sentences at all, while implication strictly depends on the meaning of sentences. For true derivation of "material implication", see Section 5.38.

$$\overline{possibility\ of\ implication} = (\overline{\rightharpoonup}) = \left\{ \begin{matrix} ((false, false), true), \\ ((false, true), true), \\ ((true, false), false), \\ ((true, true), true), \\ ((undecidable, false), false), \\ ((undecidable, true), true), \\ ((false, undecidable), true), \\ ((true, undecidable), false), \\ ((undecidable, undecidable), true) \end{matrix} \right\}$$

$$\overline{a \rightharpoonup b} = (\bar{a} \overline{\rightharpoonup} \bar{b})$$

If we have the logical function from classical logic $\overline{POSSIBILITY\ OF\ IMPLICATION}$, then:

$$\overline{POSSIBILITY\ OF\ IMPLICATION} \subset \overline{possibility\ of\ implication}$$

$$(a \rightharpoonup b) = (\sim b \rightharpoonup \sim a)$$

$$(a \Rightarrow b) \Rightarrow (a \rightharpoonup b)$$

$$\sim(a \rightharpoonup b) \Rightarrow \sim(a \Rightarrow b)$$

But:

$$\bigvee_{\{a,b\} \subset Statement^{\Uparrow}} \sim\left((a \rightharpoonup b) \Rightarrow (a \Rightarrow b)\right)$$

And:

$$\bigvee_{\{a,b\} \subset Statement^{\Uparrow}} \sim\left((a \rightharpoonup b) \equiv (a \Rightarrow b)\right)$$

$$(a \rightharpoonup b \rightharpoonup c) = \left((a \rightharpoonup b) \text{ } and \text{ } (b \rightharpoonup c)\right)$$

And we have, that $\rightharpoonup (a, b)$ is a transitive relation:

$$\left((a \rightharpoonup b) \text{ } and \text{ } (b \rightharpoonup c)\right) \Rightarrow (a \rightharpoonup c)$$

And left-distributive over “and”:

$$\left((a \rightharpoonup b) \text{ } and \text{ } (a \rightharpoonup c)\right) = \left(a \rightharpoonup (b \text{ } and \text{ } c)\right)$$

And left-distributive over “or”:

$$\left((a \rightharpoonup b) \text{ } or \text{ } (a \rightharpoonup c)\right) = \left(a \rightharpoonup (b \text{ } or \text{ } c)\right)$$

So as a consequence also:

$$\left((a \rightharpoonup c) \text{ } and \text{ } (b \rightharpoonup c)\right) = \left((a \text{ } or \text{ } b) \rightharpoonup c\right)$$

$$\left((a \rightharpoonup c) \text{ } or \text{ } (b \rightharpoonup c)\right) = \left((a \text{ } and \text{ } b) \rightharpoonup c\right)$$

Implication $(a \Rightarrow b)$ is completely different function from $(a \rightharpoonup b)$ and statements of these sentences are not equal, because implication is true only when statement, that imply other statement, really implicates this statement, while $(a \rightharpoonup b)$ is true, when arguments have appropriate logical values, that are permissible for implication. In other words, you can evaluate implication function $(a \Rightarrow b)$ to be true if and only if a really implicates b (a includes piece of information b), and you can evaluate $(a \rightharpoonup b)$ to be true for any two statements, that have appropriate logical values.

And:

$$(a \Rightarrow b) \Rightarrow (a \rightarrow b) = (a \rightharpoonup b)$$

5.41 The list of verified tautologies for possibility of implication

For any concrete statement p:

$$p \rightleftharpoons p$$

$$p \text{ } or \sim p \text{ } or \text{ } p \text{ } is \text{ } undecidable$$

$$\sim(p \textit{ and } \sim p \textit{ and } p \textit{ is decidable})$$

$$\sim\sim p \rightleftharpoons p \rightleftharpoons (p \textit{ or } p) \rightleftharpoons (p \textit{ and } p)$$

For any concrete statements p and q:

$$(p \textit{ and } q) \rightharpoonup p \rightharpoonup (p \textit{ or } q)$$

$$\sim(p \textit{ and } q) \rightleftharpoons (\sim p \textit{ or } \sim q)$$

$$(\sim p \textit{ and } \sim q) \rightleftharpoons \sim(p \textit{ or } q)$$

$$(p \rightharpoonup q) \rightleftharpoons (\sim q \rightharpoonup \sim p)$$

$$((p \rightharpoonup q) \textit{ and } p) \rightharpoonup q$$

$$((p \rightharpoonup q) \textit{ and } \sim q) \rightharpoonup \sim p$$

$$p \rightharpoonup (q \rightharpoonup p)$$

$$\sim p \rightharpoonup (p \rightharpoonup q)$$

For any concrete statements p, q and r:

$$((p \rightharpoonup q) \textit{ and } (q \rightharpoonup r)) \Rightarrow (p \rightharpoonup r)$$

$$(p \rightharpoonup q) \Rightarrow ((q \rightharpoonup r) \rightharpoonup (p \rightharpoonup r))$$

$$((p \textit{ and } q) \rightharpoonup r) = (p \rightharpoonup (q \rightharpoonup r))$$

$$((p \rightharpoonup q) \textit{ and } (p \rightharpoonup r)) = (p \rightharpoonup (q \textit{ and } r))$$

$$((p \rightharpoonup q) \textit{ or } (p \rightharpoonup r)) = (p \rightharpoonup (q \textit{ or } r))$$

$$((q \rightharpoonup p) \textit{ and } (r \rightharpoonup p)) = ((q \textit{ or } r) \rightharpoonup p)$$

$$((q \rightharpoonup p) \textit{ or } (r \rightharpoonup p)) = ((q \textit{ and } r) \rightharpoonup p)$$

And infinitely many more . . .

5.42 Possibility of implication – examples

Since a true statement can implicate only a true statement, it cannot implicate a false statement and cannot implicate an undecidable statement and, thereby, an undecidable statement cannot implicate a false statement. A true statement cannot implicate an undecidable statement, because an undecidable statement cannot be a part of a true statement, since the conjunction of a true statement and an undecidable statement is always undecidable.

All other cases of implication have the following examples (with proofs):

Example of a true and a false implication of a false statement:

$$\Lambda_\varsigma = (\Lambda_\varsigma \; and \; \varnothing_\varsigma) \Rightarrow \varnothing_\varsigma$$

$$\Lambda_\varsigma \Rightarrow \Lambda_\varsigma$$

$$(circle\ is\ a\ square) \Rightarrow (circle\ has\ a\ field)$$

$$(circle\ is\ a\ square) \Rightarrow (circle\ has\ vertexes)$$

Assume, that we have true statement t and false statement f and undecidable statement u.

Example of a true implication of an undecidable statement:

$$(\blacksquare_\varsigma \; and \; \varnothing_\varsigma) \Rightarrow \varnothing_\varsigma$$

$$(u \; and \; t) \Rightarrow t$$

Example of an undecidable implication of a false statement:

Since:

$$(\blacksquare_\varsigma \; and \; \varnothing_\varsigma) \Rightarrow \varnothing_\varsigma$$

So:

$$\sim\varnothing_\varsigma \Rightarrow \sim(\blacksquare_\varsigma \; and \; \varnothing_\varsigma) = \sim\blacksquare_\varsigma$$

So:

$$\Lambda_\varsigma \Rightarrow \sim\blacksquare_\varsigma$$

$$f \Rightarrow (f \; or \; u)$$

Examples of an undecidable implication of an undecidable statement:

$$\blacksquare_\varsigma \Rightarrow \blacksquare_\varsigma$$

$$u \Rightarrow u$$

$$(u \; and \; t) \Rightarrow u \Rightarrow (f \; or \; u)$$

5.43 When you can use contextual implication instead of implication

If we have expressions $F(t)$, $H(t)$, that uses only operators *and* and *or* and does not use negation, then:

$$F\left(\coprod_{i=1}^{n}(a_i \Rightarrow b_i)\right) \Rightarrow F\left(\coprod_{i=1}^{n}(a_i \rightarrow b_i)\right)$$

So also:

$$\left(H\left(\coprod_{i=1}^{n}(a_i \Rightarrow b_i)\right) \Rightarrow F\left(\coprod_{i=1}^{n}(a_i \Rightarrow b_i)\right)\right) \Rightarrow \left(H\left(\coprod_{i=1}^{n}(a_i \Rightarrow b_i)\right) \Rightarrow F\left(\coprod_{i=1}^{n}(a_i \rightarrow b_i)\right)\right)$$

Proof:

$$\coprod_{x}\left(x_{i,j} = \left(a_{t(i,j)} \Rightarrow b_{t(i,j)}\right)\right) \text{ by } x_{i,j}$$

$$\coprod_{y}\left(y_{i,j} = \left(a_{t(i,j)} \rightarrow b_{t(i,j)}\right)\right) \text{ by } y_{i,j}$$

Since we have:

$$x_{i,j} \Rightarrow y_{i,j}$$

Then:

$$\left(and_{j=1,\dots,k_i}\left(x_{i,j}\right) \text{ and } and_{j=1,\dots,l_i}\left(z_{i,j}\right)\right) \Rightarrow \left(and_{j=1,\dots,k_i}\left(y_{i,j}\right) \text{ and } and_{j=1,\dots,l_i}\left(z_{i,j}\right)\right)$$

$$\left(or_{j=1,\dots,k_i}\left(x_{i,j}\right) \text{ or } or_{j=1,\dots,l_i}\left(z_{i,j}\right)\right) \Rightarrow \left(or_{j=1,\dots,k_i}\left(y_{i,j}\right) \text{ or } or_{j=1,\dots,l_i}\left(z_{i,j}\right)\right)$$

Then:

$$or_{i=1,\dots,m}\left(and_{j=1,\dots,k_i}\left(x_{i,j}\right) \text{ and } and_{j=1,\dots,l_i}\left(z_{i,j}\right)\right)$$
$$\Rightarrow or_{i=1,\dots,m}\left(and_{j=1,\dots,k_i}\left(y_{i,j}\right) \text{ and } and_{j=1,\dots,l_i}\left(z_{i,j}\right)\right)$$

$$and_{i=1,\dots,m}\left(or_{j=1,\dots,k_i}\left(x_{i,j}\right) \text{ or } or_{j=1,\dots,l_i}\left(z_{i,j}\right)\right) \Rightarrow and_{i=1,\dots,m}\left(or_{j=1,\dots,k_i}\left(y_{i,j}\right) \text{ or } or_{j=1,\dots,l_i}\left(z_{i,j}\right)\right)$$

And since we can transform $H(\coprod_{i=1}^{n}(a_i \Rightarrow b_i))$ and $F(\coprod_{i=1}^{n}(a_i \Rightarrow b_i))$ to

$$or_{i=1,\dots,m}\left(and_{j=1,\dots,k_i}\left(x_{i,j}\right) \text{ and } and_{j=1,\dots,l_i}\left(z_{i,j}\right)\right)$$

or $and_{i=1,\dots,m}\left(or_{j=1,\dots,k_i}\left(x_{i,j}\right) \text{ or } or_{j=1,\dots,l_i}\left(z_{i,j}\right)\right)$, so for:

$$H\left(\coprod_{i=1}^{n}(a_i \Rightarrow b_i)\right) \Rightarrow F\left(\coprod_{i=1}^{n}(a_i \Rightarrow b_i)\right)$$

We have, that:

$$H\left(\coprod_{i=1}^{n}(a_i \Rightarrow b_i)\right) \Rightarrow F\left(\coprod_{i=1}^{n}(a_i \Rightarrow b_i)\right) \Rightarrow F\left(\coprod_{i=1}^{n}(a_i \rightarrow b_i)\right)$$

So:

$$H\left(\coprod_{i=1}^{n}(a_i \Rightarrow b_i)\right) \Rightarrow F\left(\coprod_{i=1}^{n}(a_i \rightarrow b_i)\right)$$

So, if we have rule $H(\coprod_{i=1}^{n}(a_i \Rightarrow b_i)) \Rightarrow F(\coprod_{i=1}^{n}(a_i \Rightarrow b_i))$, then it still holds, when we replace $(\Rightarrow)$ with $(\rightarrow)$ in $F(\coprod_{i=1}^{n}(a_i \Rightarrow b_i))$.

And, of course, also:

$$F\left(\coprod_{i=1}^{n}(a_i \Rightarrow b_i)\right) \Rightarrow F\left(\coprod_{i=1}^{n}(a_i \rightarrow b_i)\right)$$

So, as you can see, we can easily disprove statements of some sentences, that use implications, by disproving statements of analogous sentences, that use operator $(\rightarrow)$ instead of implication, because:

$$\sim F\left(\coprod_{i=1}^{n}(a_i \rightarrow b_i)\right) \Rightarrow \sim F\left(\coprod_{i=1}^{n}(a_i \Rightarrow b_i)\right)$$

5.44 Possibility of equality operator – equality of logical values

For any concrete statements a and b:

$$\begin{aligned}&\textit{possibility of equality}(a, b)\\&= (\textit{a and b have logical values permissible for equality of statements})\\&= (\textit{either a and b have the same logical value or both are undecidable}) = (\overline{a} = \overline{b})\\&= (a \equiv b) = (a \rightleftharpoons b)\end{aligned}$$

$$(a \rightleftharpoons b) = ((a \rightharpoonup b) \textit{ and } (a \leftharpoondown b)) = ((a \rightharpoonup b) \textit{ and } (b \rightharpoonup a))$$

$$\overline{a \rightleftharpoons b} = (\overline{a} \rightleftharpoons \overline{b})$$

$$\overline{\textit{possibility of equality}} = (\overline{\rightleftharpoons}) = \left\{\begin{array}{c}((\textit{false}, \textit{false}), \textit{true}),\\((\textit{false}, \textit{true}), \textit{false}),\\((\textit{true}, \textit{false}), \textit{false}),\\((\textit{true}, \textit{true}), \textit{true}),\\((\textit{undecidable}, \textit{false}), \textit{false}),\\((\textit{undecidable}, \textit{true}), \textit{false}),\\((\textit{false}, \textit{undecidable}), \textit{false}),\\((\textit{true}, \textit{undecidable}), \textit{false}),\\((\textit{undecidable}, \textit{undecidable}), \textit{true})\end{array}\right\}$$

$$(a \rightleftharpoons b \rightleftharpoons c) = ((a \rightleftharpoons b)\ and\ (b \rightleftharpoons c))$$

And we have, that $\rightleftharpoons (a, b)$ is transitive relation:

$$((a \rightleftharpoons b)\ and\ (b \rightleftharpoons c)) \Rightarrow (a \rightleftharpoons c)$$

$$(a \rightharpoonup a) = (a \rightleftharpoons a)$$

And:

$$\bigvee_{\{a,b\} \subset Statement^{\Uparrow}} \sim((a \rightleftharpoons b) \equiv (a = b))$$

And for any decidable statements a and b:

$$(a \rightharpoonup b) = (a \rightleftharpoons (a\, and\, b)) = ((a\, or\, b) \rightleftharpoons b)$$

$$(a \leftrightarrow b) \Rightarrow (a \rightleftharpoons b)$$

5.45 Deterministic and nondeterministic implication operator

For any concrete statements a and b:

$$deterministic\ implication(a, b) = non\, free\ will\ implication(a, b) = (\Rrightarrow (a, b)) = (a \Rrightarrow b)$$

Example: “It will be a hot day **if** the Sun will shine”

$$nondeterministic\ implication(a, b) = free\ will\ implication(a, b) = (\rightrightarrows (a, b)) = (a \rightrightarrows b)$$

Example: “I will play in the garden **if** the Sun will shine **and nothing will intervene**”

And we have:

$$(a \Rrightarrow b) = (a \Rightarrow b)$$

$$(\rightrightarrows) \neq (\Rightarrow)$$

By the way:

$$\bigvee_{\{a,b\} \subset Statement^{\Uparrow}} \sim\left(((a \rightarrow b)\ and\ a) \Rightarrow b\right)$$

But only:

$$((a \rightarrow b)\ and\ a) \rightarrow b$$

Let us notice, that:

$$\left((a \Rrightarrow b)\ and\ a\right) = \left((a \Rightarrow b)\ and\ a\right) \Rightarrow (b\ and\ a) \Rightarrow b$$

Nondeterministic inference is not just a deterministic logical consequence of the meaning of sentences, that does not depend on free will decisions, but a logical consequence that is the result of a free will decision.

For example, in the sentence "If Alice has just missed a bus and nothing intervened, then she is going on foot" the statement of the part "Alice has just missed a bus" does not implicate deterministically the statement of the part "she is going on foot", because she had also other options, for example, when she missed the bus, she could wait for another bus. In such a situation, there is the following nondeterministic implication:

$$\left((Alice\ has\ just\ missed\ a\ bus) \rightrightarrows (She\ is\ going\ on\ foot)\right) = \begin{pmatrix} She\ decided\ to\ go\ on\ foot \\ and\ nothing\ intervened \end{pmatrix}$$

For more details, see Section 6.1.

And we have the following deterministic implication:

$$\begin{pmatrix} (Alice\ has\ just\ missed\ a\ bus) \\ and \begin{pmatrix} She\ decided\ to\ go\ on\ foot \\ and\ nothing\ intervened \end{pmatrix} \end{pmatrix} \Rightarrow (She\ is\ going\ on\ foot)$$

So, if she decided to go on foot and nothing intervened, then:

$$\left(truth\ and\ (Alice\ has\ just\ missed\ a\ bus)\right) \Rightarrow \begin{pmatrix} (Alice\ has\ just\ missed\ a\ bus) \\ and \begin{pmatrix} She\ decided\ to\ go\ on\ foot \\ and\ nothing\ intervened \end{pmatrix} \end{pmatrix}$$
$$\Rightarrow (She\ is\ going\ on\ foot)$$

So:

$$(Alice\ has\ just\ missed\ a\ bus) \rightarrow (She\ is\ going\ on\ foot)$$

And that is why, we can say then, for example: "Alice has just missed a bus, **so** she is going on foot", where:

$$\left(\hat{a}\ so\ \hat{b}\right) = (a\ and\ b\ and\ (a \rightarrow b))$$

For more details, see Section 5.38.8.

Only the actions, that depend on free will decision, need the assumption, that nothing intervened, nothing intervene or nothing will intervene, because all other actions are deterministic, so there is always deterministic implication between each one of them and its every sufficient condition.

Let us see another example of nondeterministic implication:

"If [today] it will be a sunny day and nothing will intervene, then [today] Robert will be riding bike."

$$((\textit{It will be a sunny day}) \rightrightarrows (\textit{Robert will be riding bike}))$$

Let us notice, that to confirm the above implication, you have to know, what is Robert's free will decision about what he will be doing, if today will be a sunny day.

Sometimes, the same as above, an object in a sentence is implied, so implication is seemingly not a logical consequence. For example:

$$\textit{If it will rain [on the ground], then the ground will be wet.}$$

((**The Ground** is A Thing That Rain Will Fall On) $\Rrightarrow$ (**The Ground** is A Thing That Will Be Wet)) = (**A Thing That Rain Will Fall On** is A Thing That Will Be Wet)

For more details about deterministic and nondeterministic implication, see Section 6.1.

5.46 Deterministic and nondeterministic conditioning operator

For any concrete statements a and b:

$$\textit{determinisitc conditioning}(a,b) = (a \Lleftarrow\!\Rrightarrow b) = ((a \Rrightarrow b)\ \textit{and}\ (b \Rrightarrow a))$$

$$\textit{nondeterminisitc conditioning}(a,b) = (a \leftleftarrows\!\rightrightarrows b) = ((a \rightrightarrows b)\ \textit{and}\ (b \rightrightarrows a))$$

For concrete statements a and b:

$$(a \Lleftarrow\!\Rrightarrow b) = ((a \Rrightarrow b)\ \textit{and}\ (b \Rrightarrow a)) = ((a \Rightarrow b)\ \textit{and}\ (b \Rightarrow a)) = (a = b)$$

So:

$$(a \Lleftarrow\!\Rrightarrow b) = (a = b) \Rightarrow (a \leftrightarrow b)$$
$$\neq ((\leftleftarrows\!\rightrightarrows), (=), (\leftrightarrow))$$

5.47 Causation operators

And we have causation and reasoning in language.

We will assume, that in formal language for any concrete statements a and b:

$$causation(a,b) = argumentation(a,b) = (b\ is\ the\ result\ of\ a)$$
$$= (a\ is\ the\ cause\ of\ b) = (b\ is\ the\ effect\ of\ a) = (a\ is\ the\ reason\ of\ b)$$
$$= (\triangleright(a,b)) = (\triangleleft(b,a)) = (a \triangleright b) = (b \triangleleft a) = (a \therefore b) = (b \because a)$$
$$= \left(\hat{a}\ therefore\ \hat{b}\right) = \left(\hat{a}, for\ \hat{b}\right) = \left(\hat{b}\ because\ \hat{a}\right) = \left(\hat{b}\ forwhy\ \hat{a}\right)$$
$$= \left(\hat{b}\ on\ the\ grounds\ that\ \hat{a}\right)$$

$$\left(\hat{a}\ so\ that\ \hat{b}\right) = \left(\hat{a}\ because\ it\ can\ help\ to\ fulfil\ the\ condition\ that\ \hat{b}\right)$$
$$= \left(\left(it\ can\ help\ to\ fulfil\ the\ condition\ that\ \hat{b}\right) \triangleright a\right)$$

If a causation is true, then the reason and the result are both true.

$$sufficient\ causation(a,b) = (a \trianglerighteq b) = ((a \triangleright b)\ and\ (a \Rightarrow b)) = ((a \triangleright b)\ and \sim(a \gtrdot b))$$
$$\Rightarrow (a\ and\ b\ and\ (a \Rightarrow b)) = (a \rightarrowtail b)$$

For example: The water froze, be**cause** it reached 0 degrees Celsius. The plant died, be**cause** it lacked sunlight.

$$insufficient\ causation(a,b) = (a \gtrdot b) = ((a \triangleright b)\ and \sim(a \Rightarrow b)) = ((a \triangleright b)\ and \sim(a \trianglerighteq b))$$

$$necessary\ causation(a,b) = (a \succcurlyeq b) = ((a \triangleright b)\ and\ (a \Leftarrow b)) = ((a \triangleright b)\ and \sim(a \nsucceq b))$$
$$\Rightarrow (a\ and\ b\ and\ (a \Leftarrow b)) = (a \leftarrowtail b)$$

For example: The metal rusted, be**cause** there was water on it. The tree is burning, be**cause** there is oxygen.

$$unnecessary\ causation(a,b) = (a \nsucceq b) = ((a \triangleright b)\ and \sim(a \Leftarrow b)) = ((a \triangleright b)\ and \sim(a \succcurlyeq b))$$

There cannot be direct causes, that are sufficient and necessary at the same time, because if there were such a cause and its effect, then they would be equal to each other $\left(((a \Rightarrow b)\ and\ (b \Rightarrow a)) = (a = b)\right)$, but, of course, an effect cannot be its own direct cause.

And there are causes, that are neither sufficient nor necessary (e.g. He feels fine, be**cause** he ate chocolate).

Assume:

$$R \in \{\triangleright, \trianglerighteq, \gtrdot, \nsucceq, \succcurlyeq\}$$

Then:

$$(a\,R\,b) \Rightarrow (a \triangleright b)$$
$$(a\,R\,b) \Rightarrow (a\ and\ b)$$

$$(\sim a \, or \sim b) \Rightarrow \sim(a\,R\,b)$$

$$(a\,R\,b\,R\,c) = ((a\,R\,b) \, and \, (b\,R\,c))$$

We have, that some causation relations are transitive:

$$((a \triangleright b) \, and \, (b \triangleright c)) \Rightarrow (a \triangleright c)$$

$$((a \trianglerighteq b) \, and \, (b \trianglerighteq c)) \Rightarrow (a \trianglerighteq c)$$

$$((a \succcurlyeq b) \, and \, (b \succcurlyeq c)) \Rightarrow (a \succcurlyeq c)$$

And some of them are distributive over "and":

$$((a \triangleright b) \, and \, (a \triangleright c)) = (a \triangleright (b \, and \, c))$$

$$((a \trianglerighteq b) \, and \, (a \trianglerighteq c)) = (a \trianglerighteq (b \, and \, c))$$

$$((a \triangleright c) \, and \, (b \triangleright c)) = ((a \, and \, b) \triangleright c)$$

$$((a \succcurlyeq c) \, and \, (b \succcurlyeq c)) = ((a \, and \, b) \succcurlyeq c)$$

And some of them are distributive over "or":

$$((a \triangleright b) \, or \, (a \triangleright c)) = (a \triangleright (b \, or \, c))$$

$$((a \trianglerighteq b) \, or \, (a \trianglerighteq c)) \Rightarrow (a \trianglerighteq (b \, or \, c))$$

$$((a \triangleright c) \, or \, (b \triangleright c)) = ((a \, or \, b) \triangleright c)$$

$$((a \succcurlyeq c) \, or \, (b \succcurlyeq c)) \Rightarrow ((a \, or \, b) \succcurlyeq c)$$

Since there is determinism, sufficient causes must give concrete results. In such a case, *a* must precede *b* in time or happen just in the same moment, unless free will is involved because only free will combined with the consciousness of, what should happen in the future, allows the reason to come after free will and consciousness effect, e.g.: "Joana is buying schoolbooks for his son, because he will go to school this year, if nothing will intervene". In such a case, it does not matter whether consciousness or free will participate in the reason, e.g.: "Ben is working hard, because night will come soon, if nothing will intervene". The assumption, that nothing will intervene, is necessary, when the reason concerns the unpredictable future, because then we cannot be sure, whether the reason will happen in the future. The following sentences do not need such an assumption: "Joana has bought schoolbooks for his son, because he goes to school", "Ben had been working hard, because night came soon after".

And we have the following rules:

$$\bigwedge_{\{a,x,y\} \subset Statement^{\Uparrow}} \sim((a \triangleright x) \, and \, (a \triangleright y) \, and \, (x \rightarrow \sim y))$$

A cause cannot have contextually contrary effects.

And:

$$\bigwedge_{\{a,b,x\} \subset Statement^{\Uparrow}} \sim((a \triangleright x)\ and\ (b \triangleright x)\ and\ (a \rightarrow \sim b))$$

An effect cannot have contextually contrary causes.

5.48 Feedback operator

For any concrete statements a and b:

$$feedback(a,b) = (a \bowtie b) = (b \bowtie a) = ((a \triangleright b)\ and\ (b \triangleright a))$$

Feedback is also a transitive relation:

$$((a \bowtie b)\ and\ (b \bowtie c)) \Rightarrow (a \bowtie c)$$

$(a \triangleright a)$ then and only then, when there is such b different from a, that $(a \bowtie b)$.

$$(a \triangleright a) = (a \bowtie a)$$

5.49 Operator "though"

For any concrete statements a and b:

$$\left(\hat{a}\ though\ \hat{b}\right) = \left(although\ \hat{b}, \hat{a}\right) = \left(\hat{a}\ notwithstanding\ \hat{b}\right) = \left(\hat{a}\ even\ if\ \hat{b}\right)$$
$$= \left(\hat{a}\ even\ though\ \hat{b}\right) = \left(\hat{a}\ in\ spite\ of\ \hat{b}\right) = \left(\hat{a}\ despite\ \hat{b}\right) = \left(\hat{a}\ albeit\ \hat{b}\right)$$

$$\left(\hat{a}, though\ \hat{b}\right) = \left(a\ and\ b\ and\ (b \Rightarrow (\sim a\ should\ be\ or\ could\ be\ or\ would\ be\ true))\right)$$

Example: "Danna likes a kiwi fruit though it is sour", so it does not matter, that a kiwi fruit is sour (*b*), because Danna likes a kiwi fruit (*a*) even if it is sour.

5.50 Operator "but"

For any concrete statements a and b:

$$\left(\hat{a}\ but\ \hat{b}\right) = \left(\hat{a}\ but\ still\ \hat{b}\right) = \left(\hat{a}\ yet\ \hat{b}\right) = \left(\hat{a}\ and\ yet\ \hat{b}\right) = \left(\hat{a}\ only\ \hat{b}\right) = \left(\hat{a}, however, \hat{b}\right)$$
$$= \left(\hat{a}, nevertheless, \hat{b}\right) = \left(\hat{a}, nonetheless, \hat{b}\right)$$

$$\left(\hat{a}, but\, \hat{b}\right) = \bigvee_{b \Rightarrow c} \left(\begin{matrix} a\, and\, b\, and\, c \\ and \left(a \Rightarrow (\sim c\, should\, be\, or\, could\, be\, or\, would\, be\, true)\right) \end{matrix} \right)$$

$$= \bigvee_{b \Rightarrow c} \left(\begin{matrix} a\, and\, b \\ and \left(a \Rightarrow (\sim c\, should\, be\, or\, could\, be\, or\, would\, be\, true)\right) \end{matrix} \right)$$

Example: "Jane can dance very well, but she has broken leg", so it is not enough, that she is very good at dancing (*a*), and she cannot dance now (*c*), because she has broken leg (*b*).

And here can be clearly seen the only difference between "though" and "but", that rely on the fact, that fact, that Jane can dance very well does not necessarily implicate, that she should not or could not or would not have broken leg, as would be in case of the statement of the sentence "Jane has broken leg, though she can dance very well" (remember, that this is an artificial sentence). But the fact, that Jane can dance very well, implicates, that something else unexpressed directly in the sentence should be or could be or would be true, which in this case, is the statement, that she can dance now, but she cannot do it now, because she has broken leg.

As we already know:

$$\left(\hat{a},\, but\, \hat{b}\right) = \bigvee_{b \Rightarrow c} \left(\begin{matrix} a\, and\, b \\ and \left(a \Rightarrow (\sim c\, should\, be\, or\, could\, be\, or\, would\, be\, true)\right) \end{matrix} \right)$$

And:

$$\left(\hat{b},\, though\, \hat{a}\right) = \left(b\, and\, a\, and\, (a \Rightarrow (\sim b\, should\, be\, or\, could\, be\, or\, would\, be\, true))\right)$$

It can be, that $c = b$, for example: "You are mean to Mary recently, but she still likes you". In such a case:

$$\left(\hat{b}\, though\, \hat{a}\right) = \left(\hat{a}\, but\, \hat{b}\right)$$

And in general:

$$\left(\hat{b}\, though\, \hat{a}\right) \Rightarrow \left(\hat{a}\, but\, \hat{b}\right)$$

By this example, you can clearly see, how complex logic you use intuitively, whenever you use these words correctly. So be aware, that all this is knowledge, that is already in your head, and you only need to realize these rules, that you already use, to think even more efficiently.

5.51 The negation of a set

Now, we can return to sets. So, we have for any thing Y and any set X:

$$(Y \notin X) = \sim(Y \in X) = \sim(Y \text{ is an element of set } X) = \sim(Y \text{ belongs to set } X)$$

$\sim X$ is called the negation of set X or the complement of set X such, that:

$$\sim\sim X = X$$

$$(X = \sim Y) = (\sim X = Y)$$

For any thing Y and any set X:

$$((Y \in X) \equiv true) = ((Y \in \sim X) \equiv false)$$

$$(Y \in X) = \sim(Y \in \sim X) = (Y \notin \sim X)$$

$$(Y \notin X) = \sim(Y \in X) = (Y \in \sim X)$$

5.52 Elementary operations and operators

And we have four elementary operations on sets:

$$((A \cup B) = (B \cup A) = X) = \Big((Y \in X) = ((Y \in A) \text{ } or \text{ } (Y \in B))\Big)$$

$$((A \cap B) = (B \cap A) = X) = \Big((Y \in X) = ((Y \in A) \text{ } and \text{ } (Y \in B))\Big)$$

$$((A - B) = X) = \Big((Y \in X) = ((Y \in A) \text{ } and \sim(Y \in B))\Big)$$
$$= \Big((Y \in X) = ((Y \in A) \text{ } and \text{ } (Y \in \sim B))\Big) = ((A \cap \sim B) = X)$$

$$((A -_{\cap} B) = X) = \Big((Y \in X) = ((Y \in A) \text{ } or \sim(Y \in B))\Big)$$
$$= \Big((Y \in X) = ((Y \in A) \text{ } or \text{ } (Y \in \sim B))\Big) = ((A \cup \sim B) = X)$$

And three logical operators:

$$(A = B) = (B = A) = ((Y \in A) = (Y \in B))$$

$$(A \neq B) = \sim(A = B)$$

$$(A \subset B) = (B \supset A) = ((Y \in A) \Rightarrow (Y \in B))$$

5.53 The empty set and the set of everything

The empty set $\varnothing$ (set of nothing) is defined, since we can define $Y \in \varnothing$ for every thing Y: for any thing Y: $\overline{Y \in \varnothing = false}$.

The content of $\varnothing$ is the general negation of a thing – is nothing.

A set is not, what it contains, taken together, but a new being with its own properties. That is, why $\varnothing$ exists, because it is not equal to nothing. But since the empty set contains nothing, we have:

$$\varnothing = \Phi@ = \{\Phi\} \neq \Phi = @\varnothing$$

A set is a conjunction of atomic sets.

The set Ω of all things (set of everything) is defined, since we can define $Y \in \Omega$ for every thing Y: for any thing Y: $\overline{Y \in \Omega} = \overline{true}$.

The conjunction of all elements of Ω is equal to totality, since every thing belongs to Ω.

$$\varnothing \in \Omega$$

$$\Omega \in \Omega$$

$$\sim\varnothing = \Omega$$

$$\sim\Omega = \varnothing$$

For any thing x:

$$(x \in \mathcal{B}) = (x \subset Reality)$$

$$(x \in \Omega) = (x \subset Totality)$$

For any set X:

$$\varnothing \subset X \subset \Omega$$

$$(\Omega - X) = \sim X$$

$$(X - \varnothing) = (X \cup \varnothing) = (X \cap \Omega) = (X \cup X) = (X \cap X) = X$$

$$(X - X) = \varnothing$$

$$(X \cup \sim X) = (X \cup \Omega) = \Omega$$

$$(X \cap \sim X) = (X \cap \varnothing) = \varnothing$$

$$(X - \sim X) = X$$

5.54 More about things and nothing

$$nothing = \Phi = (opposite\ of\ \omega) = \wr\omega = \text{Not}\ \omega$$

$$Thing = \omega$$

$$true \equiv \sim \bigvee_{x \in \Omega} (\Phi \text{ is } x) = \bigwedge_{x \in \Omega} \sim(\Phi \text{ is } x) = \bigwedge_{x \in \Omega} \sim(Everything\ is\ not\ x) = \bigwedge_{x \in \Omega} (Something\ is\ x)$$

$$true \equiv \sim \bigvee_{x \in \Omega} (x \text{ is } \Phi) = \bigwedge_{x \in \Omega} \sim (x \text{ is } \Phi) = \bigwedge_{x \in \Omega} \sim (x \text{ is not a thing}) = \bigwedge_{x \in \Omega} (x \text{ is something})$$

And:

$$true \equiv (\text{Nothing is nothing}) = (\text{Nothing is no thing}) = (\text{Nothing is not a thing})$$

$$true \equiv (\text{Nothing is the opposite of a thing}) = (\text{Nothing is the general negation of a thing})$$
$$= (\text{Nothing is not a thing})$$

The statements of the sentences "Nothing can exist" and "Nothing cannot exist" are just false, because the statements, that are, respectively, equal to them, of the sentences "Everything cannot exist" and "Everything can exist" are false. So you cannot prove, that it is impossible, that nothing would exist, using argument, that nothing cannot exist, since it is a false statement. You should use different statement, that "There cannot exist/be nothing".

The rest of this section is for advanced readers.

In the goodness of our hearts we might think that it is impossible, that nothing would exist, for example, because something, that cannot be created from nothing, exists and cannot be turned into nothing. But reality is not so simple.

Since the word "nothing" is just a quantifier, so it has not the independent meaning, the sentences, that you may have on mind, does not mean, what you probably think. Here they are:

$$(\textbf{\textit{everything cannot be turned into nothing}})$$
$$= \bigwedge_{x} \sim \bigvee_{y} (x \text{ cannot be turned into } y)$$
$$= \bigwedge_{x} \sim \bigvee_{y} \sim (x \text{ can be turned into } y)$$
$$= \bigwedge_{x} \bigwedge_{y} (\boldsymbol{x} \textbf{ can be turned into } \boldsymbol{y})$$
$$= \sim \bigvee_{x} \sim \bigwedge_{y} (x \text{ can be turned into } y) =$$
$$= \sim (\textbf{\textit{something can be turned into not anything}})$$
$$\neq (\textbf{\textit{nothing can be turned into nothing}})$$
$$= \sim \bigvee_{x} \sim \bigvee_{y} (x \text{ can be turned into } y)$$
$$= \bigwedge_{x} \bigvee_{y} (\boldsymbol{x} \textbf{ can be turned into } \boldsymbol{y})$$

$$(\textbf{\textit{everything can be turned into nothing}}) = \bigwedge_x \sim \bigvee_y (\textit{x can be turned into y})$$

$$= \sim \bigvee_x \bigvee_y (\textbf{\textit{x can be turned into y}}) =$$

$$= \sim (\textbf{\textit{something cannot be turned into not anything}})$$

$$\neq (\textbf{\textit{nothing cannot be turned into nothing}})$$

$$= \sim \bigvee_x \sim \bigvee_y \sim (\textit{x can be turned into y})$$

$$= \bigwedge_x \sim \bigwedge_y (\textbf{\textit{x can be turned into y}})$$

$$(\textbf{\textit{something cannot be turned into nothing}})$$

$$= \bigvee_x \sim \bigvee_y \sim (\textit{x can be turned into y})$$

$$= \bigvee_x \bigwedge_y (\textbf{\textit{x can be turned into y}})$$

$$(\textbf{\textit{nothing cannot be turned into something}})$$

$$= \sim \bigvee_x \bigvee_y \sim (\textit{x can be turned into y})$$

$$= \bigwedge_x \bigwedge_y (\textbf{\textit{x can be turned into y}})$$

$$= (\textbf{\textit{everything cannot be turned into nothing}})$$

$$\neq \sim (\textbf{\textit{nothing can be turned into something}})$$

$$= \sim\sim \bigvee_x \bigvee_y (\textbf{\textit{x can be turned into y}})$$

$$= \bigvee_x \bigvee_y (\textbf{\textit{x can be turned into y}})$$

$$= \sim (\textbf{\textit{everything can be turned into nothing}})$$

Summing, up:

$$\textbf{\textit{false}} \equiv (\textbf{\textit{everything cannot be turned into nothing}})$$

$$= \bigwedge_x \bigwedge_y (\textbf{\textit{x can be turned into y}})$$

$$= \sim (\textbf{\textit{something can be turned into not anything}})$$

$$\neq \bigwedge_x \bigvee_y (\textbf{\textit{x can be turned into y}})$$

$$= (\textbf{\textit{nothing can be turned into nothing}}) \equiv \textbf{\textit{true}}$$

$$\textbf{\textit{false}} \equiv (\textbf{\textit{everything can be turned into nothing}})$$

$$= \sim \bigvee_{x} \bigvee_{y} (\textbf{\textit{x can be turned into y}}) =$$

$$= \sim (\textbf{\textit{something cannot be turned into not anything}})$$

$$\neq \bigwedge_{x} \sim \bigwedge_{y} (\textbf{\textit{x can be turned into y}})$$

$$= (\textbf{\textit{nothing cannot be turned into nothing}}) = \textbf{\textit{true}}$$

$$\textbf{\textit{false}} \equiv (\textbf{\textit{something cannot be turned into nothing}})$$

$$= \bigvee_{x} \bigwedge_{y} (\textbf{\textit{x can be turned into y}})$$

$$\textbf{\textit{false}} \equiv (\textbf{\textit{nothing cannot be turned into something}})$$

$$= \bigwedge_{x} \bigwedge_{y} (\textbf{\textit{x can be turned into y}})$$

$$= (\textbf{\textit{everything cannot be turned into nothing}})$$

$$\neq \sim\sim \bigvee_{x} \bigvee_{y} (\textbf{\textit{x can be turned into y}})$$

$$= \sim (\textbf{\textit{nothing can be turned into something}}) =$$

$$= \sim (\textbf{\textit{everything can be turned into nothing}}) \equiv \textbf{\textit{true}}$$

The most important conclusion is, that counterintuitively the statement, that nothing cannot be turned into something, is equal to the statement, that everything cannot be turned into nothing, and both are false, but, as you can see, fortunately, it does not mean respectively, that the statement, that nothing can be turned into something, and the statement, that is equal to it, that everything can be turned into nothing, are also both false. And you cannot be satisfied with any of these statements, because they do not mean, what you probably think. The statement, that nothing cannot be turned into something, the same as the statement, that everything cannot be turned into nothing, means, that every thing can be turned to every thing, which is not true, because non-physical things cannot be turned into physical things and vice versa, and this is not a statement, that would answer our question, why there is rather something than nothing. And the statement, that nothing can be turned into something, the same as the statement, that everything can be turned into nothing, means, that there is not any pair of things such that the first one of them can be turned to the second one of them, which is also not true, because there are physical things, that can be turned into other physical things, so this is also not a statement, that would answer that question. This is a proof, that we cannot even state, what we want to state about this problem.

It is very easy to prove, that if there is nothing, then nothing cannot be created, because when nothing exists, then everything does not exist, so it does not mean, that something, such as 'nothing', exists, because 'nothing' is not something, but the word "nothing" has the same meaning as the phrase "neither of things", so it cannot be treated as a symbol of the subject of a sentence. So, there is no possibility of creation of anything, so also there is no possibility of the creation of the world. And also, since time does not exist, nothing can happen, because every action takes place in time. So nothing can be created, so the world cannot be created.

But this is not yet the full answer to the question, why there is rather something than nothing, because even if we proved, that there never was and never will be nothing, then we still would not prove, that it is logically impossible, that there would be nothing.

Nothingness is nothing trying to be a thing defined as "the general state of nonexistence" (English Wikipedia, 10-10-2018), but unfortunately nothing cannot be a thing (something), so it has self-contrary definition, that nothingness is there, if and only if nothing is there, so nothingness defined such a way has failed attempt to define it, so it is impossible thing and a nobeing. So, nothingness cannot be just equal to the meaning of the word "nothing", because the word "nothingness" is a noun, so it must be a thing.

The correct definition of what logicians have really on mind could be, that it is there, if and only if nothing exists, which is a weaker assumption about it than above self-contrary assumption, that it is there, when nothing is there, and another self-contrary assumption, that it exists, if and only if nothing exists.

Here is a proof for this attempt to define it:

$$\left(\boldsymbol{X \text{ is there if and only if nothing exists}}\right)$$
$$= (X \textit{ is there if and only if there are no beings})$$
$$\Rightarrow (X \textit{ is not a being}) = (X \textit{ is a nobeing}) = (X \textit{ has failed attempt to define it})$$

So, X is never as it was meant to be, so X is always impossible.

You may think, that we should say about nobeingness instead of nothingness and then, of course, it has nothing to do with the meaning of the word "nothing", because it is a thing. But unfortunately, as long as there is something, the single element set, that contains it, exists, so nobeingness does not differ from nothingness, so it is not a solution to the problem.

Finally, even if we do not demand from 'nothing' to be a thing, that is there, then we have:

$$A = (\textit{there is nothing}) \Rightarrow (\Omega = \varnothing) \Rightarrow \bigwedge_{x \in \Omega} \left((x \in \varnothing)\, xor \sim(x \in \varnothing)\right)$$
$$= (\varnothing \textit{ exists}) \Rightarrow (\textit{something exists}) \Rightarrow (\textit{there is something}) \Rightarrow \sim(\textit{there is nothing}) = \sim A$$

For more details, see Section 5.2.

So:

$$A \Rightarrow \sim A$$

So, *A* is false, so there is always something. What is more, always something exists. The empty set is enough to define arithmetic:

$$0 = |\varnothing|$$

$$1 = |\{0\}|$$

$$2 = |\{0,1\}|$$

and so on.

So, always at least, arithmetic exists. And, finally, only this is a proof, that it is logically impossible, that there would be nothing. For more details about arithmetic, see Section 16.2.

Q.E.D.

Thing is a wider notion, than *being*. Every**thing**, that you can give a standalone symbol to, have to be able to be referred by word "it" to say something about **it** without repeating its symbol, so everything is some**thing**, but not every**thing**, that has a standalone symbol, is a being, because there are standalone symbols, that have attempts to define them, that failed, so they are all nobeings, but you can still distinguish them by their different symbols, that are connected to different attempts to define them. For example, a square circle and the set from Russell's paradox are both nobeings, but they are not the same thing, since they have different attempts to define them.

So, you have to be able to use symbols of things, that have failed attempts to define them, that distinguish them, for example, to express about each one of them the true statement, that it does not exist.

Thanks to this, you can say something, for example, about impossible things and distinguish them by their failed attempts to define them, while they all do not exist, so they are all nobeings. And you can use word "every**thing**" to name all possible and impossible things. That is, why, for example, the statement of the sentence "all things are possible for god" means, that also impossible things are possible for god, because there are things, that are impossible.

So:

$$(\textit{Nothing is impossible}) = (\textit{Neither of things is impossible})$$
$$= (\textit{Everything is not impossible}) = (\textit{Everything is possible})$$

5.55 Cardinalities

$--n$ and $n-1$ are the cardinalities of a set, that has one element less than a set, that has cardinality n.

$++n$ and $n+1$ are the cardinalities of a set, that has one element more than a set, that has cardinality n.

The cardinalities of finite sets are defined as:

$$0 = |\{\Phi\}| = |\varnothing|, successor(0) = 1$$

$$1 = |\{0\}|, successor(1) = 2, predecessor(1) = 0$$

$$2 = 1+1 = |\{0,1\}|, successor(2) = 3, predecessor(2) = 1$$

$$3 = 2+1 = 1+1+1 = |\{0,1,2\}|, successor(3) = 4, predecessor(3) = 2$$

And so on.

$$n = predecessor(n) + 1 = \overbrace{1 + \ldots + 1}^{n} = |S_n|$$

$$= \left|S_{predecessor(n)} \cup \left\{\left|S_{predecessor(n)}\right|\right\}\right|,\ successor(n) = n+1,\ predecessor(n) = n-1$$

Where $S_0 = \varnothing$.

Any set, that has cardinality 2, is unordered pair.

The cardinality of each finite set is a natural number. So, a natural number is not a set, but the cardinality of a finite set.

From this, we have infinite set of natural numbers $N = \{0,1,2,3,\ldots\}$, that is the first and the smallest infinite set. The largest infinite set is Ω, that has the greatest cardinality.

$$|\Omega| = |truth@| = |(truth\,and\,untruth)@| = |\Lambda_\varsigma @|$$

Here is a proof of above equality:

For every thing x, we have true statement $(x\,is\,a\,thing)$ and false statement $(x\,is\,not\,a\,thing)$.

And every statement is a thing.

So, we have bijection between these sets, which proves, that they have the same cardinality.

5.56 Lists and tuples

A finite tuple is defined as:

$$(a_1, \ldots, a_n) = \left\{\{1, \{a_1\}\}, \ldots, \{n, \{a_n\}\}\right\}$$

Definition $(a_1, \ldots, a_n) = \{\{1, a_1\}, \ldots, \{n, a_n\}\}$ would not be correct, because then for example:

$$(2,3,1) = \{\{1,2\},\{2,3\},\{3,1\}\} = \{\{1,3\},\{2,1\},\{3,2\}\} = (3,1,2)$$

The empty tuple $\varnothing_\gamma$ can be defined as:

$$\varnothing_\gamma = () = (\Phi) = \varnothing$$

$$1_\times = \{()\} = \{(\Phi)\} = \{\varnothing\}$$

A tuple can be also defined as a function:

$$(a_1, \ldots, a_n) = \{((1), a_1), \ldots, ((n), a_n)\}$$

A tuple can be also defined without numbers:

$$(\Phi) = \varnothing$$

$$x = (a, b, c, \ldots, d) = \{\{a\}, (b, c, \ldots, d)\} = \left\{\{a\}, \left\{\{b\}, \left\{\{c\}, \ldots \{\{d\}, (\Phi)\}\right\}\right\}\right\}$$

Then:

$$head(x) = a$$

$$tail(x) = (b, c, \ldots, d)$$

And so on.

And we have:

$$x_i = x(i) = head\left(tail^{i-1}(x)\right)$$

$$length\left((\Phi)\right) = 0$$

$$length(x) = 1 + length\left(tail(x)\right)$$

The symbol of a sum, for $A(x)$ decidable for any x, can be used the following different ways:

$$\sum_{A(x)} f(x) = \sum_{x \in \{y:A(y)\}} f(x) = \sum_{\{y:A(y)\}} f(x) = \int_{\{y:A(y)\}} f(x)\, for\, x$$

A tuple can be generalized with the use of the following big symbol of a list to any number of elements:

$$\left(\coprod_A a\, by\, a\right)$$

Where $\coprod_X f(x)\ by\ x$ is a generalized list (that does not have to have countable number of elements) composed from elements of X transformed by function f. The list is in the order of elements of X, when X is a tuple, and in any order, when X is a set.

And we have the following double square brackets to express a tuple, that allow to incrementally build a tuple:

$$\left[\!\left[\left[\!\left[\left[\!\left[\ldots\left[\!\left[[\![a_1]\!],a_2\right]\!\right]\ldots\right]\!\right],a_{n-1}\right]\!\right],a_n\right]\!\right]=(a_1,\ldots,a_n)$$

Where we can convert a tuple to recursive double square bracket:

$$(a_1,a_2,\ldots,a_n)=[\![(a_1,a_2,\ldots,a_{n-1}),a_n]\!]$$

$$(a_1)=[\![a_1]\!]$$

And vice versa:

$$\left[\!\left[[\![a_1,a_2,\ldots,a_{n-1}]\!],a_n\right]\!\right]=[\![a_1,a_2,\ldots,a_n]\!]=(a_1,a_2,\ldots,a_n)$$

Where, of course, every list of elements in round brackets is a tuple of these elements.

A generalized tuple must fulfill only the condition, that the set of its indexes is in form X^Y, where (X is a countable set of numbers or X is in the same form) and (Y is a countable set of numbers or Y is in the same form), so that the set of its indexes has defined a total order. For example, $\{0,1\}^N$ is such a set.

And we have new form of an integral:

$$\int_A f(a)\ by\ a=\int_A f(a)\ da$$

And we have weighted integral:

$$\int_A f(a)\ by\ a_{\ avg(weight)}=\frac{\int_A weight(a)*f(a)\ by\ a}{\int_A weight(a)\ by\ a}$$

$$\int_A f(a)\ by\ a_{\ avg}=\int_A f(a)\ by\ a_{\ avg((1)\leftarrow(t))}$$

Whenever you expect an expression, (a), that otherwise would be understood as a single element tuple, is treated as the result of the expression a, so you have to evaluate it before you will place the result in the context of the external expression, e.g.:

$$((1+2)*3)=(3*3)=9\neq(1+2*3)=7$$

$$((\mathit{Horses\ are\ running})\ \mathit{is\ true})$$
$$=(\mathit{The\ statement\ of\ the\ sentence\ “Horses\ are\ running”\ is\ true})$$

So $(a) \in A$ for single element a will be treated as $a \in A$, but (a) will be treated as a tuple for multielements list a. This is important, when we define a function and the exponentiation of sets.

So, use $[\![a]\!]$ instead of (a) in expressions, when you have on mind the tuple, that has single element a.

Of course, 'nothing' cannot be the result of an expression, because if nothing is the result of an expression, then it means, that there is no result of this expression.

And we have the operation of the transformation of the elements of list x by function f:

$$\overset{f}{\widehat{x}} = \coprod_{x} f(x_i) \, by \, x_i$$

5.57 Cartesian product and exponentiation

And we have the Cartesian product operation:

$$((A \times B) = X) = \left(([\![Y, Z]\!] \in X) = ((Y \in A) \; and \; (Z \in B)) \right)$$

$$(A^n \times A) = A^{n+1}$$

$$A^0 = \frac{A}{A} = 1_\times = \{(\Phi)\}$$

$$A \times 0_\times = 0_\times = \varnothing = \{\Phi\}$$

And we have exponentiation operation:

$$A^B = \{f\colon (([\![b]\!], a) \in f) \; and \; (a \in A) \; and \; ((b) \in B)\}$$
$$= \{f\colon (f(b) = a) \; and \; (a \in A) \; and \; ((b) \in B)\}$$

So, A^B is the set of all functions, that has domain B and codomain A.

For example:

R^{R^n} for any natural number n greater than 0 is the set of all n-ary real functions.

For:

$$A = \{false, true\}$$

$$B = \left\{ \begin{array}{c} (false, \; false), \\ (false, \; true), \\ (true, \; false), \\ (true, \; true) \end{array} \right\}$$

A^B is the set of all classical binary logical functions:

$$A^B = \left\{ \left\{ \begin{array}{c} ((false, false), false), \\ ((false, true), false), \\ ((true, false), false), \\ ((true, true), false) \end{array} \right\}, \ldots, \overline{AND}, \ldots, \overline{XOR}, \overline{OR}, \ldots, \left\{ \begin{array}{c} ((false, false), true), \\ ((false, true), true), \\ ((true, false), true), \\ ((true, true), true) \end{array} \right\} \right\}$$

5.58 Definitions

Unlike in case of true statements, that have their realizations, there is no intermediate being between defined thing itself and a complex symbol, that defines it, because the meaning of the symbol is this defined thing itself. In case of statements, any declarence is a complex symbol (sentence built from words, that are simpler symbols, under rules of some grammar), that describes action, that either (in case of an affirmative declarence) takes or (in case of a negative declarence) does not take place in reality, that do not have simpler symbol yet. So, you can treat the statement of any either affirmative or negative declarence as equal to the statement, that action described by it, respectively, takes or does not take place in reality. So, the difference is, that definitions of things determine things, while declarences describe states of things – what and when happens (where in this book, we assume, that in case of a negative declarence, just the opposite action happens) with and to some things. Of course, such a state of a thing (state of affairs) as a part of totality is also a thing.

When we define a thing, then we equate the meaning of a complex symbol, that can be understood by recipient of the definition and usually describes unambiguously the thing, with the meaning of some new simpler symbol, which means, that they are interchangeable from now in any context, where their meaning is taken into account. One symbol cannot represent two different things each at once – this is a mistake in logic. Equal definitions define the same thing.

An intensional definition of a being can have one of two forms:

$$SomeBeing = (Something,\ about\ which\ the\ following\ facts\ are\ true\colon SomeFacts)$$

$$SomeBeing = (Something,\ that\ has\ the\ following\ attributes\colon SomeAttributes)$$

Additionally, every abstract thing can be defined two ways:

a.) by a set of facts about it or a set of its attributes, e.g.: the dog is a species that . . .

b.) by the set of the common facts about or the common attributes of all members of the class of it, e.g.: a dog is an animal that . . .

All facts about a thing specify all its attributes. This means, that you can define every thing two ways, either by some facts, from which you can infer all facts about and all attributes of this thing, or by some set of attributes, from which you can infer the same. So, the definition of a thing can be expressed in a language in many ways, that do not have to recite all the attributes of the thing, but have to point out the thing unambiguously, so that all the attributes are determined and can be inferred from this definition (assuming, of course, that you know also definitions of all other things used in these attributes), e.g.:

$$(\textit{the Sun})$$
$$= (\textit{Something, about which the following fact is true: it is a star of the Solar System})$$
$$= (\textit{Something, that has the following attribute: being a star of the Solar System})$$

Which simplifies to:

$$(\textit{the Sun}) = (\textit{Something, that is a star of the Solar System})$$
$$= (\textit{star of the Solar System})$$

All things have physical attributes, that come from their presence or existence in time.

Things, that are meant to be nonphysical, abstracted from their existence in time, can be defined by definitions abstracted from physical totality.

Things, that are meant to be physical, even abstracted from their existence in time, cannot be ever defined by definitions abstracted from physical totality. So, such things, even abstracted from their existence in time, have attributes, that cannot be inferred from definitions abstracted from physical totality, but can be inferred only from empirical knowledge we have about these things. That is why their definitions must always have reference to physical totality, from which we infer their empirical attributes, e.g.: in above definition of the Sun, Solar System is this physical thing, that is the reference to physical totality.

On the other hand, nonphysical things do not need physical totality to be there, because they do not depend on physical totality, so their definitions can reference physical totality (e.g.: A circle is something Juliet has just thought about), but do not need such a reference (A circle is the conjunction of all points equally distant from its center). And all nonphysical attributes of nonphysical things can be inferred from their abstract definitions.

5.59 Recurrent set definition

A set can belong to itself, because it is a simple recurrence. We can define a recurrence, though a recurrence seems to need the definition of itself to be defined, because in a recurrence, we reference some being, that we are guaranteed, that it exists, if the attempt to define this being is not self-contrary. So, if this reference refers to the definition, the attempt of which did not fail for any other reason, then this reference is also not a reason for this attempt to fail, so there is no reason at all it should fail, so the recurrence is successfully defined. This is also a proof, that axiom of regularity is wrong, because this axiom implicates, that a set cannot belong to itself.

The same concerns more complicated cases, e.g.:

$$X_1 = \{X_2, X_3, \ldots, X_n\}$$

$$X_2 = \{X_1, X_3, \ldots, X_n\}$$

$$\ldots$$

$$X_n = \{X_1, \ldots, X_{n-1}\}$$

For example, that set, the attempt to define of which is in “Russell’s paradox”, does not exist not because a set cannot contain itself, but because the attempt to define the set failed, because it is self-contrary. For more details, see Section 15.6.

5.60 Set specification

Since everything can be an element of a set, straight from the definition of a set we have, that we can define a set X of all elements, that have any property P, under condition, that $P(x)$ is decidable for every x, for which it is defined, as follows:

$$\begin{aligned}(X = \{x : P(x)\}) &= (X = \{x : x \in \Omega,\ P(x)\}) \\ &= \bigwedge_{x \in \Omega} \big((x \in X) = (P(x) \textit{ is defined and } P(x))\big)\end{aligned}$$

And for any e:

$$(X = \{x : P(x)\}) = \big((e \in X) = (P(e) \textit{ is defined and } P(e))\big)$$

Where Ω is the set of all things. For more details, see Section 5.53.

In general, for any relation $P(x)$ decidable for any tuple (x) for which it is defined and for any tuple (e) that fits into tuple (x) and for any function $Q(x)$:

$$(X = \{Q(x) : P(x)\}) = \big((Q(e) \in X) = (Q(e) \textit{ and } P(e) \textit{ are defined and } P(e))\big)$$

And we have:

$$(X = \{Q(x) : {\sim}P(x)\}) = (X = {\sim}_{Q\langle[\textit{domainOf}(Q) \cap \textit{domainOf}(P)]\rangle}\{Q(x) : P(x)\})$$

Where $Q\langle[S]\rangle$ transforms set S of tuples of arguments to the set of the appropriate results.

So only if $P(x)$ and $Q(x)$ are defined for any x, then:

$$(X = \{Q(x) : {\sim}P(x)\}) = (X = {\sim}\{Q(x) : P(x)\})$$

Remember, that we cannot write $\{x : P(x)\}$ the following way $\{x \in \Omega : P(x)\}$, because $(x \in \Omega)$ is a statement, that can be an element of a set.

Set X of all sets, that do not belong to themselves, does not exist not because is naïvely defined, but because this property is undecidable for this set. For more details, see Section 15.6.

But there, of course, exists the set of all sets, that do not belong to themselves, except this set.

The statement, that the set X of all von Neumann's ordinal numbers is an ordinal number, is false, so the von Neumann's definition of ordinal numbers is self-contrary, if it implicates this statement. Here is a proof:

$$(X \textit{ is ordinal}) \rightarrow (X \in X) = (X < X) = \big((X < X) \textit{ or } (X > X)\big) = (X \neq X) \equiv \textit{false}$$

So the set of all ordinal numbers is not an ordinal number.

5.61 Completeness of truth – logical causality

There is, of course, **logical causality**, because if some statement was, for example, true without a sufficient reason (cause), then it could also be false or undecidable without a sufficient reason (cause) as well (because why could not it be, since there is no sufficient reason (cause) it has to be true?), which is impossible, because these possibilities are mutually exclusive. So, to prove any true or false statement, you only need to point out that sufficient reason (cause), why it is true or false, because it is obviously sufficient for the proof.

And, of course, every provable statement is true.

Q.E.D.

For another proof of logical causality, see Section 5.14.

And yet another proof:

Assume, that statement x is decidable.

First of all:

$$(\textit{there is a sufficient reason, why x must be true})$$

$$= \begin{pmatrix} \textit{there is a sufficient reason, why it cannot be truthfully said, that x is false,} \\ \textit{if it can be said, that x is false} \end{pmatrix}$$

So:

$$(\textit{there is not any sufficient reason, why x must be true})$$
$$= \sim(\textit{there is a sufficient reason, why x must be true})$$
$$= \sim\left(\begin{array}{c}\textit{there is a sufficient reason, why it cannot be truthfully said, that x is false,}\\ \textit{if it can be said, that x is false}\end{array}\right)$$
$$= \left(\begin{array}{c}\textit{there is not any sufficient reason, why it cannot be truthfully said, that x is false,}\\ \textit{if it can be said, that x is false}\end{array}\right)$$

So:

$$(\textit{there is not a proof, that x is true})$$
$$= (\textit{there is not any sufficient reason, why x must be true})$$
$$= \left(\begin{array}{c}\textit{there is not any sufficient reason, why it cannot be truthfully said, that x is false,}\\ \textit{if it can be said, that x is false}\end{array}\right)$$
$$= (\textit{it can be truthfully said, that x is false, if it can be said, that x is false})$$
$$= (\textit{x is false}) = \sim x$$

So:

$$\sim x = (\textit{there is not a proof, that x is true}) = (\textit{x is not provable})$$

So:

$$x = (\textit{there is a proof, that x is true}) = (\textit{x is provable})$$

So every decidable statement has a proof, that it is true or false.

Q.E.D.

This means, that every true or false statement has a proof, that it is true or false. So any statement has a proof, that it is true, if only it is true, and any statement has a proof, that it is false, if only it is false. It is undecidable whether an undecidable statement has a proof, that it is true, and a proof, that it is false, and, of course, it has a proof, that it is undecidable, whether the statement is true or false.

So:

$$x = (\textit{x is true}) = (\textit{x is provable}) = (\sim x \textit{ is not provable})$$
$$\sim x = (\textit{x is false}) = (\sim x \textit{ is provable}) = (\textit{x is not provable})$$

In other words, the statement of the Gödel sentence is the simplest example of an undecidable statement (the Liar statement), because:

$$x = (x\ is\ not\ provable) = (\sim x\ is\ true) = (x\ is\ false) = \sim x$$

Which would not be fulfilled, if the statement was decidable.

And in Section 15.12, you will find more about that.

5.62 Definitions instead of standalone axioms

The definition of a thing determines all attributes of the thing. So, the definitions of things are enough to determine all relations between these things, because any relation between any subset of them makes corresponding attributes of these things, from which we can obtain this relation.

5.63 Definitions based systems

All axioms of, for example, set theory can be proved or disproved using this definitions-based definition of a set, that you can find in Section 5.2. Every statement (that is either true or false) is not true or false for no sufficient reason, so to prove any statement, it is enough to indicate this sufficient reason. This is indeed a proof, that all standalone axioms are unnecessary. Such axioms as some basic and trivial statements are all very easily provable. To be able to prove any statement, you need to understand it, so you have to know the definitions of the terms used in this statement and the definitions of the terms used in these definitions and so on. And every term used in any axiom has its meaning, since an axiom is always a meaningful declarative sentence. So, standalone axioms are only supplements and elements of incapable or missing definitions. Of course, definitions can use axioms, but you have to be sure, that these axioms define a term completely. So, by a definitions-based system, I mean also a set of axioms divided into separate definitions of terms. In such a system, axioms are not just some statements assumed to be true without proofs and for no reason, because the validity of assumed definitions is a proof of them.

You always need to define something symbolically only to the point, when someone, that unambiguously understands the terms used by you, can understand what it is about. Indeed, any statements, that taken together are true only for (or can be attributed only to) some thing, in case of a definition of some concrete thing (e.g. the Sun, 1, etc.) or are true only for (or can be attributed only to) all representatives of some kind in case of a definition of an abstract thing (e.g. a star, a natural number, etc.), are a correct definition of this thing. That is, why you can define something completely by a set of axioms. Every relation between things, among which, one is a variable, is a correct definition of some thing, if it is true only for this variable being equal to this thing, because then it unambiguously indicates this thing. For example: "X is the star of the Solar System" is true only for $X = (the\,Sun)$, so the Sun can be defined as the star of the Solar System by the sentence "It is the star of the Solar System". In real life, you often indicate unambiguously a being, partially symbolically and partially non-symbolically, e.g.: you can point some door with your finger and say, that the being is behind that door. Then such a statement alone will be understandable only in the concrete non-symbolical context of the situation, in which it is used.

A definition implicates all attributes of a defined thing. For example, the definition of the Sun as the star of the Solar System implicates, that the Sun is hot, because the star of the Solar System is hot and the Sun is this star. Definitions of nonphysical things (and of all nonphysical relations) implicate all nonphysical relations between these things, that specify all their nonphysical attributes, while you have to also have the whole empirical knowledge about things, that are meant to be physical, to implicate all relations between them, that specify all their attributes, because all of them are physical (reference physical totality).

For example, proving the axiom of choice is trivial. You can prove it the simplest way, defining needed set C:

Since every set S from set T of nonempty sets has at least one element, then by definition of a set, every element of this set is a thing. So, by definition of a set, you can define for exactly one such element x relation $(x \in C) \equiv true$. And, you can do it for all sets from T, and for every other thing y you can define $(y \in C) \equiv false$. That is, how you have defined C. So, such a set always exists for any T.

Q.E.D.

5.64 The set of all concrete sets

The set Ψ, that contains only all concrete sets, is defined, since we can define $Y \in \Psi$ for every thing Y: for any thing Y, that is a concrete set, $\overline{Y \in \Psi} = true$, and for all other things $\overline{Y \in \Psi} = false$.

$$\Psi = Set^{\Uparrow} = \Omega_{\Rightarrow Set}$$

Where operation $S_{\Rightarrow K}$ returns the set, that contains every element of set S, that is a concrete K. For more details, see Chapter 10.

$$\Omega \in \Psi$$

$$\Psi \in \Omega$$

$$\Psi \in \Psi$$

$$\Omega \in \Omega$$

$$\varnothing \in \Psi$$

$$\Psi \subset \Omega$$

$$(x \in \Psi) = (x \subset \Omega) = (x \in_{\downarrow Set} \Omega) = (x \in \Omega_{\Rightarrow Set}) \Rightarrow (x \in \Omega)$$

$$(\Psi \in \Psi) = (\Psi \subset \Omega) \equiv true$$

$$\big((x \in \Omega) = (x \subset \Omega)\big) = (x \in \Psi)$$

It is not true, that the set of all concrete sets has cardinality smaller than its power set, because the set of all concrete sets includes its power set and power set of this power set and so on. Every set, that includes the set of all concrete sets, is a special case, for which Cantor's diagonal argument does not work, because such a set has the greatest possible cardinality by definition, because any thing counts in the cardinality of this set as the concrete single element set, that contains this thing, so there is not greater set at all. So, the same concerns the set of everything, which includes the set of all concrete sets. In other words, it is not true, that the power set of any set is greater than this set. So, Cantor's paradox does not exist. For more details, see Section 15.10.

So, we have:

$$\left(\{0,1\}^{\Psi} \cup \{0,1\}^{\{0,1\}^{\Psi}} \cup \ldots \cup \{0,1\}^{\Omega} \cup \{0,1\}^{\{0,1\}^{\Omega}} \cup \ldots\right) \subset \Psi \subset \Omega$$

But:

$$\Psi \not\supset \Omega$$

And:

$$|\Omega| = 2^{|\Omega|} = 2^{2^{|\Omega|}} = \ldots = |\Psi| = 2^{|\Psi|} = 2^{2^{|\Psi|}} = \ldots$$

And it does not mean that there is no infinite sequence of greater and greater cardinalities, because we have:

$$X_0 = |N| = \left|\bigcup_{k=1}^{\infty} \{\varnothing, \Omega\}^k\right| < X_1 = 2^{|N|} < X_2 = 2^{X_1} < \ldots < X_i = 2^{X_{i-1}} < \ldots < |\Psi| = |\Omega|$$

Which means only, that we cannot put in order all cardinalities in a countable list.

5.65 Quantifiers

Now, come back to logic one more time.

First of all, we have for any unary relations $R(x)$ and $S(x)$:

$$\bigwedge_{S(x)} R(x) = \left(\widehat{R(x)}\, for\, any\, x, for\, which\, S(x)\, is\, defined\, and\, \widehat{S(x)}\right)$$

$$\bigwedge_{x} R(x) = \left(\widehat{R(x)}\, for\, any\, x\right) = \bigwedge_{x \in \Omega} R(x)$$

$$\bigvee_{S(x)} R(x) = \sim \bigwedge_{S(x)} \sim R(x)$$

$$\bigvee_{x} R(x) = \bigvee_{x \in \Omega} R(x) = \sim \bigwedge_{x} \sim R(x) = \sim \bigwedge_{x \in \Omega} \sim R(x)$$

Let $\coprod_X x$ be a generalized list constructed from set X. Then:

$$\bigwedge_{X} F(x)\, by\, x\, of\, kind\, K = \bigwedge_{x \in_K X} F(x) = and\left(\coprod_{x \in_K X} F(x)\right)$$

$$\bigvee_{X} F(x)\, by\, x\, of\, kind\, K = \bigvee_{x \in_K X} F(x) = or\left(\coprod_{x \in_K X} F(x)\right)$$

For any $F(x)$, that is some statement for any x:

$$\Lambda_\varsigma = \sim\varnothing_\varsigma = \bigvee_{\varnothing} F(x)\, by\, x \equiv false$$

Because, then there is not any element of the above disjunction, that is true.

So also:

$$\varnothing_\varsigma = \bigwedge_{\varnothing} F(x)\, by\, x = \sim \bigvee_{\varnothing} \sim F(x)\, by\, x = \sim\Lambda_\varsigma \equiv true$$

Where $\varnothing_\varsigma$ is the empty statement and Λ_ς is the totality for the negation of a statement. For more details, see Section 9.2.4.

For any tuple (a), that fits into tuple (x) of variables, and any two relations $A(x)$ and $B(x)$, each of which as the restriction of a quantifier makes (x) a tuple of bound variables and is decidable for any (x), we have:

$$\bigwedge_{\boldsymbol{A}(\boldsymbol{x})} \boldsymbol{B}(\boldsymbol{x}) = \bigwedge_{(\boldsymbol{x}) \in \{(\boldsymbol{y}):\boldsymbol{A}(\boldsymbol{y})\}} ((\boldsymbol{x}) \in \{(\boldsymbol{y}): \boldsymbol{B}(\boldsymbol{y})\})$$

$$= (\{(\boldsymbol{y}): \boldsymbol{A}(\boldsymbol{y})\} \subset \{(\boldsymbol{y}): \boldsymbol{B}(\boldsymbol{y})\}) = \Big(((\boldsymbol{a}) \in \{(\boldsymbol{y}): \boldsymbol{A}(\boldsymbol{y})\})$$

$$\Rightarrow ((\boldsymbol{a}) \in \{(\boldsymbol{y}): \boldsymbol{B}(\boldsymbol{y})\})\Big) = (\boldsymbol{A}(\boldsymbol{a}) \Rightarrow \boldsymbol{B}(\boldsymbol{a}))$$

$$\bigvee_{A(x)} \boldsymbol{B}(\boldsymbol{x}) = \sim \bigwedge_{A(x)} \sim B(x) = \sim(\{(\boldsymbol{y}): \boldsymbol{A}(\boldsymbol{y})\} \subset \{(\boldsymbol{y}): \sim\boldsymbol{B}(\boldsymbol{y})\})$$

$$= \sim\Big(\big((\boldsymbol{a}) \in \{(\boldsymbol{y}): \boldsymbol{A}(\boldsymbol{y})\}\big) \Rightarrow \big((\boldsymbol{a}) \in \{(\boldsymbol{y}): \sim\boldsymbol{B}(\boldsymbol{y})\}\big)\Big) = \sim(\boldsymbol{A}(\boldsymbol{a}) \Rightarrow \sim\boldsymbol{B}(\boldsymbol{a}))$$

$$= \sim(\boldsymbol{B}(\boldsymbol{a}) \Rightarrow \sim\boldsymbol{A}(\boldsymbol{a}))$$

Q.E.D.

For more details, see Sections 6.3 and 5.38.3.

So, very often, you do not need quantifiers at all.

For example, for continuity of real function f at point x_0, we have that for any ε, δ and x:

$$((\varepsilon \in R)\ and\ (\varepsilon > 0))$$

$$\Rightarrow \bigg(\sim\Big(((\delta \in R)\ and\ (\delta > 0)) \Rightarrow \sim\big((x \in R) \Rightarrow ((|x - x_0| < \delta) \Rightarrow (|f(x) - f(x_0)| < \varepsilon))\big)\Big)\bigg)$$

So, for any relations $A(x)$ and $B(x)$, each of which as the restriction of a quantifier makes (x) a tuple of bound variables and is decidable for any x, and any tuples (a) and (b), that fit into tuple (x):

$$\left(\bigwedge_{A(x)} B(x)\ and \bigwedge_{x(a)} x(b)\right) = \Big((A(a) \Rightarrow B(a))\ and\ (B(a) \Rightarrow B(b))\Big) \Rightarrow (A(a) \Rightarrow B(b))$$

So, you can use these rules to prove or disprove any simple implication, because:

$$\left(Something, that\ \widehat{Y(\_)}\right) = \left(Something, for\ which\ it\ is\ true, that\ \widehat{Y(it)}\right)$$

$$\left(X\ is\ Something, that\ \widehat{A(\_)}\right) = A(X) \Rightarrow B(X) = \left(X\ is\ Something, that\ \widehat{B(\_)}\right)$$

$$\left(X\ is\ Something, that\ \widehat{B(\_)}\right) = B(X) \Rightarrow B(Y) = \left(Y\ is\ Something, that\ \widehat{B(\_)}\right)$$

For more details, see Chapter 6.

So, we have two interchangeable logical quantifiers.

For any tuple (x) of variables and any two relations $A(x)$ and $B(x)$, each of which as the restriction of a quantifier makes (x) a tuple of bound variables and is decidable for any (x), and any tuple (p) that fits into tuple (x), we have:

$$\bigwedge_{A(x)} B(x) = @\{B(x):A(x)\} = (\{(x):A(x),B(x)\} = \{(x):A(x)\}) = \bigwedge_{\sim B(x)} \sim A(x)$$

$$= \sim \bigvee_{A(x)} \sim B(x) = \sim \bigvee_{\sim B(x)} A(x) = \sim \bigvee_{x \in \{y:A(y)\}} \sim B(x)$$

$$= \sim \bigvee_{\Omega} A(x)\; and \sim B(x)\; by\, x = \bigwedge_{\Omega} \sim A(x)\; or\, B(x)\; by\, x$$

$$= \left(\widehat{B(x)}\; for\; any\; x\; such,\; that\; \widehat{A(x)}\right)$$

$$= \left(\widetilde{A(x)}\; or\; \widehat{B(x)}\; for\; any\; x\right) = \sim\left(\widehat{A(x)}\; and\; \widetilde{B(x)}\; for\; some\; x\right)$$

$$= (A(p) \Rightarrow B(p))$$

$$\bigvee_{A(x)} B(x) = \sim @\{\sim B(x):A(x)\} = (\{(x):A(x),B(x)\} \neq \varnothing) = \bigvee_{B(x)} A(x) = \sim \bigwedge_{A(x)} \sim B(x)$$

$$= \sim \bigwedge_{B(x)} \sim A(x) = \sim \bigwedge_{x \in \{y:A(y)\}} \sim B(x) = \bigvee_{\Omega} A(x)\; and\, B(x) by\, x$$

$$= \sim \bigwedge_{\Omega} \sim A(x)\; or \sim B(x)\; by\, x = \left(\widehat{B(x)}\; for\; some\; x\; such,\; that\; \widehat{A(x)}\right)$$

$$= \left(\widehat{A(x)}\; and\; \widehat{B(x)}\; for\; some\; x\right) = \sim\left(\widetilde{A(x)}\; or\; \widetilde{B(x)}\; for\; any\; x\right)$$

$$= \sim(A(p) \Rightarrow \sim B(p)) = \sim(B(p) \Rightarrow \sim A(p))$$

A conjunction of some conditions is equal to the disjunction of the same conditions if and only if all these conditions are equal to each other.

Notice, that an existential quantifier cannot be interpreted as the phrase "at least one exists", because "everything" includes also things, that are nobeings, so an existential quantifier can be interpreted as the phrase "there is at least one". The same universal quantifier cannot be interpreted as the phrase "for all beings", but it can be interpreted as the phrase "for all things". Of course, you can get such quantifiers ("at least one exists", "for all beings") the following way:

$$\bigvee_{x\, is\, a\, being} R(x) = (at\; least\; one\; being\; x\; exists, for\; which\; R(x))$$

$$\bigwedge_{x\, is\, a\, being} R(x) = (R(x)\; for\; every\; being\; x)$$

For any a and b and any relations $A(x)$, $B(y)$ and $C(x,y)$, each of which as the restriction of a quantifier makes respectively x, y and x and y bound variables and is decidable for any x and y, we have the following logical rules:

$$\left(\boldsymbol{A}(\boldsymbol{a}) \Rightarrow (\boldsymbol{B}(\boldsymbol{b}) \Rightarrow \boldsymbol{C}(\boldsymbol{a},\boldsymbol{b}))\right) = \bigwedge_{A(x)} \bigwedge_{B(y)} C(x,y) = \bigwedge_{B(y)} \bigwedge_{A(x)} C(x,y)$$

$$= \left(\boldsymbol{B}(\boldsymbol{b}) \Rightarrow (\boldsymbol{A}(\boldsymbol{a}) \Rightarrow \boldsymbol{C}(\boldsymbol{a},\boldsymbol{b}))\right)$$

$$= \bigwedge_{A(x)\, and\, B(y)} C(x,y) = \left((\boldsymbol{A}(\boldsymbol{a})\ \boldsymbol{and}\ \boldsymbol{B}(\boldsymbol{b})) \Rightarrow \boldsymbol{C}(\boldsymbol{a},\boldsymbol{b})\right)$$

$$= \sim \bigvee_{A(x)} \bigvee_{B(y)} \sim C(x,y) = \left(A(a) \Rightarrow (B(b) \Rightarrow C(a,b))\right)$$

So:

$$\left(\boldsymbol{A}(\boldsymbol{a}) \Rightarrow (\boldsymbol{A}(\boldsymbol{a}) \Rightarrow \boldsymbol{C}(\boldsymbol{a}))\right) = \left((\boldsymbol{A}(\boldsymbol{a})\ \boldsymbol{and}\ \boldsymbol{A}(\boldsymbol{a})) \Rightarrow \boldsymbol{C}(\boldsymbol{a})\right) = (\boldsymbol{A}(\boldsymbol{a}) \Rightarrow \boldsymbol{C}(\boldsymbol{a}))$$

So:

$$\left(\boldsymbol{A}(\boldsymbol{a}) \Rightarrow \left(\boldsymbol{A}(\boldsymbol{a}) \Rightarrow (\boldsymbol{A}(\boldsymbol{a}) \Rightarrow \boldsymbol{C}(\boldsymbol{a}))\right)\right) = \left(\boldsymbol{A}(\boldsymbol{a}) \Rightarrow (\boldsymbol{A}(\boldsymbol{a}) \Rightarrow \boldsymbol{C}(\boldsymbol{a}))\right) = (\boldsymbol{A}(\boldsymbol{a}) \Rightarrow \boldsymbol{C}(\boldsymbol{a}))$$

And so on.

And:

$$\bigvee_{A(x)} \bigvee_{B(y)} C(x,y) = \sim\left(A(a) \Rightarrow (B(b) \Rightarrow \sim C(a,b))\right) = \bigvee_{B(y)} \bigvee_{A(x)} C(x,y)$$

$$= \sim\left(B(b) \Rightarrow (A(a) \Rightarrow \sim C(a,b))\right) = \bigvee_{A(x)\, and\, B(y)} C(x,y) = \sim\left((\boldsymbol{A}(\boldsymbol{a})\ \boldsymbol{and}\ \boldsymbol{B}(\boldsymbol{b})) \Rightarrow \sim \boldsymbol{C}(\boldsymbol{a},\boldsymbol{b})\right)$$

And we have:

$$\left(\bigvee_{A(x)} \bigwedge_{B(y)} C(x,y) \Rightarrow \bigwedge_{B(y)} \bigvee_{A(x)} C(x,y)\right)$$

$$= \left(\sim\left(\boldsymbol{A}(\boldsymbol{a}) \Rightarrow \sim(\boldsymbol{B}(\boldsymbol{b}) \Rightarrow \boldsymbol{C}(\boldsymbol{a},\boldsymbol{b}))\right) \Rightarrow \left(\boldsymbol{B}(\boldsymbol{b}) \Rightarrow \sim(\boldsymbol{A}(\boldsymbol{a}) \Rightarrow \sim \boldsymbol{C}(\boldsymbol{a},\boldsymbol{b}))\right)\right)$$

And we also have exclusive-or quantifier for any relations $A(x)$ and $B(x)$, each of which as the restriction of a quantifier makes (x) a tuple of bound variables and is decidable for any (x), and any tuples (p) and (q) that fit into tuple (x):

$$\bigvee_{A(x)}^{\oplus} B(x) = xor\left(\coprod_{A(x)} B(x)\right)$$

$$= \bigvee_{A(x)}\left(B(x)\ and \sim \bigvee_{y \in (\Omega - \{x\})\ and\ A(y)} B(y)\right)$$

$$= \left(\widehat{B(x)}\ for\ exactly\ one\ x\ such,\ that\ \widehat{A(x)}\right)$$

$$= \bigwedge_{A(x)\ and\ B(x)} \bigwedge_{A(y)\ and\ B(y)} ((x) = (y))$$

$$= \left((A(p)\ and\ B(p)) \Rightarrow \left((A(q)\ and\ B(q)) \Rightarrow ((p) = (q))\right)\right)$$

for almost every and *for almost none* for $B(The\ \omega)$:

$$\left(((\varepsilon \in R)\ and\ (\varepsilon > 0)) \Rightarrow \left(0 \neq P\left(\sim B\left(The\ \{x{:}\, A(x)\}_{\downarrow}\right)\right) < \varepsilon\right)\right)$$

$$= \left(\widehat{B(x)}\ for\ almost\ every\ x\ such,\ that\ \widehat{A(x)}\right)$$

$$\left(((\varepsilon \in R)\ and\ (\varepsilon > 0)) \Rightarrow \left(0 \neq P\left(B\left(The\ \{x{:}\, A(x)\}_{\downarrow}\right)\right) < \varepsilon\right)\right)$$

$$= \left(\widehat{B(x)}\ for\ almost\ none\ x\ such,\ that\ \widehat{A(x)}\right)$$

$$= \left(\widetilde{B(x)}\ for\ almost\ every\ x\ such,\ that\ \widehat{A(x)}\right)$$

The above probabilities are smaller than any positive real number but not equal to zero, which means that they are some differentials. For more details see Section 16.1.

6 Inference rules

6.1 The difference between deterministic and nondeterministic implication

This chapter, except this subsection and subsection *Elementary notions*, (Section 6.2) is for people, that are really interested to know everything about logic, and it will teach you how to infer new knowledge from information you already have.

All methods of evaluation of implication included in this section are for deterministic implications ("⇛"), in case of which, there is deterministic relation between a condition and the effect of the fulfillment of the condition. It is exactly the same implication as referenced by the symbol "⇒" and has different symbol in Chapter 6 only to highlight its deterministic property. So, whenever in given implication the fulfillment of the implicated statement is unpredictable on the basis of fulfillment of the implicating statement, which means that free will decisions are involved in implication, we deal with nondeterministic implication (⇉), that does not have any rules of implication, because any being, that has free will, can by a pure caprice join anything, that depends on his or her will, with any condition without any deterministic logical consequence.

For example, we have the following true deterministic implication:

$$\left(\left(\begin{matrix}\textit{Mary will dance at midnight}\\ \textit{and}\left(\begin{matrix}\textit{John decided to dance with her then}\\ \textit{and nothing will intervene}\end{matrix}\right)\end{matrix}\right)\Rightarrow(\textit{John will dance with her then})\right)$$

And if John decided, that he will dance with Mary at midnight, if she will dance at midnight, then:

$$\left(\left(\begin{matrix}\textit{truth}\\ \textit{and Mary will dance at midnight}\\ \textit{and nothing will intervene}\end{matrix}\right)\Rightarrow\left(\begin{matrix}\textit{Mary will dance at midnight}\\ \textit{and}\left(\begin{matrix}\textit{John decided to dance with her then}\\ \textit{and nothing will intervene}\end{matrix}\right)\end{matrix}\right)\right.$$

$$\left.\Rightarrow(\textit{John will dance with her then})\right)$$

$$\Rightarrow((\textit{Mary will dance at midnight and nothing will intervene}).$$

$$\rightarrow(\textit{John will dance with her then}))$$

https://doi.org/10.1515/9783111441382-006

And if nothing will intervene:

$$\left(\left(\begin{array}{c}\textit{truth}\\ \textit{and Mary will dance at midnight}\end{array}\right)\Rightarrow\left(\begin{array}{c}\textit{Mary will dance at midnight}\\ \textit{and}\left(\begin{array}{c}\textit{John decided to dance with her then}\\ \textit{and nothing will intervene}\end{array}\right)\end{array}\right)\right.$$

$$\left.\Rightarrow(\textit{John will dance with her then})\right)$$

$$\Rightarrow((\textit{Mary will dance at midnight})\rightarrow(\textit{John will dance with her then}))$$

And we have the following equality:

$$((\textit{Mary will dance at midnight})\rightrightarrows(\textit{John will dance with her then}))$$
$$=\left(\begin{array}{c}\textit{John decided to dance with her then}\\ \textit{and nothing will intervene}\end{array}\right)\equiv\textit{true}$$

So, every nondeterministic implication is true or false due to the appropriate free will decisions and what happened, happens or will happen.

And we have the following inequality:

$$\textit{true}\equiv\left(\begin{array}{c}(\textit{Mary will dance at mid night})\\ \rightrightarrows(\textit{John will dance with her then})\end{array}\right)$$
$$=\left(\begin{array}{c}(\textbf{\textit{Midnight}}\ \textit{is a}\ \textbf{\textit{Moment When Mary Will Dance}})\\ \rightrightarrows(\textbf{\textit{Midnight}}\ \textit{is a}\ \textbf{\textit{Moment When John Will Dance With Her}})\end{array}\right)$$
$$\neq\left(\begin{array}{c}(\textbf{\textit{Midnight}}\ \textit{is a}\ \textbf{\textit{Moment When Mary Will Dance}})\\ \Rightarrow(\textbf{\textit{Midnight}}\ \textit{is a}\ \textbf{\textit{Moment When John Will Dance With Her}})\end{array}\right)$$
$$=\left(\begin{array}{c}\textit{a}\ \textbf{\textit{Moment When Mary Will Dance}}\\ \textit{is}\\ \textit{a}\ \textbf{\textit{Moment When John Will Dance With Her}}\end{array}\right)\equiv\textit{false}$$

So, there is no rule of implication, that joins fulfillment of this condition with such an effect, other than free will decision of John. So his decision does not have any deterministic cause. So this nondeterministic implication is true, though we have:

$$\left(\begin{array}{c}\textit{a}\ \textbf{\textit{Moment When Mary Will Dance}}\\ \textit{is}\\ \textit{a}\ \textbf{\textit{Moment When John Will Dance With Her}}\end{array}\right)\equiv\textit{false}$$

Remember, that the realization of any free will decision is unpredictable, so you must add the statement of the sentence "nothing intervened" or the sentence "nothing intervenes" or the sentence "nothing will intervene", because a free will decision does not guarantee, that it will be realized in reality, since something can get in the way of the realization of this decision, for example, a free will being can change its mind.

There is no conflict between nondeterministic implication and deterministic rules, because the conditions, that someone made some decision and that nothing intervened, intervenes or will intervene, are necessary, since the realization of a free will decision is unpredictable, and it changes these statements used in implication, so then this is implication between other statements.

Of course, someone, that has free will, can do something determined as the consequence of fulfillment of some condition. But then it is not free will, because in such a case, he must do it as the logical consequence, so he has no choice.

Here we have another example: let us assume, that John cannot keep balance, when he trips on the roof, and that everyone who cannot keep the balance, when he trips on the roof, will fall of the roof, if nothing will intervene. Then we have:

$$((\textit{John will trip on the roof}) \Rrightarrow (\textit{John will fall of the roof}))$$
$$= ((\boldsymbol{\textit{John}}\ \text{is}\ \boldsymbol{\textit{Someone that will trip on the roof}})$$
$$\Rrightarrow (\boldsymbol{\textit{John}}\ \text{is}\ \boldsymbol{\textit{Someone that will fall of the roof}}))$$
$$= (\boldsymbol{\textit{Someone that will trip on the roof}}\ \text{is}\ \boldsymbol{\textit{Someone that will fall of the roof}})$$

Which is true if and only if everyone cannot keep balance, when he trips on the roof, so it is not true, so something can intervene, so that John does not fall of the roof. For example, someone that has free will and is unpredictable due to this can help him keep the balance. That is why, there is no deterministic implication. So John is not determined to fall of the roof, even if he himself cannot keep balance, when he trips on the roof. And only the statement, that John will trip on the roof and he cannot keep balance, when he trips on the roof, and nothing will intervene, will implicate the statement, that John will fall of the roof.

So, nondeterministic implication takes place only when someone had a choice, what to do, and it is still possible that something can get in the way of the realization of this decision, so he does not have to do, what he chose, as deterministic logical consequence of fulfillment of some condition. That is the difference.

6.2 Elementary notions

A **monothematic statement about *X*** is a statement, that references *X* and does not have parts, that do not refer to *X*, and its negation does not have parts, that do not refer to *X*. An **attribute** can be obtained from a monothematic statement. A true decidable monothematic statement about *X* is equal to a **fact about *X***. And an attribute of X is the attribute obtained from a fact about X. The negation of an attribute is an attribute.

Function $\boldsymbol{facts}(\boldsymbol{X})$ returns the set of all facts about X.

Function $\boldsymbol{attr}(\boldsymbol{X})$ returns the set of all attributes of X.

Every **attempt of definition** is a set of attributes.

A **necessary attribute** of a thing in an attempt of definition is an attribute of the thing, that cannot be derived from the conjunction of all other attributes of that attempt.

Remember, that every attempt of definition of X must contain at least all its necessary attributes, but on the other hand, any set of attributes leads to a definition of some thing. So, you should always ensure, that you know, what you want to define.

An attempt of definition of X is **complete** if and only if it contains all directly **intended attributes** of X and all their consequences (attributes that are implicated by the conjunction of these intended attributes; they are also intended, but indirectly).

An attempt of definition of X is **successful** if and only if it defines, what was intended to be defined. So, when it **fails**, when it is an attempt to define a being, its result is not the intended being, but the appropriate nobeing, and when it is an attempt to define a nobeing, its result is not the intended nobeing, but nothing. So if you want to define a nobeing, you must ensure that this attempt will be successful, because otherwise it will be simply incorrect. When it is successful, you will get the intended nobeing, but if it fails, then such an attempt is just incorrect and you will get nothing, because we allow to define nobeings alternatively only when they already have definitions, that come from the attempts to define them as beings, and, of course, totality is consistent, so the definition of every thing is not contrary to itself and is not contrary to any other definition of a thing. In other words, the attempt to define a nobeing must not fail (cannot be self-contrary and cannot be contrary to (truth of) reality) in order to be considered correct.

By the examination of the intended attributes of a thing, it can be determined whether it is an attempt to define a being or an attempt to define a nobeing, so you do not need to declare it directly, but since it can be difficult, it will be more reader friendly, if you declare it in the attempt of a definition.

For example, you can define a square circle two ways:

$$(a\ \boldsymbol{square\ circle}) = (a\ \boldsymbol{being}\ \textit{that is a square and a cricle at the same time})$$

And directly as a nobeing:

$$\begin{aligned}&(a\ \boldsymbol{square\ circle})\\&= (a\ \boldsymbol{nobeing}\ \textit{that was meant to be a square and a cricle at the same time})\end{aligned}$$

In both situations, the result of these attempts will be the same.

The following definition of a being will fail:

$$(a\ \boldsymbol{squareNotSquare\ circle}) = \begin{pmatrix} a\ \boldsymbol{being}\ \textit{that was meant to be a square cricle} \\ \textit{and was not meant to be a square} \end{pmatrix}$$

And you will get the following nobeing:

$$(a\ \textbf{squareNotSquare circle}) = \begin{pmatrix} a\ \textbf{nobeing}\ that\ was\ meant\ to\ be\ meant\ to\ be\ a\ square\ cricle \\ and\ was\ meant\ to\ be\ not\ meant\ to\ be\ a\ square \end{pmatrix}$$

Function $thing(A)$ returns the result of attempt A of definition (the thing defined by set A of attributes), so if it is an attempt to define a being and these attributes are self-contrary or contrary to (truth of) reality, then it returns the appropriate nobeing.

So:

$$thing(attr(X)) = X$$

But for the set A of attributes, that is not a complete attempt to define something, as well as when it is an attempt to define a being and these attributes are self-contrary or contrary to (truth of) reality:

$$attr(thing(A)) \neq A$$

Function $attr(X, F)$ will return set of attributes of X, that can be derived from set of facts F.

And we have for any x:

$$(x \in \Omega) = \bigvee_{A \subset Attribute^{\Uparrow}} (x = thing(A))$$

Which means, that any thing has the appropriate set of attributes, that define it.

Any set of attributes leads to the definition of something.

$$(x \in \mathcal{B}) = \bigvee_{A \subset Attribute^{\Uparrow}} \big((x = thing(A))\ and\ (primaryAttr(x) = A) \big)$$

Which means, that a thing is a being if and only if it has its all primary attributes (defined below).

A **primary attribute** of X is an attribute, that was meant to be an attribute of X. For example, fulfilling, that it can bark, is a primary attribute of a dog and fulfilling, that it is a square and a circle at the same time, is a primary attribute of a square circle. Remember, that If you define a being, then all its intended attributes are primary, but If you define a nobeing, then all its intended attributes are not primary.

Function $primaryAttr(X)$ returns the set of all primary attributes of X.

Function $primaryFacts(X)$ returns all facts about X, that can be obtained from $primaryAttr(X)$.

For every X:

$$\boldsymbol{attrp(X)} = \bigcap_{\boldsymbol{e \in X^{\uparrow}}} \boldsymbol{attr(e)}$$

Always, at least, the attribute from the fact $(e \in X^{\uparrow})$ is common. For example, if X is equal to *Dog*, then the set of all common attributes of its all representatives contains the attribute 'fulfilling, that it was consistently meant to be concrete Dog' and all attributes that follow from it.

To simplify the rules in this section, let us assume, that $attrx \in \{attr, attrp\}$.

Function $facts(X)$ returns all facts about X, that can be obtained from $attr(X)$.

For any thing X and any concrete attribute x and unary relation T:

$$\boldsymbol{attrToMonoStat(x, X) = (X\ has\ an\ attribute\ x)}$$

$$\boldsymbol{monoStatToAttr(T(X), X) = (fulfilling\ T(it))}$$

$$\boldsymbol{monoStatToAttr(attrToMonoStat(x, X), X) = x}$$

$$\boldsymbol{attrToMonoStat(monoStatToAttr(T(X), X), X) = T(X)}$$

Where $T(X)$, when expressed in a language, cannot contain words "it" and "its", so you have to repeat symbols of the appropriate referents instead. Remember, that function $attrToMonoStat(x, X)$ returns a monothematic statement about X, which means that its result does not have to be true. And function $monoStatToAttr(T(X), X)$ transforms a monothematic statement about X to an attribute, so it can transform also a false monothematic statement.

For any things X and Y and any concrete attribute a:

$$((a \in attrp(X)) \Rightarrow (a \in attrp(Y))) = (attrp(X) \subset attrp(Y)) = (Y^{\uparrow} \subset X^{\uparrow})$$

For any thing X and any concrete attribute x:

$$(x \in attr(X)) = (attrToMonoStat(x, X) \in facts(X)) = attrToMonoStat(x, X)$$

For any thing X and any concrete monothematic statement x:

$$(x \in facts(X)) = (monoStatToAttr(x, X) \in attr(X))$$

For any unary relations A and B and any thing X that belongs to the intersection of their domains:

$$\begin{aligned}&((\{A(X), B(X)\} \subset facts(X))\ and\ (A(X) \Rightarrow B(X)))\\&= ((\{(fulfilling\ A(it)), (fulfilling\ B(it))\}\\&\subset attr(X))\ and\ ((fulfilling\ A(it)) \in attr(X))\\&\Rightarrow ((fulfilling\ B(it)) \in attr(X)))\end{aligned}$$

If $A(it)$ in some context is meaningful and its statement is decidable, then in this context:

$$\sim(\mathit{fulfilling}\ A(it)) = (\mathit{not\ fulfilling}\ A(it)) = (\mathit{fulfilling} \sim A(it))$$

For any concrete attributes x and y:

$$(x=y) = \bigvee_{R\ \mathit{is\ unary\ relation}} \bigwedge_{\{X,Y\}\subset\Omega} \big((\mathit{attrToMonoStat}(x,X)=R(X)) \\ \mathit{and}\ (\mathit{attrToMonoStat}(y,Y)=R(Y))\big)$$

For any Z and any concrete attributes x and y and every concrete binary operation op, that is defined for concrete statements:

$$op(x,y)=(x\, op\, y)=\mathit{monoStatToAttr}\big(op(\mathit{attrToMonoStat}(x,Z), \mathit{attrToMonoStat}(y,Z)),Z\big)$$

And for every concrete unary operation op:

$$op(x) = \mathit{monoStatToAttr}\big(op(\mathit{attrToMonoStat}(x,Z)),\ Z\big)$$

For example:

$$\sim x = \mathit{monoStatToAttr}(\sim \mathit{attrToMonoStat}(x,Z),\ Z)$$

So all axioms for statements as attributes of any thing are also true for attributes in place of statements. Therefore all rules, that applies to statements, should apply also to attributes, but remember that, for example, the analog of *truth* for attributes, which is equal to $@attr(X)$, is different for every thing X, to which we apply attributes.

And we have for any concrete attributes a and b:

$$((a\ \mathit{and}\ b) \in \mathit{attrx}(X)) \\ = \big((a \in \mathit{attrx}(X))\ \mathit{and}\ (b \in \mathit{attrx}(X))\big) \\ \Rightarrow ((a\ \mathit{or}\ b) \in \mathit{attrx}(X))$$

$$(a \in \mathit{attrx}(X)) \Rightarrow \bigwedge_{a\Rightarrow b} (b \in \mathit{attrx}(X))$$

For any thing X and any concrete attribute a:

$$(a \in \mathit{attrp}(X)) \Rightarrow \sim(\sim a \in \mathit{attrp}(X)) = (\sim a \in \sim \mathit{attrp}(X))$$

For any thing X and any concrete attribute a, for which $\mathit{attrToMonoStat}(a,X)$ is decidable:

$$(a \in \mathit{attr}(X)) = \sim(\sim a \in \mathit{attr}(X)) = (\sim a \in \sim \mathit{attr}(X))$$

And for any thing X, we have:

$$\wr X = (Not\, X) = (Something\ That\ Is\ Not\, X)$$

And for any thing X, we have, that:

$$\left(\sim \bigwedge_{x \in attrp(X)} x \right) = \left(\bigvee_{x \in attrp(X)} \sim x \right) \in attrp(\wr X)$$

Let us define positive form $R(X, Y)$ of a simple declarence:

$$\boldsymbol{R_{singular}}(\boldsymbol{X}, \boldsymbol{Y}) = (X\ isA\ Y) = \begin{pmatrix} Any\ member\ of\ the\ class\ of\ X \\ is\ some\ member\ of\ the\ class\ of\ Y \end{pmatrix}$$

$$\boldsymbol{R_{plural}}(\boldsymbol{X}, \boldsymbol{Y}) = (X\ areA\ Y) = \begin{pmatrix} Any\ members\ of\ the\ class\ of\ X \\ are\ some\ members\ of\ the\ class\ of\ Y \end{pmatrix}$$

Where:

$$(X\ isA \backslash areA\ Y) = \left(X^{\uparrow} \subset Y^{\uparrow}\right)$$

For more details about relations *isA, isp* and *areA, arep,* see Section 10.9.

$$\begin{aligned} \boldsymbol{R}(\boldsymbol{X}, \boldsymbol{Y}) &= \left(the\ class\ of\ X\ is\ a\ subclass\ of\ (the\ class\ of)\ Y\right) \\ &= \left(the\ class\ of\ Y\ is\ super\ class\ of\ (the\ class\ of)\ X\right) = \Subset (\boldsymbol{X}, \boldsymbol{Y}) \\ &= (\boldsymbol{X} \Subset \boldsymbol{Y}) = \left(\boldsymbol{X}^{\uparrow} \subset \boldsymbol{Y}^{\uparrow}\right) = \left(\boldsymbol{attrp}(\boldsymbol{Y}) \subset \boldsymbol{attrp}(\boldsymbol{X})\right) \\ &= \left(\boldsymbol{R_{singular}}(\boldsymbol{X}, \boldsymbol{Y})\ \boldsymbol{or}\ \boldsymbol{R_{plural}}(\boldsymbol{X}, \boldsymbol{Y})\right) \end{aligned}$$

Where $X^{\uparrow}$ is the set of all representatives of the kind of X. For more details, see Section 10.9.

And we have:

$$(\$\boldsymbol{X} \Subset \boldsymbol{Y}) = \left((\$\boldsymbol{X})^{\uparrow} \subset \boldsymbol{Y}^{\uparrow}\right) = \left(\boldsymbol{X} \in \boldsymbol{Y}^{\uparrow}\right) = (\boldsymbol{X}\ \boldsymbol{isBp} \backslash \boldsymbol{areBp}\ \boldsymbol{Y})$$

And for any things X and Y:

$$\Subset (\boldsymbol{X}, \boldsymbol{Y}) = \left(\boldsymbol{attrp}(\boldsymbol{Y}) \subset \boldsymbol{attrp}(\boldsymbol{X})\right)$$

Remember, that all the rules for implication from Chapter 6 will also be correct for the following definition of relation $\Subset (X, Y)$:

$$\Subset (X, Y) = (Any\, X\ is\ Some\, Y) = \left(X^{\Uparrow} \subset Y^{\Uparrow}\right)$$

Additionally, for classifications, we have:

$$\Subset (Classification\, A\, of\, C, Classification\, B\, of\, D)$$
$$= \Subset (Classifications\, A\, of\, C, Classifications\, B\, of\, D) =$$
$$= \left(R_{singular}(A, B)\, and\, R_{singular}(C, D)\right)$$

And for any X and Y:

$$(X\, is\backslash isp\, Y) = (Y\, is\backslash isp\, Something\, That\, X\, is/isp)$$
$$(X\, are\backslash arep\, Y) = (Y\, are\backslash arep\, Some\, things\, That\, X\, are/arep)$$
$$(X \Subset Y) = (\wr Y \Subset \wr X)$$
$$(X \Subset X) \equiv true$$
$$(X \Subset \wr X) \equiv false$$

And, of course, from transitivity of inclusion of sets, we have:

$$\left((A \Subset B)\, and\, (B \Subset C)\right) \Rightarrow (A \Subset C)$$

6.3 Basic properties of inclusion of things

6.3.1 Rules

For any concrete a, b and x:

$$((\boldsymbol{a} \subset \boldsymbol{x}) \Rightarrow (\boldsymbol{b} \subset \boldsymbol{x})) = \Big((a \subset x) = ((a \subset x)\, and\, (b \subset x)) = ((a \cup b) \subset x)\Big)$$
$$= \Big((a \subset x) = ((a \cup b) \subset x)\Big) = (a = (a \cup b)) = (\boldsymbol{b} \subset \boldsymbol{a})$$

Where we have obviously true:

$$\Big((a \subset x) = ((a \cup b) \subset x)\Big) \Leftarrow (b \subset a)$$

And we have:

$$((a \cup b) \subset x) \Rightarrow \left(\sim(b \subset a) \rightarrow \bigvee_{y \subset b, \sim(y \subset a)} (y \subset x)\right)$$

$$\sim\left((a \subset x) \Rightarrow \left(\sim(b \subset a) \rightarrow \bigvee_{y \subset b, \sim(y \subset a)} (y \subset x)\right)\right)$$

So, for $\sim(b \subset a)$, both statements $\big((a \cup b) \subset x\big)$ and $(a \subset x)$ have different necessary conditions, so $\big((a \cup b) \subset x\big) \neq (a \subset x)$. So:

$$\sim\Big((a \subset x) = ((a \cup b) \subset x)\Big) \Leftarrow \sim(b \subset a)$$

So:

$$\Big((a \subset x) = ((a \cup b) \subset x)\Big) \Rightarrow (b \subset a)$$

So, finally:

$$\Big((\boldsymbol{a} \subset \boldsymbol{x}) = ((\boldsymbol{a} \cup \boldsymbol{b}) \subset \boldsymbol{x})\Big) = (\boldsymbol{b} \subset \boldsymbol{a})$$

Q.E.D.

So:

$$((\boldsymbol{a} \subset \boldsymbol{x}) \Rightarrow (\boldsymbol{b} \subset \boldsymbol{x})) = (\boldsymbol{b} \subset \boldsymbol{a})$$

And this proof does not fail, when we use an assumption, that two different things are equal, like, for example: $1=2$. It may seem like assumption $(1=2)$ implies, that all numbers are equal, so, for example:

$$((\{\boldsymbol{1}\} \subset \{\boldsymbol{2,3,4}\}) \Rightarrow (\{5\} \subset \{2,3,4\})) \equiv \mathit{true}$$

Though $\sim(\{5\} \subset \{1\})$.

But this is not the case. For more details, see Section 6.3.2.2.

Q.E.D.

So, using the following rule:

$$((\boldsymbol{a} \subset \boldsymbol{x}) \Rightarrow (\boldsymbol{b} \subset \boldsymbol{x})) = (\boldsymbol{b} \subset \boldsymbol{a})$$

We have, that for any plain concrete a, b and x such that $(a \cup b \cup x) \subset \Lambda_A^{\mho}$:

$$((\boldsymbol{x} \subset \boldsymbol{a}) \Rightarrow (\boldsymbol{x} \subset \boldsymbol{b})) = \Big((\sim_A^{\mho} a \subset \sim_A^{\mho} x) \Rightarrow (\sim_A^{\mho} b \subset \sim_A^{\mho} x)\Big) = (\sim_A^{\mho} b \subset \sim_A^{\mho} a) = (\boldsymbol{a} \subset \boldsymbol{b})$$

And for any concrete statements a, b and x:

$$((\boldsymbol{x} \subset \boldsymbol{a}) \Rightarrow (\boldsymbol{x} \subset \boldsymbol{b})) = ((\sim a \subset \sim x) \Rightarrow (\sim b \subset \sim x)) = (\sim b \subset \sim a) = (\boldsymbol{a} \subset \boldsymbol{b})$$

So, for any concrete statements or plain things a, b and x:

$$((\boldsymbol{x} \subset \boldsymbol{a}) \Rightarrow (\boldsymbol{x} \subset \boldsymbol{b})) = (\boldsymbol{a} \subset \boldsymbol{b})$$

For more details about the negation, see Section 9.2.4.

Q.E.D.

So also for any concrete x, y, a, b:

$$((\boldsymbol{a} \subset \boldsymbol{x}) \Rightarrow (\boldsymbol{b} \subset \boldsymbol{y})) \Leftarrow ((\boldsymbol{x} \subset \boldsymbol{y})\ \boldsymbol{and}\ (\boldsymbol{b} \subset \boldsymbol{a}))$$

Here is the following proof of it:

$$((a \subset x) \Rightarrow (b \subset y)) \Leftarrow \left(((a \subset x) \Rightarrow (a \subset y)) \; and \; ((a \subset y) \Rightarrow (b \subset y))\right)$$
$$= ((x \subset y) \; and \; (b \subset a))$$

Q.E.D.

And since implication is the inclusion of information, for any concrete statements x, a, b:

$$((\boldsymbol{a} \Rightarrow x) \Rightarrow (\boldsymbol{b} \Rightarrow x)) = (\boldsymbol{b} \Rightarrow \boldsymbol{a})$$

$$((x \Rightarrow \boldsymbol{a}) \Rightarrow (x \Rightarrow \boldsymbol{b})) = (\boldsymbol{a} \Rightarrow \boldsymbol{b})$$

$$((\boldsymbol{a} \Rightarrow x) \Rightarrow (\boldsymbol{b} \Rightarrow y)) \Leftarrow ((\boldsymbol{b} \Rightarrow \boldsymbol{a}) \; \boldsymbol{and} \; (x \Rightarrow y))$$

And for any a and any sets X and Y:

$$((\boldsymbol{a} \in \boldsymbol{X}) \Rightarrow (\boldsymbol{a} \in \boldsymbol{Y})) = ((\{a\} \subset X) \Rightarrow (\{a\} \subset Y)) = (\boldsymbol{X} \subset \boldsymbol{Y})$$

So:

$$((\boldsymbol{a} \in \boldsymbol{X}) \Rightarrow (\boldsymbol{a} \in \boldsymbol{Y})) = (\boldsymbol{X} \subset \boldsymbol{Y})$$

6.3.2 No exceptions

6.3.2.1 Introduction

We have, that:

$$((a \Rightarrow b) \Rightarrow a) \Rightarrow ((a \Rightarrow b) \Rightarrow b)$$

Here is the following proof of it:

$$((a \Rightarrow b) \Rightarrow a) = \left((a \Rightarrow b) = (a \; and \; (a \Rightarrow b)) \Rightarrow b\right) \Rightarrow ((a \Rightarrow b) \Rightarrow b)$$

Q.E.D.

And we have that:

$$(a \Rightarrow (a \Rightarrow b)) \Rightarrow (a \Rightarrow b)$$

Here is the following proof of it:

$$(a \Rightarrow (a \Rightarrow b)) = \left(a = (a \; and \; (a \Rightarrow b)) \Rightarrow b\right) \Rightarrow (a \Rightarrow b)$$

Q.E.D.

6.3.2.2 Not an exception

It may seem like we have the following exception:

If for any a, b and x, we assume:

$$\sim(b \subset a) \text{ and } ((a \subset x) \Rightarrow (b \subset a))$$

Then:

$$(a \subset x) \Rightarrow ((a \subset x) \text{ and } (b \subset a)) \Rightarrow (b \subset x)$$

So:

$$(a \subset x) \Rightarrow (b \subset x)$$

And we know that:

$$((a \subset x) \Rightarrow (b \subset x)) = (b \subset a)$$

So:

$$(b \subset a)$$

So, we have a contradiction because we assumed that $\sim(b \subset a)$.

So, our assumption must be false:

$$\sim(b \subset a) \text{ and } ((a \subset x) \Rightarrow (b \subset a))$$

So, above statement is never true.

It can also be proved the following way:

$$((a \Rightarrow b) \Rightarrow a) = \Big((a \Rightarrow b) = (a \text{ and } (a \Rightarrow b)) \Rightarrow b\Big) \Rightarrow ((a \Rightarrow b) \Rightarrow b)$$

$$((a \subset x) \Rightarrow (b \subset a))$$

$$= \Big(((b \subset a) \Rightarrow (b \subset x)) \Rightarrow (b \subset a)\Big) \Rightarrow \Big(((b \subset a) \Rightarrow (b \subset x)) \Rightarrow (b \subset x)\Big)$$

$$\Rightarrow ((a \subset x) \Rightarrow (b \subset x)) = (b \subset a)$$

And the following way:

$$(a \Rightarrow (a \Rightarrow b)) = \Big(a = (a \text{ and } (a \Rightarrow b)) \Rightarrow b\Big) \Rightarrow (a \Rightarrow b)$$

$$((a \subset x) \Rightarrow (b \subset a)) = \Big((a \subset x) \Rightarrow ((a \subset x) \Rightarrow (b \subset x))\Big) \Rightarrow ((a \subset x) \Rightarrow (b \subset x))$$

$$= (b \subset a)$$

So:

$$((a \subset x) \Rightarrow (b \subset a)) \Rightarrow (b \subset a)$$

So, the following assumption is never true:

$$\sim(b \subset a) \text{ and } ((a \subset x) \Rightarrow (b \subset a))$$

Q.E.D.

This may seem to concern, for example, numbers and sets of numbers, e.g. the following statements may seem to be true:

$$((2 \subset 1) \Rightarrow (3 \subset 2)) \leftrightarrow \big((2=1) \Rightarrow (2=1+(1)=1+(2)=3)=(3=2)\big) \equiv true$$

And:

$$((2 \subset 1) \Rightarrow (3 \subset 1)) \leftrightarrow \big((2=1) \Rightarrow (1=2=1+(1)=1+(2)=3)=(3=1)\big) \equiv true$$

But $\sim(3 \subset 2)$.

It may seem, that it concerns numbers, because for numbers, any assumption of equality of two different numbers seems to imply, that there is only one number (since any number is equal to any number), so sets of numbers are single element sets, that are atomic things equal to each other.

But here is a proof, that it is not the case:

First of all, for any different a and b there is function $h_{a,b}$, that fulfills the following conditions:

$$h_{a,b}(x) = a$$

$$h_{a,b}(y) = b$$

Let us assume, that $x=y$ for some different x and y.

Then for any different things a and b:

$$a = h_{a,b}(x) = h_{a,b}(y) = b$$

e.g. for $h_{2,3}(t) = (1+t)$ if $1=2$, then:

$$2 = h_{2,3}(1) = h_{2,3}(2) = 3$$

So anything is anything, so there is only one thing. But that function is defined only for different a and b, so then for any a and b function $h_{a,b}$ does not exist.

In other words, if everything is equal to everything then everything is equal to some nobeing x, so nothing exists. Then:

$$\bigwedge_{a \neq b}(h_{a,b}\ exists) \to \bigwedge_{a,b}(a=b) \to \bigwedge_{a}(a=x) \to \bigwedge_{a \neq b} \sim(h_{a,b}\ exists) \to \sim \bigwedge_{a \neq b}(h_{a,b}\ exists)$$

$$\bigwedge_{a \neq b}(h_{a,b}\ exists) \to \sim \bigwedge_{a \neq b}(h_{a,b}\ exists)$$

So $\sim\bigwedge_{a \neq b}(h_{a,b}\ exists)$.

So if $x=y$ for some different x and y, then the conclusion, that anything is anything, is unjustified.

Q.E.D.

And any analogous proof for any possible functions h, that seems to prove, that if $x = y$ for some different x and y, then anything is anything, will fail the same way.

6.4 Minimal step of a simple implication

A simple implication occurs only due to inheritance of members by a class from all its subclasses.

From the previous sections, we know, that:

For any concrete sets a, b and x:

$$((\boldsymbol{a} \subset \boldsymbol{x}) \Rightarrow (\boldsymbol{b} \subset \boldsymbol{x})) = (\boldsymbol{b} \subset \boldsymbol{a})$$

And for any concrete sets a, b and x:

$$((\boldsymbol{x} \subset \boldsymbol{a}) \Rightarrow (\boldsymbol{x} \subset \boldsymbol{b})) = (\boldsymbol{a} \subset \boldsymbol{b})$$

And:

$$(X \Subset Y) = (X^{\uparrow} \subset Y^{\uparrow}) = (attrp(X) \supset attrp(Y))$$

So, we have two ways, that:

For any X, Y and Z:

$$((X \Subset Y) \Rrightarrow (X \Subset Z)) = (Y \Subset Z)$$

This forward minimal step of implication has also the following special case:

$$((X\ isBp\backslash areBp\ Y) \Rrightarrow (X\ isBp\backslash areBp\ Z)) = ((\$X \Subset Y) \Rrightarrow (\$X \Subset Z)) = (Y \Subset Z)$$

Where:

$$(X\ isBp\backslash areBp\ Y) = (X \in Y^{\uparrow})$$

For more details about relations *isBp* and *areBp*, see Section 10.8.

In the following examples, relations *is* and *are* have the same meaning as respectively relations *isA* and *areA*:

$$((Rose\ is\ Flowering\ Plant) \Rrightarrow (Rose\ is\ Plant)) = (Flowering\ Plant\ is\ Plant) \equiv true$$

$$((Sky\ is\ Being\ That\ Is\ Blue) \Rrightarrow (Sky\ is\ Being\ That\ Has\ A\ Color))$$
$$= (Being\ That\ is\ Blue\ is\ Being\ That\ Has\ A\ Color) \equiv true$$

$$((Pterosaur\ is\ Being\ That\ Can\ Fly) \Rrightarrow (Pterosaur\ is\ Being\ That\ Do\ Not\ Only\ Walk))$$
$$= (Being\ That\ Can\ Fly\ is\ Being\ That\ Do\ Not\ Only\ Walk) \equiv true$$

$$((AI\ is\ Intelligent\ Machine) \Rrightarrow (AI\ is\ Something\ That\ Can\ Solve\ Problems))$$
$$= (Intelligent\ Machine\ is\ Something\ That\ Can\ Solve\ Problems) \equiv true$$

$$((Elpehant\ is\ Something\ That\ is\ As\ Small\ as\ Ant) \Rrightarrow (Elephant\ is\ Something\ Small))$$
$$= (Something\ That\ is\ As\ Small\ as\ Ant\ is\ Something\ Small) \equiv true$$

$$((Batman\ is\ Bat) \Rrightarrow (Batman\ is\ Something\ That\ Does\ Not\ See))$$
$$= (Bat\ is\ Something\ That\ Does\ Not\ See) \equiv true$$

$$((Instrumental\ Piece\ is\ Something\ You\ Can\ Dance\ To)$$
$$\Rrightarrow (Instrumental\ Piece\ is\ Something\ You\ Can\ Sing))$$
$$= (Something\ You\ Can\ Dance\ To\ is\ Something\ You\ Can\ Sing) \equiv false$$

$$((Song\ is\ Something\ You\ Can\ Listen) \Rrightarrow (Song\ is\ Something\ You\ Can\ Dance\ To))$$
$$= (Something\ You\ Can\ Listen\ is\ Something\ You\ Can\ Dance\ To) \equiv false$$

And for any X, Y and Z:

$$((X \Subset Y) \Rrightarrow (Z \Subset Y)) = (Z \Subset X)$$

In the following examples, relations *is* and *are* have the same meaning as respectively relations *isA* and *areA*:

$$((Rose\ is\ Flowering\ Plant) \Rrightarrow (French\ Rose\ is\ Flowering\ Plant))$$
$$= (French\ Rose\ is\ Rose) \equiv true$$

$$((Flightless\ Bird\ is\ Being\ That\ Cannot\ Fly) \Rrightarrow (Penguine\ is\ Being\ That\ Cannot\ Fly))$$
$$= (Penguine\ is\ Flightless\ Bird) \equiv true$$

$$((Star\ is\ Being\ That\ Is\ Heavier\ Than\ Its\ Planet)$$
$$\Rrightarrow (The\ Sun\ is\ Being\ That\ Is\ Heavier\ Than\ Its\ Planet)) = (The\ Sun\ is\ Star) \equiv true$$

$$((Tool\ That\ Allow\ to\ Solve\ Problems\ is\ Useful\ Tool) \Rrightarrow (Computer\ is\ Useful\ Tool))$$
$$= (Computer\ is\ Tool\ That\ Allow\ to\ Solve\ Problems) \equiv true$$

$$((Mammal\ is\ Something\ That\ Lay\ Eggs) \Rrightarrow (Lampart\ is\ Something\ That\ Lay\ Eggs))$$
$$= (Lampart\ is\ Mammal) \equiv true$$

$$((Superhero\ is\ Something\ That\ Cannot\ Fly) \Rrightarrow (Superman\ is\ Something\ That\ Cannot\ Fly))$$
$$= (Supermen\ is\ Superhero) \equiv true$$

$$((Movie\ is\ Something\ That\ Can\ Be\ Played) \Rrightarrow (Book\ is\ Something\ That\ Can\ Be\ Played))$$
$$= (Book\ is\ Movie) \equiv false$$

$$((Human\ is\ Something\ That\ Can\ Think) \Rrightarrow (Animal\ is\ Something\ That\ Can\ Think))$$
$$= (Animal\ is\ Human) \equiv false$$

And:

$$((X \Subset Y)\ and\ (Y \Subset X)) = \left((X^{\uparrow} \subset Y^{\uparrow})\ and\ (Y^{\uparrow} \subset X^{\uparrow})\right) = (X^{\uparrow} = Y^{\uparrow}) = (X = Y)$$

And we have:

$$(X \Subset Y) = (X\ isA \backslash areA\ Y) \Rrightarrow (X\ isp \backslash arep\ Y)$$

Relations *isp* and *arep* are defined in Section 10.8.

So, finally:

$$((\boldsymbol{X} \Subset \boldsymbol{Y}) \Rrightarrow (\boldsymbol{X} \Subset \boldsymbol{Z})) = (\boldsymbol{Y} \Subset \boldsymbol{Z}) = (\boldsymbol{Y}\ \boldsymbol{isA} \backslash \boldsymbol{areA}\ \boldsymbol{Z}) \Rrightarrow (\boldsymbol{Y}\ \boldsymbol{isp} \backslash \boldsymbol{arep}\ \boldsymbol{Z})$$
$$((\boldsymbol{X} \Subset \boldsymbol{Y}) \Rrightarrow (\boldsymbol{Z} \Subset \boldsymbol{Y})) = (\boldsymbol{Z} \Subset \boldsymbol{X}) = (\boldsymbol{Z}\ \boldsymbol{isA} \backslash \boldsymbol{areA}\ \boldsymbol{X}) \Rrightarrow (\boldsymbol{Z}\ \boldsymbol{isp} \backslash \boldsymbol{arep}\ \boldsymbol{X})$$
$$((\$\boldsymbol{X} \Subset \boldsymbol{Y}) \Rrightarrow (\$\boldsymbol{X} \Subset \boldsymbol{Z})) = (\boldsymbol{Y} \Subset \boldsymbol{Z}) = (\boldsymbol{Y}\ \boldsymbol{isA} \backslash \boldsymbol{areA}\ \boldsymbol{Z}) \Rrightarrow (\boldsymbol{Y}\ \boldsymbol{isp} \backslash \boldsymbol{arep}\ \boldsymbol{Z})$$

So, we can turn every minimal step of a simple implication into form $(A \Subset B)$.

So, for any K for $F(x) = (K \Subset x)$ we have:

$$(F(Y) \Rrightarrow F(Z)) = (Y \Subset Z)$$

And for any K for $F(x) = (x \Subset K)$ we have:

$$(F(Y) \Rrightarrow F(X)) = (X \Subset Y)$$

Now we can define minimal step of implication:

$$\left(V \underbrace{\Rrightarrow}_{minimal} Z\right) = \left(\begin{array}{l} \bigwedge\limits_{V=(X \Subset Y), Z=(X \Subset A)} (Y \Subset A) \\ and \bigwedge\limits_{V=(X \Subset Y), Z=(A \Subset Y)} (A \Subset X) \end{array}\right) = \left(\begin{array}{l} \bigvee\limits_{V=(X \Subset Y), Z=(X \Subset A)} (Y \Subset A) \\ or \bigvee\limits_{V=(X \Subset Y), Z=(A \Subset Y)} (A \Subset X) \end{array}\right)$$

All these conditions are equal to each other, because the conjunction of these conditions is equal to the disjunction of these conditions.

Where for $(X^{\uparrow} \cup Y^{\uparrow}) \subset A^{\uparrow}$ in the first case and $A^{\uparrow} \subset (X^{\uparrow} \cap Y^{\uparrow})$ in the second case V implicates true statements and for $(Y^{\uparrow} \subset A^{\uparrow})\ and \sim((X^{\uparrow} \cup Y^{\uparrow}) \subset A^{\uparrow})$ in the first case and $(A^{\uparrow} \subset X^{\uparrow})\ and \sim\left(A^{\uparrow} \subset (X^{\uparrow} \cap Y^{\uparrow})\right)$ in the second case V implicates false statements. So from this, we have, that the true statement of the sentence $(X \Subset Y)$ cannot implicate not true statement, because $(X^{\uparrow} \cup Y^{\uparrow}) = Y^{\uparrow}$ and $(X^{\uparrow} \cap Y^{\uparrow}) = X^{\uparrow}$ for $X^{\uparrow} \subset Y^{\uparrow}$.

The same can be also expressed the following way:

Where for $\left(attrp(A) \subset (attrp(X) \cap attrp(Y))\right)$ in the first case and $\left((attrp(X) \cup attrp(Y)) \subset attrp(A)\right)$ in the second case, V implicates true statements and for $(attrp(A) \subset attrp(Y))$ $and \sim\left(attrp(A) \subset (attrp(X) \cap attrp(Y))\right)$ in the first case and $(attrp(X) \subset attrp(A))$ $and \sim\left((attrp(X) \cup attrp(Y)) \subset attrp(A)\right)$ in the second case, V implicates false statements. So, from this we have, that the true statement of the sentence $(X \Subset Y)$ cannot implicate not true statement, because $(attrp(X) \cup attrp(Y)) = attrp(X)$ and $(attrp(X) \cap attrp(Y)) = attrp(Y)$ for $attrp(Y) \subset attrp(X)$.

6.5 Simple implication

A **basic implication** is an implication expressed as implication between two basic declarences.

A **simple implication not by negation** is in the form $((A \Subset B) \Rrightarrow (X \Subset Y))$.

From two minimal steps of a simple implication, every other simple implication not by negation can be constructed with a use of transitivity of implication, e.g. using these two elementary steps for sequence A, B, C, D, that fulfills $(A \Subset B)$ and $(C \Subset D)$, we have $(B \Subset C) \Rrightarrow (A \Subset D)$:

$$(\boldsymbol{B} \Subset \boldsymbol{C}) \Rrightarrow (B \Subset D) \Rrightarrow (\boldsymbol{A} \Subset \boldsymbol{D})$$

In addition, every **simple implication by negation** can be constructed using rule $(A \Subset B) \Rrightarrow \sim(A \Subset \wr B)$ and two simple implications:

$$\begin{aligned}&\Big(((A \Subset B) \Rrightarrow (C \Subset D))\ and\ ((C \Subset D) \Rrightarrow \sim(C \Subset \wr D))\ and\ ((X \Subset Y) \Rrightarrow (C \Subset \wr D))\Big)\\ &\quad = ((A \Subset B) \Rrightarrow (C \Subset D) \Rrightarrow \sim(C \Subset \wr D) \Rrightarrow \sim(X \Subset Y))\\ &\quad \Rrightarrow ((\boldsymbol{A} \Subset \boldsymbol{B}) \Rrightarrow \sim(\boldsymbol{X} \Subset \boldsymbol{Y}))\end{aligned}$$

So, we have the following hypothesis about the definition of any simple implication:

$$\begin{aligned}&\Big(S \underbrace{\Rrightarrow}_{\text{simple}} T\Big) = \Big(\sim T \underbrace{\Rrightarrow}_{\text{simple}} \sim S\Big)\\ &= \left(\begin{array}{l}\bigvee_{\{A,B,X,Y\} \subset \Omega} \left(\begin{array}{c}(S = (A \Subset B))\ and\ (T = (X \Subset Y))\\ and\ (X \Subset A)\ and\ (B \Subset Y)\end{array}\right)\\ or \bigvee_{\{A,B,C,D,X,Y\} \subset \Omega} \left(\begin{array}{c}(S = (A \Subset B))\ and\ (T = \sim(X \Subset Y))\\ and\ (C \Subset A)\ and\ (B \Subset D)\ and\ (C \Subset X)\ and\ (Y \Subset \sim D)\end{array}\right)\end{array}\right)\end{aligned}$$

In the following examples, relations *is* and *are* have the same meaning as respectively relations *isA* and *areA*:

$$((Rose\ is\ Flowering\ Plant) \Rrightarrow (French\ Rose\ is\ Plant))$$
$$\Lleftarrow ((French\ Rose\ is\ Rose)\ and\ (Flowering\ Plant\ is\ Plant)) \equiv true$$

$$((Star\ is\ Being\ That\ Is\ Greater\ Than\ Its\ Planet)$$
$$\Rrightarrow (The\ Sun\ is\ Being\ That\ Is\ Greater\ Than\ Earth))$$
$$\Lleftarrow \left(\begin{array}{c} (The\ Sun\ is\ Star)\ and \\ \left(\begin{array}{c} Being\ That\ Is\ Greater\ Than\ Solar\ System\ Planet\ is\ Being \\ That\ Is\ Greater\ Than\ Earth \end{array} \right) \end{array} \right)$$
$$\equiv true$$

$$((Vehicle\ is\ Transportation\ Machine) \Rrightarrow (Car\ is\ Machine))$$
$$\Lleftarrow ((Car\ is\ Vehicle)\ and\ (Transportation\ Machine\ is\ Machine)) \equiv true$$

For sure, we only know, that:

$$\boldsymbol{((X \Subset A)\ and\ (B \Subset Y)) \Rrightarrow ((A \Subset B) \Rrightarrow (X \Subset Y))}$$

$$\boldsymbol{((C \Subset A)\ and\ (B \Subset D)\ and\ (C \Subset X)\ and\ (Y \Subset \wr D))}$$
$$\Rrightarrow ((A \Subset B) \Rrightarrow (C \Subset D) \Rrightarrow \sim(C \Subset \wr D) \Rrightarrow \sim(X \Subset Y))$$
$$\Rrightarrow \boldsymbol{((A \Subset B) \Rrightarrow \sim(X \Subset Y))}$$

And we know, that:

$$\Big(((A \Subset B) \Rrightarrow (X \Subset B))\ and\ ((A \Subset B) \Rrightarrow (A \Subset Y))\Big) = ((X \Subset A)\ and\ (B \Subset Y))$$
$$\Rrightarrow ((A \Subset B) \Rrightarrow (X \Subset Y))$$

$$\Big(((A \Subset B) \Rrightarrow (X \Subset B))\ and\ ((A \Subset B) \Rrightarrow (A \Subset \wr Y) \Rrightarrow \sim(A \Subset Y))\Big)$$
$$= ((X \Subset A)\ and\ (B \Subset \wr Y)) \Rrightarrow ((A \Subset B) \Rrightarrow \sim(X \Subset Y))$$

And for any two classifications *Classification A of C* and *Classification B of D* and any *X* we have the following rule, that can be extended only to the right, but in two dimensions:

$$\Big(((a \backslash the) X\ is \backslash are\ (a)\ Classification(s)\ A\ of\ C)$$
$$\Rrightarrow ((a \backslash the)\ X\ is \backslash are\ (a)\ Classification(s)\ B\ of\ D)\Big)$$
$$= (R(A, B)\ and\ R(C, D))$$

For example:

$$(X\ is\ Warm-blooded\ Species\ of\ Domesticated\ Animal) \Rightarrow (X\ is\ Species\ of\ Animal)$$

$$(X\ is\ Thoroughbred\ of\ Horse) \Rightarrow (X\ is\ Breed\ of\ Animal)$$

There are also mixed rules for A not being a subclass of $Classification(s)\ B\ of\ C$:

$$((A\ isBp \backslash areBp\ Classification(s)\ B\ of\ C) \Rightarrow (A \Subset D)) = (C \Subset D)$$
$$((A \Subset D) \Rightarrow (A\ isBp \backslash areBp\ Classification(s)\ B\ of\ C)) = (D \Subset C)$$

And in addition for X not being a subclass of $Classification(s)\ A\ of\ Y$:

$$(X\ isBp \backslash areBp\ Classification(s)\ A\ of\ Y) \Rightarrow (X \Subset Y)$$

6.6 A simple declarence

A **basic declarence** is every declarence, that does not use the logical operators on declarences.

A **simple declarence** is a declarence, that can be presented in the positive form $\Subset (A,B)$ or the negative form $(! \Subset)(A,B)\ (= \sim(A \Subset B))$, e.g.:

$$(He\ knows\ that\ she\ likes\ him)$$
$$= (\mathbf{He\ is\ Someone\ Who\ Knows\ That\ (She\ is\ Someone\ Who\ Likes\ Him)}) =$$
$$\Subset (\mathbf{He, Someone\ Who\ Knows\ That} \Subset (\mathbf{She, Someone\ Who\ Likes\ Him}))$$

$$(A\ Horse\ runs\ sometimes) = (\mathbf{A\ Horse\ is\ Something\ That\ Runs\ Sometimes}) =$$
$$\Subset (\mathbf{Horse,\ Something\ That\ Runs\ Sometimes})$$

$$(Some\ Horse\ runs\ sometimes) = \sim(Neither\ of\ Horses\ runs\ sometimes)$$
$$= \sim(\mathbf{Horse\ is\ Something\ That\ Does\ Not\ Ever\ Run}) = \sim$$
$$\Subset (\mathbf{Horse,\ Something\ That\ Does\ Not\ Ever\ Run})$$
$$= \sim \bigwedge_{X \in Horse^{\Uparrow}} (X\ is\ Something\ That\ Does\ Not\ Ever\ Run)$$
$$= \sim \bigwedge_{X \in Horse^{\Uparrow}} \bigvee_{Y \in (Something\ That\ Does\ Not\ Ever\ Run)^{\Uparrow}} (X = Y)$$

So for any things X and Y, we have the following rules:

$$\sim(Every\ X\ is\ Some\ Y) = \sim(Any\ X\ is\ Some\ Y) = (Some\ X\ is\ not\ Any\ Y)$$
$$= (Some\ X\ is\ Some\ Not\ Y) \neq (Every\ X\ is\ Some\ Not\ Y)$$
$$= (Any\ X\ is\ not\ Any\ Y) = (Neither\ of\ \&X\ is\ Some\ Y)$$
$$(Not\ Not\ X) = X$$

Let us assume:

$$X_\beta = \downarrow X = \left(X^{\Uparrow}\right)_\downarrow$$

Then we have:

$$(X \Subset Y) \Rightarrow (X_\beta \Subset Y_\beta)$$

$$X_\beta^\uparrow = \left(\left(X^{\Uparrow}\right)_\downarrow\right)^\uparrow = X^{\Uparrow}$$

$$(X_\beta \Subset Y_\beta) = (Any\, X\, is\, Some\, Y) \Rrightarrow (Some\, X\, is\, Some\, Y)$$
$$= \sim(X_\beta \Subset \wr Y_\beta) = \sim(X_\beta \Subset Not\, Y_\beta)$$

$$\sim(X_\beta \Subset Y_\beta) = \sim(Any\, X\, is\, Some\, Y) = (Some\, X\, is\, not\, Any\, Y)$$
$$= (Some\, X\, is\, Some\, Not\, Y)$$

And for concrete thing X:

$$(X \Subset Y) = \sim(X \Subset \wr Y) = \sim(X \Subset Not\, Y)$$

And again for any things X and Y, we have the following rules:

$$(\boldsymbol{Only\, Y\, is\, X}) = \sim(\boldsymbol{Some\, X\, is\, not\, Any\, Y}) = (\boldsymbol{Any\, X\, is\, Some\, Y}) = (\boldsymbol{X_\beta \Subset Y_\beta})$$
$$= \bigwedge_{Z \in X_\beta^\uparrow} (Z\, is\, Y_\beta) = \bigwedge_{Z \in X_\beta^\uparrow} (Z \Subset Y_\beta) = \bigwedge_{Z \in X_\beta^\uparrow} \bigvee_{V \in Y_\beta^\uparrow} (Z = V) = \bigwedge_{Z \in X_\beta^\uparrow} \sim \bigvee_{V \in \sim Y_\beta^\uparrow} (Z = V)$$
$$= \bigwedge_{Z \in X_\beta^\uparrow} \bigwedge_{V \in \sim Y_\beta^\uparrow} [Z \neq V]$$

$$(\boldsymbol{Not\, Only\, Y\, is\, X}) = (\boldsymbol{Some\, X\, is\, not\, Any\, Y}) = \sim(\boldsymbol{Any\, X\, is\, Some\, Y})$$
$$= \sim\sim(Some\, X\, is\, not\, Any\, Y) = (\boldsymbol{Not\, Any\, X\, is\, Some\, Y})$$
$$= \sim(\boldsymbol{X_\beta \Subset Y_\beta}) = \sim \bigwedge_{Z \in X_\beta^\uparrow} \bigvee_{V \in Y_\beta^\uparrow} (Z = V) = \bigvee_{Z \in X_\beta^\uparrow} \sim \bigvee_{V \in Y_\beta^\uparrow} (Z = V)$$
$$= \bigvee_{Z \in X_\beta^\uparrow} \bigvee_{V \in \sim Y_\beta^\uparrow} (Z = V) = \bigvee_{Z \in X_\beta^\uparrow} \bigwedge_{V \in Y_\beta^\uparrow} \sim(Z = V)$$
$$= \bigwedge_{W \in \times_{Z \in X_\beta^\uparrow} \left\{ \coprod_{V \in \sim Y_\beta^\uparrow} (Z,V) \right\}} \left[\bigvee_{\{i, \{(Z,V)\}\} \in W} (Z \neq V)\, by\, i \right]$$

$$(\textbf{\textit{Only Not Y is X}}) = \sim(\textbf{\textit{Some X is Some Y}}) = (\textbf{\textit{Any X is not Any Y}})$$

$$= (\textbf{\textit{Any X is Some Not }}\boldsymbol{Y_\beta}) = (\boldsymbol{X_\beta} \Subset \textbf{\textit{Not }}\boldsymbol{Y_\beta}) = \bigwedge_{Z \in X_\beta^\uparrow} (Z \textit{ is Not } Y_\beta)$$

$$= \bigwedge_{Z \in X_\beta^\uparrow} (Z \Subset Not\ Y_\beta) = \bigwedge_{Z \in X_\beta^\uparrow} \bigvee_{V \in \sim Y_\beta^\uparrow} (Z = V) = \bigwedge_{Z \in X_\beta^\uparrow} \bigwedge_{V \in Y_\beta^\uparrow} [Z \neq V]$$

$$= \bigwedge_{Z \in X_\beta^\uparrow} \sim \bigvee_{V \in Y_\beta^\uparrow} (Z = V) = \bigwedge_{Z \in X_\beta^\uparrow} \bigvee_{V \in \sim Y_\beta^\uparrow} (Z = V)$$

$$(\textbf{\textit{Not Only Not Y is X}}) = (\textbf{\textit{Some X is Some Y}}) = \sim(\textbf{\textit{Any X is not Any Y}})$$

$$= \sim\sim(\textit{Some X is Some Y}) = (\textbf{\textit{Not Any X is Some Not }}\boldsymbol{Y_\beta})$$

$$= \sim(\boldsymbol{X_\beta} \Subset \textbf{\textit{Not }}\boldsymbol{Y_\beta}) = \sim \bigwedge_{Z \in X_\beta^\uparrow} \sim \bigvee_{V \in Y_\beta^\uparrow} (Z = V) = \sim \bigwedge_{Z \in X_\beta^\uparrow} \bigvee_{V \in \sim Y_\beta^\uparrow} (Z = V)$$

$$= \bigvee_{Z \in X_\beta^\uparrow} \bigvee_{V \in Y_\beta^\uparrow} (Z = V) = \bigvee_{Z \in X_\beta^\uparrow} \bigwedge_{V \in \sim Y_\beta^\uparrow} \sim(Z = V)$$

$$= \bigwedge_{W \in \times_{Z \in X_\beta^\uparrow} \left\{ \coprod_{V \in \sim Y_\beta^\uparrow} (Z,V) \right\}} \left[\bigvee_{\{i,\{(Z,V)\}\} \in W} (Z \neq V)\ by\ i \right]$$

Where $X^\uparrow$ is the set of all representatives of the kind of X. For more details, see Section 10.18.

Where operation $\&X$ makes plural form from singular form. For more details, see Chapter 10.

As you can see, you can even go to the level of the representatives of kinds and make calculations there.

6.7 Turning a plural form into its singular form

Sometimes you can turn a plural form *are* into its singular form *is* using the following rule:

For not concrete thing X:

$$(\&X\ are\ \&Y) = (X\ is\ Y)$$

Where operation $\&X$ makes the plural form from singular form X. For more details, see Chapter 10.

For example:

$$(Dogs\ are\ Animals) = (Dog\ is\ Animal)$$

$$(Dogs\ run) = (Dogs\ are\ Those\ Things\ That\ Run) = (Dog\ is\ Something\ That\ Runs)$$

$$(Some\ Dogs\ are\ Rottweilers) = (More\ Than\ One\ Dog\ is\ Rottweiler)$$
$$= ((Some\ Dog\ is\ Rottweiler)\ and\ (Some\ Other\ Dog\ is\ Rottweiler))$$
$$= \bigvee_{x \in Dog^{\Uparrow}} \left((x\ is\ Rottweiler)\ and \bigvee_{y \in (Dog^{\Uparrow} - \{x\})} ((y\ is\ Rottweiler)) \right)$$

6.8 Turning a simple declarence into all its forms

Turning into form $(X \Subset Y)$ or $\sim(X \Subset Y)$ can be done only for every thing X referenced at least implicitly by given declarence.

For full normalized relation $R(Any\,A_{)(})$ (more about this type of relation, you will find in Section 10.12), it can be done, for example, using the following rules:

$$R\left(Any\,(A_1, \ldots, A_n)_{)(}\right)$$
$$= \left(A_i \Subset Something\ for\ which\ R \left(Any \begin{pmatrix} Any\,(A_1, \ldots, A_{i-1})_{)(}, \\ repr\ it, \\ Any\,(A_{i+1}, \ldots, A_n)_{)(} \end{pmatrix}_{)(}, \right) \right)$$
$$= \left(A_i \Subset \wr \left(Something\ for\ which \sim R \begin{pmatrix} Any\,(A_1, \ldots, A_{i-1})_{)(}, \\ repr\ it, \\ Any\,(A_{i+1}, \ldots, A_n)_{)(} \end{pmatrix} \right) \right)$$
$$= \left(A_i \Subset \wr \left(Something\ for\ which\ (!R) \begin{pmatrix} Some\,(A_1, \ldots, A_{i-1})_{)(}, \\ repr\ it, \\ Some\,(A_{i+1}, \ldots, A_n)_{)(} \end{pmatrix} \right) \right)$$

So, for full normalized relation $R(Some\,A_{)(})$:

$$R\left(Some\,(A_1,\,\ldots,A_n)_{)(}\right)$$

$$= \sim \left(A_i \Subset Something\,for\,which \sim\!R\left(\begin{array}{c} Some\,(A_1,\,\ldots,A_{i-1})_{)(},\\ repr\,it,\\ Some\,(A_i,\,\ldots,A_n)_{)(}\end{array}\right)\right)$$

$$= \sim \left(A_i \Subset \wr \left(Something\,for\,which\,R\left(\begin{array}{c} Some\,(A_1,\,\ldots,A_{i-1})_{)(},\\ repr\,it,\\ Some\,(A_i,\,\ldots,A_n)_{)(}\end{array}\right)\right)\right)$$

$$= \sim \left(A_i \Subset Something\,for\,which\,(!R)\left(\begin{array}{c} Any\,(A_1,\,\ldots,A_{i-1})_{)(},\\ repr\,it,\\ Any\,(A_i,\,\ldots,A_n)_{)(}\end{array}\right)\right)$$

And:

$$R\left(A_{)(}\right) = \left(R \Subset \left(Something\,for\,which\,(it)\left(A_{)(}\right)\right)\right)$$

The sentence "The Children plays in The Garden today" can be turned into, for example:

"**The Children** are <u>Those That Are Playing In The Garden Today</u>"
"**The Garden** is <u>The Place Where The Children Plays Today</u>"
"**Playing** is <u>Something That The Children Do Today In The Garden</u>"
"**Playing Today** is <u>Something That The Children Do In The Garden</u>"
"**Playing In The Garden** is <u>Something That The Children Do Today</u>"
"**Playing In The Garden Today** is <u>Something That The Children Do</u>"
"**Today** is <u>The Day When The Children Plays In The Garden</u>"
"**Today In The Garden** is <u>The Day When The Children Plays</u>"

And so on.

If a relation can be undecidable, then it cannot be transformed to such a form $(X \Subset Y)$, because the first or the second argument would not be a thing, since the class of it would not have its representation, e.g. for any concrete statement x:

$$(x\,is\,false) \neq (x\,is\,a\,Thing\,That\,Is\,False) = \left(\{x\} \subset (Thing\,That\,Is\,False)^{\Uparrow}\right)$$

$$(x\,is\,false) \neq (x\,is\,a\,False\,Statement) = \left(\{x\} \subset (False\,Statement)^{\Uparrow}\right)$$

But:

$$(x\ \textit{is decidable and} \sim x) = (x\ \textit{is a False Decidable Statement})$$

$$= \left(\{x\} \subset (\textit{False Decidable Statement})^{\Uparrow}\right)$$

There are not things *Thing That Is False* and *False Statement*, because $(x\ \textit{is false})$ is undecidable for undecidable statement x, so the representations of such things do not exist. But there is a thing *False Decidable Statement*, because a decidable statement is always either true or false, so the class of false decidable statement exists.

Remember, that, for example:

$$(\textit{A decidable statement is either true or false})$$

$$= (\textit{A decidable statement is a statement, that is decidable and either true or false})$$

$$\neq (\textit{A decidable statement is a statement, that is either true or false})$$

Important note: Whenever in the text (not in expressions) of this book we use the phrases “true/false statement”, we have on mind a true/false decidable statement, if not stated otherwise. In all expressions in this book, except the above three statements about x, we use the phrases “true/false statement” only if we have on mind a true/false statement, if not stated otherwise.

If thing X is referenced by the sentence more than once, then you can reference it by the words “it”/“they”/“them”, “its”/ “their”, “itself”/ “themselves”, e.g.:

$$(\textit{My cat loves my cat}) = (\textbf{\textit{My Cat}}\ \textit{is Something That Loves Itself})$$

Or:

$$(\textit{My cat loves my cat}) = (\textbf{\textit{My Cat}}\ \textit{is Something That Loves My Cat})$$

Although both above declarences are different simple declarences, they have, of course, the same statement.

Another example:

$$(\text{Some Cat is eating Some Mouse})$$

$$= \sim(\textit{A}\backslash\textit{Any}\ \textbf{Cat}\ \textit{is}\ \underline{\textit{Something That Is Not Eating Any Mouse}})$$

$$= \sim(\textit{A}\backslash\textit{Any}\ \textbf{Mouse}\ \textit{is}\ \underline{\textit{Something That Is Not Being Eaten By Any Cat}})$$

6.9 Inferring an implication from the expression of simple implications

And you can turn:

$$\left(\bigvee_I a_i \, by\, i \Rightarrow b\right) = \bigwedge_I (a_i \Rightarrow b)\, by\, i$$

$$\left(a \Rightarrow \bigwedge_I b_i \, by\, i\right) = \bigwedge_I (a \Rightarrow b_i)\, by\, i$$

$$\bigvee_I (a_i \Rightarrow b)\, by\, i \Rightarrow \left(\bigwedge_I a_i \, by\, i \Rightarrow b\right)$$

$$\bigvee_I (a \Rightarrow b_i)\, by\, i \Rightarrow \left(a \Rightarrow \bigvee_I b_i \, by\, i\right)$$

And, of course, you can turn every implication of two statements into form:

$$or_{i=1,\ldots,n}\left(and_{i,j=1,\ldots,n_i}(A_{i,j})\right) \Rightarrow and_{i=1,\ldots,m}\left(or_{i,j=1,\ldots,m_i}(B_{i,j})\right)$$

Where each $A_{i,j}$ and $B_{i,j}$ is the statement of a basic declarence or the negation of the statement of a basic declarence.

So, you can infer each such implication from a conjunction of disjunctions of simple implications:

$$and_{(i,k)\in\{1,\ldots,n\}\times\{1,\ldots,m\}}\left(or_{(j,l)\in\{1,\ldots,n_i\}\times\{1,\ldots,m_k\}}(A_{i,j} \Rightarrow B_{k,l})\right)$$

So, you can infer any implication from such an expression of simple implications and you can state whether simple implications are true or false, so when such an expression is true, you can confirm, that given implication is true. To state whether any implication is true or false is more difficult.

6.10 Additional useful rules for composition and decomposition of implication

The following rules will help you in analysis of implication:

For any concrete statements *a*, *b*, *c*, *x* and *y*:

$$((a\, and\, b) \Rightarrow \sim c) = (a \Rightarrow \sim(c\, and\, b)) = ((c\, and\, b) \Rightarrow \sim a)$$

[not verified]

$$(a \Rightarrow b) \Rightarrow ((a\, and\, x) \Rightarrow (b\, or\, y))$$

For a proof of both above rules, see Section 9.2.12.

For any X, Y, A, B:

$$((X \Subset A)\ and\ (X \Subset B)) = \cancel{(X \Subset (A\ and\ B))} = \left(X \Subset \left(A \hat{\cup} B\right)\right)$$

$$((X \Subset A)\ and\ (Y \Subset A)) = \cancel{((X\ and\ Y) \Subset \&A)} = (([X]\ and\ [Y]) \Subset \&A) = \left(\left(X \hat{\cap} Y\right) \Subset A\right)$$

Where $\&A$ is the plural form of A and $A\&$ is the singular form of A.

In this section, sentences in the informal language are crossed out. Do not use them in the formal reasoning so as not to mix different kinds of operators.

In the informal language, a conjunction of things in the left argument will be always treated as a plural form, so you cannot expand any of these things to other expressions of things or you should use the formal language and write the conjunction in the form of a plural form, e.g.: $(([X_1]\ and \ldots and\ [X_n]) \Subset Y)$ instead of $((X_1\ and \ldots and\ X_n) \Subset Y)$ for abstract X_i for $i = 1, .., n$.

A conjunction of things in the right argument will be treated in the informal language as a fusion of these things $\left(\text{operator "}\hat{\cup}\text{"}\right)$ only if the singular/plural property of the elements of the conjunction is the same as of the left argument, so then you cannot also expand any of its arguments to other expressions of things, because, for example, two different kinds of a conjunction cannot be mixed or you should use the formal language and write the conjunction in the form of a fusion: $\left(Y \Subset \left(X_1 \hat{\cup} \ldots \hat{\cup} X_n\right)\right)$ instead of $(Y \Subset (X_1\ and \ldots and\ X_n))$.

$$(A \Subset B) \Rightarrow \left(\left(X \Subset \left(A \hat{\cup} B\right)\right) = (X \Subset A)\right)$$

$$(X \Subset Y) \Rightarrow \left(\left(\left(X \hat{\cap} Y\right) \Subset A\right) = (Y \Subset (A\ turned\ to\ form\ of\ Y))\right)$$

$$\bigvee_{A \in T}^{\oplus} (X \Subset A) = \left(X \Subset \bigvee_{A \in T}^{\oplus} A\right)$$

$$\bigvee_{X \in T}^{\oplus} (X \Subset A) = \left(\bigvee_{X \in T}^{\oplus} X \Subset A\right)$$

$$((X \Subset A)\ or\ (X \Subset B)) = \cancel{(X \Subset (A\ or\ B))} = \left(X \Subset \left(A \hat{\times} B\right)\right)$$

$$((X \Subset A)\ or\ (Y \Subset A)) = \cancel{((X\ or\ Y) \Subset A)}$$

$$((X \Subset A)\ and\ (Y \Subset B)) \Rrightarrow \left(\left(\left(X \hat{\cap} Y\right) \Subset \left(A \hat{\cap} B\right)\right)\ and\ \left(\left(X \hat{\cup} Y\right) \Subset \left(A \hat{\cup} B\right)\right)\right)$$

$$((X \Subset A)\ or\ (Y \Subset B)) \Rrightarrow \left(\left(X \hat{\cup} Y\right) \Subset \left(A \hat{\cap} B\right)\right)$$

$$attrp\left(X \breve{\times} Y\right) = \{x\ or\ y\colon (x,y) \in (attrp(X) \times attrp(Y))\}$$

$$attrp\left(X \hat{\cup} Y\right) = (attrp(X) \cup attrp(Y))$$

$$attrp\left(X \hat{\cap} Y\right) = (attrp(X) \cap attrp(Y))$$

$$\left(X \hat{\cup} Y\right) = \left(X^{\uparrow} \cap Y^{\uparrow}\right)_{\downarrow}$$

$$\left(X \hat{\cap} Y\right) = \left(X^{\uparrow} \cup Y^{\uparrow}\right)_{\downarrow}$$

Where $X^{\uparrow}$ is the set of all members of the class of X. $X_{\downarrow}$ is the thing, the class of which is the class of an element of set X. For more details, see Section 10.18.

A simple implication can be constructed from simple declarences using the following rules:

$$((X \Subset A)\ and\ (B \Subset Y)) \Rrightarrow ((A \Subset B) \Rrightarrow (X \Subset Y))$$

$$((X \Subset A)\ and\ (B \Subset Y)\ and\ (A \Subset B)) \Rrightarrow ((X \Subset B)\ and\ (A \Subset Y)) \Rrightarrow ((B \Subset A) \Rrightarrow (X \Subset Y))$$

$$((X \Subset A)\ and\ (B \Subset Y)\ and\ (Y \Subset X)) \Rrightarrow ((Y \Subset A)\ and\ (B \Subset X)) \Rrightarrow ((A \Subset B) \Rrightarrow (Y \Subset X))$$

$$\left(((A \Subset B) \Rrightarrow (C \Subset D))\ and\ ((X \Subset Y) \Rrightarrow (C \Subset \wr D))\right) \Rrightarrow ((A \Subset B) \Rrightarrow {\sim}(X \Subset Y))$$

Important note: The statement of such a simple declarence, that can be presented only in the negative form, does not ever implicate the statement of such a simple declarence, that can be presented only in the positive form, because you cannot deduce any inclusion of sets of attributes from the negation of the inclusion of sets of attributes. And, of course, we can turn every implication in the form ${\sim}(X \Subset Y) \Rrightarrow {\sim}(A \Subset B)$ into a simple implication:

$$({\sim}(X \Subset Y) \Rrightarrow {\sim}(A \Subset B)) = ((A \Subset B) \Rrightarrow (X \Subset Y))$$

The following implication is the only interesting implication of such a simple declarence, that can be presented only in the negative form:

$${\sim}(X \Subset Y) \Rrightarrow {\sim}((X \Subset Y)\ and\ (Y \Subset X)) = (X \neq Y)$$

6.11 An implication as a simple declarence

First of all, an implication can be equal to the statement of a sentence in the form $(A \Subset B)$:

$$(A \Subset B) = ((X \Subset A) \Rrightarrow (X \Subset B)) = ((B \Subset Y) \Rrightarrow (A \Subset Y))$$

So, above implications implicate whatever $(A \Subset B)$ implicates and can be implicated by whatever $(A \Subset B)$ is implicated by.

Remember also, that $(a \Rightarrow b)$ is also a basic declarence, that can be turned into form $(X \Subset Y)$ or $\sim(X \Subset Y)$ only three ways:

$$\begin{aligned}(a \Rightarrow b) &= \Subset (a,\ \textit{statement that implicates}\ b)\\ &= \Subset (b,\ \textit{statement that is implicated by}\ a)\\ &= \Subset (\textit{implication, binary relation between}\ a\ \textit{and}\ b)\end{aligned}$$

Where practically, nothing interesting comes from the last form.

And a and b are concrete beings, so the only implications are by the forward and backward minimal step:

$$\begin{aligned}&((a \Rightarrow b) \Rrightarrow (a \Rightarrow d))\\ &\quad = (\Subset (a,\ \textit{statement that implicates}\ b) \Rightarrow \Subset (a, \textit{statement that implicates}\ d))\\ &\quad = \Subset (\textit{statement that implicates}\ b,\ \textit{statement that implicates}\ d)\\ &\quad = (b \Rightarrow d)\end{aligned}$$

A proof of it is very simple:

If $\sim(b \Rightarrow d)$, then b is an example of a statement, that implicates b and does not implicate d, so:

$$\sim \Subset (\textit{statement that implicates}\ b,\ \textit{statement that implicates}\ d)$$

So:

$$\sim(b \Rightarrow d) \Rightarrow \sim \Subset (\textit{statement that implicates}\ b,\ \textit{statement that implicates}\ d)$$

$$\Subset (\textit{statement that implicates}\ b,\ \textit{statement that implicates}\ d) \Rightarrow (b \Rightarrow d)$$

And we have the following trivial implication:

$$(b \Rightarrow d) \Rightarrow \Subset (\textit{statement that implicates}\ b,\ \textit{statement that implicates}\ d)$$

So, putting both together we have:

$$(b \Rightarrow d) = \Subset (\textit{statement that implicates}\ b,\ \textit{statement that implicates}\ d)$$

Q.E.D.

And from backward minimal step, we have:

$$\begin{aligned}
&((a \Rightarrow d) \Rrightarrow (c \Rightarrow d))\\
&= (\Subset (d, \textit{statement that is implicated by } a) \Rightarrow \Subset (d, \textit{statement that is implicated by } c))\\
&= \Subset (\textit{statement that is implicated by } a, \textit{ statement that is implicated by } c)\\
&= (c \Rightarrow a)
\end{aligned}$$

Again, a proof of it is very simple:

If $\sim(c \Rightarrow a)$, then a is an example of a statement, that is implicated by a and is not implicated by c, so:

$$\sim \Subset (\textit{statement that is implicated by } a, \textit{ statement that is implicated by } c)$$

So:

$$\sim(c \Rightarrow a) \Rightarrow \sim \Subset (\textit{statement that is implicated by } a, \textit{ statement that is implicated by } c)$$

$$\Subset (\textit{statement that is implicated by } a, \textit{ statement that is implicated by } c) \Rightarrow (c \Rightarrow a)$$

And we have the following trivial implication:

$$(c \Rightarrow a) \Rightarrow \Subset (\textit{statement that is implicated by } a, \textit{ statement that is implicated by } c)$$

So, putting both together, we have:

$$(c \Rightarrow a) = \Subset (\textit{statement that is implicated by } a, \textit{ statement that is implicated by } c)$$

Q.E.D.

Summing up:

$$\begin{aligned}
((a \Rightarrow b) \Rrightarrow (c \Rightarrow d)) &\Leftarrow \begin{pmatrix} ((a \Rightarrow b) \Rrightarrow (a \Rightarrow d) \Rrightarrow (c \Rightarrow d)) \\ \textit{or } ((a \Rightarrow b) \Rrightarrow (c \Rightarrow b) \Rrightarrow (c \Rightarrow d)) \end{pmatrix}\\
&= \begin{pmatrix} (((a \Rightarrow b) \Rrightarrow (a \Rightarrow d)) \textit{ and } ((a \Rightarrow d) \Rrightarrow (c \Rightarrow d))) \\ \textit{or } (((a \Rightarrow b) \Rrightarrow (c \Rightarrow b)) \textit{ and } ((c \Rightarrow b) \Rrightarrow (c \Rightarrow d))) \end{pmatrix}\\
&= \begin{pmatrix} ((b \Rightarrow d) \textit{ and } (c \Rightarrow a)) \\ \textit{or } ((c \Rightarrow a) \textit{ and } (b \Rightarrow d)) \end{pmatrix} = ((b \Rightarrow d) \textit{ and } (c \Rightarrow a))
\end{aligned}$$

So, finally:

$$((a \Rightarrow b) \Rrightarrow (c \Rightarrow d)) \Leftarrow ((c \Rightarrow a) \textit{ and } (b \Rightarrow d))$$

And for any not necessarily concrete statements A and B such that $(A^{\to\infty}\ and\ B^{\to\infty}) \subset Statement^{\Uparrow}$ (where $X^{\to}$ is the expansion of X and $X^{\to\infty}$ is the infinite expansion (concretization) of X; for more detail, see Section 10.11):

$$(A \Rightarrow B) = ((Any\ A\ implicates\ Some\ B)\ and\ (Any\ B\ is\ implicated\ by\ Some\ A))$$

$$= \left(\bigwedge_{a \in A^{\uparrow}} \bigvee_{b \in B^{\uparrow}} (a \Rightarrow b)\ and \bigwedge_{b \in A^{\uparrow}} \bigvee_{a \in B^{\uparrow}} (a \Rightarrow b) \right)$$

Remember also, that you can treat an implication like inclusion of information, since:

$$(a \Rightarrow b) = (a\ includes\ all\ information\ that\ b\ includes) = (b \subset a)$$

From this, you have above reasoning in very simplified form:

$$((a \supset b) \Rrightarrow (a \supset d)) = (b \supset d)$$

$$((a \supset b) \Rrightarrow (c \supset b)) = (c \supset a)$$

For more details, see Section 6.3.

So:

$$((a \supset b) \Rrightarrow (c \supset d)) \Leftarrow \begin{pmatrix} ((a \supset b) \Rrightarrow (a \supset d) \Rrightarrow (c \supset d)) \\ or\ ((a \supset b) \Rrightarrow (c \supset b) \Rrightarrow (c \supset d)) \end{pmatrix}$$

$$= \begin{pmatrix} \left(((a \supset b) \Rrightarrow (a \supset d))\ and\ ((a \supset d) \Rrightarrow (c \supset d))\right) \\ or\ \left(((a \supset b) \Rrightarrow (c \supset b))\ and\ ((c \supset b) \Rrightarrow (c \supset d))\right) \end{pmatrix}$$

$$= ((b \supset d)\ and\ (c \supset a))$$

6.12 Some simple implications

Some simple implications, where the same statement implicates some statement and its negation, cannot be both true:

$$\left(((A \Subset B) \Rrightarrow (X \Subset Y))\ and\ ((A \Subset B) \Rrightarrow \sim(A \Subset Not\,B) \Rrightarrow \sim(X \Subset Y))\right)$$

$$\Rrightarrow ((A \Subset B) \Rrightarrow (X \Subset Y) \Rrightarrow (A \Subset Not\,B)) \Rrightarrow ((A \Subset B) \Rrightarrow (A \Subset Not\,B))$$

$$= ((B \Subset Not\,B)) \equiv false$$

Next example:

$$\left(((A \Subset B) \Rrightarrow (A \Subset Y))\ and\ ((A \Subset B) \Rrightarrow (A \Subset Not\,Y))\right) = ((B \Subset Y)\ and\ (B \Subset Not\,Y)) \equiv false$$

6.13 Additional deduction rules

In this section, we can also use operator $\Uparrow$ instead of $\uparrow$.

We have, for example, the following additional deduction rule:

$$\left(\left(\boldsymbol{A}^{\uparrow}\subset\boldsymbol{B}^{\uparrow}\right)\ \boldsymbol{and}\ \left(\left(\boldsymbol{X}^{\uparrow}\subset\boldsymbol{B}^{\uparrow}\right)\Rightarrow\boldsymbol{F}\right)\right)\Rightarrow\left(\left(\boldsymbol{X}^{\uparrow}\subset\boldsymbol{A}^{\uparrow}\right)\Rightarrow\boldsymbol{F}\right)$$

Proof:

$$\left(\left(A^{\uparrow}\subset B^{\uparrow}\right)\ and\ \left(\left(X^{\uparrow}\subset B^{\uparrow}\right)\Rightarrow F\right)\right)\Rightarrow\left(\left(X^{\uparrow}\subset A^{\uparrow}\right)\Rightarrow\left(X^{\uparrow}\subset B^{\uparrow}\right)\Rightarrow F\right)$$
$$\Rightarrow\left(\left(X^{\uparrow}\subset A^{\uparrow}\right)\Rightarrow F\right)$$

Q.E.D.

And we have analogous rule:

$$\left(\left(\boldsymbol{A}^{\uparrow}\subset\boldsymbol{B}^{\uparrow}\right)\ \boldsymbol{and}\ \left(\left(\boldsymbol{A}^{\uparrow}\subset\boldsymbol{X}^{\uparrow}\right)\Rightarrow\boldsymbol{F}\right)\right)\Rightarrow\left(\left(\boldsymbol{B}^{\uparrow}\subset\boldsymbol{X}^{\uparrow}\right)\Rightarrow\boldsymbol{F}\right)$$

Which has analogous proof.

Proof:

$$\left(\left(A^{\uparrow}\subset B^{\uparrow}\right)\ and\ \left(\left(A^{\uparrow}\subset X^{\uparrow}\right)\Rightarrow F\right)\right)\Rightarrow\left(\left(B^{\uparrow}\subset X^{\uparrow}\right)\Rightarrow\left(A^{\uparrow}\subset X^{\uparrow}\right)\Rightarrow F\right)$$
$$\Rightarrow\left(\left(B^{\uparrow}\subset X^{\uparrow}\right)\Rightarrow F\right)$$

Q.E.D.

Another deduction rule:

$$(the\ A\ is\ Classification\ B\ of\ C)\Rightarrow((a\ A\ is\ a\ C)\ and\ (the\ A\ is\ Classification\ B))$$

For example: the Rottweiler is a Breed of Dog, so a Rottweiler is a Dog and the Rottweiler is a Breed.

7 Proving and disproving statements

7.1 All kinds of logical proofs

A proof can be a subjective thing, because, what a proof is for us, depends on what we know. If, for example, we are not hallucinating and we see the wet ground, then it is a proof for us that the ground is wet, so we do not need any other proof of it. And we cannot prove a physical fact from the axioms of logic only, so we always have to know something about physical reality to prove a physical fact. For example, if we know only that it is raining and we do not see the ground, then we need a proof, that the ground is wet, which in this case is very simple, because the fact that it is raining implicates, that the ground is wet, so the proof will be as follows:

$$\Big((\{0\}, \textit{It is raining}), (\{0\}, (1) \Rightarrow (\textit{The ground is wet})), (\{1,2\}, \textit{The ground is wet})\Big)$$

The above form of a proof will be explained in detail further in this section. For now you only need to know, that each element of above tuple is the ordered pair of some premises and their conclusion, where numbers are the indexes of the elements of the above tuple and 0 refers to what you already know.

In other words, before we will recognize any thing as a proof for us, we have to determine, what we already know as an observer.

For any observer s:

$$(\textit{the known reality for observer } s) = \mathrm{K}_s \subset \mathcal{R}$$

$$(\textit{the known truth for observer } s) = @\sphericalangle_s \Leftarrow (\textit{the achievable truth for observer } s)$$

$$= \kappa_s \Leftarrow \textit{truth}$$

Where the symbol "$\sphericalangle_s$" is the symbol of the set of all true statements known to s and should resemble opened eye and means everything that he sees – what he knows, understands and is aware of. So $\sphericalangle_s$ is his true theory of the world he knows (K_s) and κ_s is the fullness of that theory, where $\kappa_s@$ is the true, consistent, full and exhaustive theory of K_s.

But remember, that if you, for example, know, that some bike is standing there, but you do not know its engine and the rest of its interior, then you only know the exterior of this bike, because its interior in such a case is not automatically included. The same way, if you know only the definition of some object, then you know only its external attributes and all other attributes remain unknown for you.

And we have:

$$(\textit{in the strict sense we know that statement } x \textit{ is true}) = (\textit{we have a proof of statement } x)$$

In the broad sense, we know something, if someone, in who we trust, said it, but in the strict sense, it is only a belief. The simplest consequence of that, is that if you in your proof invoke some other proof you do not know, then you can only believe that

https://doi.org/10.1515/9783111441382-007

your proof is correct. Fortunately, the basics of mathematics are quite simple to be understood and are brought home to you in this book. For example, this is not a dishonor to ask about proofs of the simplest things, like, for example, $1+1=2$, because it can teach you a lot. A proof that $1+1=2$ is very simple and does not need thousand pages at all. You will find this proof in Section 16.2.

The body of a **logical proof**, which is a rooted directed acyclic graph (rooted DAG), in which every node (except for a leaf) is a conclusion and (except for the root) a premise to another conclusion, can be sometimes (if we are lucky) flattened into and sometimes even (if we are even luckier) expressed as a generalized tuple of pairs of a premise and its conclusion, that fulfills the condition, that every conclusion in that tuple has a proof for us, so either it is a statement, that is already proved for us (the premise is equal to $\{0\}$), if the main proof is abbreviated, or it is implicated by a conjunction of some previous conclusions in that tuple (the premise is equal to the set of their numbers) and the last statement t is our final conclusion, which is our statement that needs to be proved. And if only a proof is in this form, then no matter how much steps it has, it obviously proves the last statement. Remember, that not every generalized tuple of pairs of a premise and its conclusion can the body of a logical proof for us, because if it does not fulfill the above conditions for us then at most it was meant to be the body of a logical proof for us and does not exist as the body of a logical proof for us, since such an attempt of a logical proof fails for us.

So, each statement in the tuple is implicated by the conjunction of all the previous statements in the tuple. Since axioms are universal, they are always assumed before a proof the same as what we, as an observer, already know ("$\sphericalangle_{observer}$"), but are not in the tuple of the proof. All beings exist due to the fact, that they have their definitions, that are not contrary, which means that they are not self-contrary and are not contrary to (truth of) reality, so they are not contrary to all other definitions of beings. All beings you know form the part of reality you know (K_{you}) and the fact of their existence have some implications, that form what you know ($\sphericalangle_{you}$) from what you can know ($\kappa_{you}@$) from the part of reality you know.

Assumption a assumed to prove, that it has implication b, has the following form:

$$\left\{\begin{matrix} \left(1,\left\{\left(\{0\},x_1\right)\right\}\right), \\ \ldots, \\ \left(n,\left\{\left(\{\ldots\},x_n\right)\right\}\right), \\ \left(n+1,\left\{\left(\{0\},a=a_0\Rightarrow a_1\right)\right\}\right), \\ \ldots, \\ \left(n+k,\left\{\left(\{0\},a_{k-1}\Rightarrow a_k=b\right)\right\}\right), \\ \left(n+k+1,\left\{\left(\{n+1,\ldots,n+k\},a\Rightarrow b\right)\right\}\right), \\ \ldots, \\ \left(m,\left\{\left(\{\ldots\},t\right)\right\}\right) \end{matrix}\right\}$$

Counterfactual assumption a assumed to prove its negation has the following form:

$$\left\{\begin{array}{c} \left(1, \left\{\left(\{0\}, x_1\right)\right\}\right), \\ \ldots, \\ \left(n, \left\{\left(\{\ldots\}, x_n\right)\right\}\right), \\ \left(n+1, \left\{\left(\{0\}, a = a_0 \rightarrow a_1\right)\right\}\right), \\ \ldots, \\ \left(n+k, \left\{\left(\{0\}, a_{k-1} \rightarrow a_k = b\right)\right\}\right), \\ \left(n+k+1, \left\{\left(\{n+1, \ldots, n+k\}, a \rightarrow b\right)\right\}\right), \\ \left(n+k+2, \left\{\left(\{0\}, \sim b\right)\right\}\right), \\ \left(n+k+3, \left\{\left(\{0\}, (a \rightarrow b) = (\sim b \rightarrow \sim a)\right)\right\}\right), \\ \left(n+k+4, \left\{\left(\{n+k+1, n+k+3\}, \sim b \rightarrow \sim a\right)\right\}\right), \\ \left(n+k+5, \left\{\left(\{n+k+2, n+k+4\}, \sim a\right)\right\}\right), \\ \ldots, \\ \left(m, \left\{\left(\{\ldots\}, t\right)\right\}\right) \end{array}\right\}$$

Only proofs of nonphysical facts can be unabbreviated, so only they can be objective. So, objective proofs of physical facts do not exist.

In other words, every segment of a proof, that starts at the beginning of the proof, is also a proof of some intermediate statement. So, we have a sequence of implications of statements, each of which is not stronger than the conjunction of all the previous statements in that sequence and of what we know. In the simplified form for any observer s:

$$\begin{aligned} &(\varnothing_{ç}\ and\ (\mathrm{K}_s \in \mathcal{B})) \\ &\qquad = (\mathrm{K}_s \in \mathcal{B}) \Rightarrow @\sphericalangle_s \Rightarrow (@\sphericalangle_s\ and\ x_1) \Rightarrow \left((@\sphericalangle_s\ and\ x_1)\ and\ x_2\right) \\ &\qquad \Rightarrow \left((@\sphericalangle_s\ and\ x_1\ and\ x_2)\ and\ x_3\right) \Rightarrow \ldots \\ &\qquad \Rightarrow \left((@\sphericalangle_s\ and\ x_1\ and \ldots and\ x_{n-1})\ and\ x_n\right) \Rightarrow x_n = t \subset \kappa_s \end{aligned}$$

Where:

$$(\mathrm{K}_s \in \mathcal{B}) = \bigwedge_{k \subset \mathrm{K}_s} (k \in \mathcal{B}) \Rightarrow \kappa_s = @\sphericalangle_s$$

$$\sphericalangle_s \subset \kappa_s @$$

And after we discover proof p of true statement t:

$$true \equiv \left(\mathrm{K}_s = (\mathrm{K}_s\ and\ p)\right) = (p \subset \mathrm{K}_s)$$

$$true \equiv \left(\sphericalangle_s = (\sphericalangle_s\ and\ \{t\})\right) = (t \in \sphericalangle_s)$$

And also an intermediate statement is proved at each step of a proof:

$$true \equiv \left(\mathrm{K}_s = (\mathrm{K}_s \ and\ p_i)\right) = (p_i \subset \mathrm{K}_s)$$

$$true \equiv \left(\sphericalangle_s = (\sphericalangle_s \ and\ \{x_i\})\right) = (x_i \in \sphericalangle_s)$$

So, such a proof looks as follows:

$$\begin{aligned}&(\varnothing_\varsigma \ and\ (\mathrm{K}_s \in \mathcal{B}))\\ &\quad = (\mathrm{K}_s \in \mathcal{B}) \Rightarrow @\sphericalangle_s \Rightarrow (@\sphericalangle_s \ and\ x_1) \Rightarrow (@\sphericalangle_s \ and\ x_2) \Rightarrow (@\sphericalangle_s \ and\ x_3)\\ &\quad \Rightarrow \ldots \Rightarrow (@\sphericalangle_s \ and\ x_n) \Rightarrow x_n = t \subset \kappa_s\end{aligned}$$

And can be reduced to a single step:

$$@\sphericalangle_s \Rightarrow x_n = t \subset \kappa_s$$

And no matter how many steps a proof has, it can be done in one moment.

A logical proof expressed in some language is correct only as long as all intermediate statements of sentences are true. Since, sentences can have different statements in different situations, such an expression of a proof can become obsolete.

Do not forget, that as long as your $\sphericalangle$ is not equal to your $\kappa@$, you can gain new knowledge almost for free – only at the cost of thinking. In other words, proving a statement relies on discovering all truth implied by the part of reality you know (your K). You should be able to discover every part of your κ, knowing at start only, what you already know (your $\sphericalangle$). So, if you truly cannot prove some statement, then it means, that you need to get to know better the appropriate part of reality to expand your K before proving that statement. And, this is dangerous to shut yourself in your K and be restricted to $\kappa@$ that arises out of it, so, better get out of your K and experience sometimes something different to widen your perspective, so that you can $\sphericalangle$ more.

In everyday practice, we also have a collective observer and authority, in which we trust more or less. So, when we say not strictly, that something is already proved we rely on and invoke the knowledge of a collective observer or some authority. In the ideal situation, the collective observer in science should depend on authorities and vice versa. Sometimes, a single fresh idea from one person can be enough to break through a period of long-standing stagnation. However, the scientific community needs to be open to new ideas, courageous and self-critical.

And we have:

$$\mathcal{R} = \bigwedge_{b \in \mathcal{B}} b$$

$$(\mathcal{R} \in \mathcal{B}) = \bigwedge_{r \subset \mathcal{R}} (r \in \mathcal{B})$$

$$true \equiv ((\mathcal{R} \in \mathcal{B}) \Rightarrow truth) = \bigwedge_{x \text{ is a true decidable statement}} ((\mathcal{R} \in \mathcal{B}) \Rightarrow x)$$

A proof of it is as follows:

$$\bigwedge_{x \text{ is a statement}} \left(x = \bigvee_{r \in [x]} (r \in \mathcal{B}) \right)$$

So:

$$true \equiv \bigwedge_{x \text{ is a true decidable statement}} \bigvee_{r \subset \mathcal{R}} ((r \in \mathcal{B}) \Rightarrow x)$$

$$\Rightarrow \bigwedge_{x \text{ is a true decidable statement}} \left(\bigwedge_{r \subset \mathcal{R}} (r \in \mathcal{B}) = (\mathcal{R} \in \mathcal{B}) \Rightarrow x \right) = ((\mathcal{R} \in \mathcal{B}) \Rightarrow truth)$$

Q.E.D.

So, the only way to prove anything is by implication, that is defined as:

$$(\boldsymbol{a} \Rightarrow \boldsymbol{b}) = (\sim\boldsymbol{b} \Rightarrow \sim\boldsymbol{a}) = (\textbf{\textit{any proof of a proves b}})$$

So, to find a proof of statement Y_i, we have to always find statement Y, that is true by some definition or is already proved, that implicates Y_i, or statement Y_{i+1}, that implicates Y_i and still needs a proof and so on to the moment when $Y_i = \sphericalangle_S$.

So, in general, each step of a proof goes this way:

$$\textbf{\textit{a proof of Y and of }} (\boldsymbol{Y} \Rightarrow \boldsymbol{X}) \textbf{\textit{ proves X}}$$

And since we cannot know, that a statement is true, without a proof:

$$\textbf{\textit{if we know that Y is true and that }} (\boldsymbol{Y} \Rightarrow \boldsymbol{X}) \textbf{\textit{ then it proves X}}$$

Because:

$$(\boldsymbol{Y} \textbf{\textit{ and }} (\boldsymbol{Y} \Rightarrow \boldsymbol{X})) \Rightarrow \boldsymbol{X}$$

So, if we know, that Y is true, then it is a proof of every statement implicated by Y. So, Y is a statement, not weaker than X, expressed in such a form, that can be evaluated to be true with use of our current knowledge.

And that is, why a proof of some statement X is always a proof of any special case Z of this statement. For example, for $X \Rightarrow Z$:

$$(Y \textit{ and } (Y \Rightarrow X \Rightarrow Z)) \Rightarrow X \Rightarrow Z$$

$$(Y \textit{ and } (Y \Rightarrow Z)) \Rightarrow Z$$

A sufficient reason for a statement is a true statement, that implicates this statement. The existence of a proof of a statement is a sufficient reason for this statement. And the implication of this statement by a sufficient reason is a proof of this statement. So, a proof of a statement exists if and only if a sufficient reason for the statement exists.

A contextual sufficient reason for a statement is a true statement, that contextually implicates this statement. The existence of a contextual proof of a statement is a contextual sufficient reason for this statement. And the implication of this statement by a contextual sufficient reason is a contextual proof of this statement. So, a contextual proof of a statement exists if and only if a contextual sufficient reason for the statement exists.

A contextual proof proves a statement if and only if all assumptions, that are made in our context, turns out to be true.

We can prove statement X by two ways:

- top-down ($\boldsymbol{Y \Rightarrow X}$) – **direct proof** – if we know statement Y, that is true by some definition or already proved and implicates X, then X is true by virtue of the fact, that true statement can implicate only true statements, since every statement is the conjunction of all statements implicated by it, so the statement is true if and only if all its implications are also true.
- down-top ($\boldsymbol{\sim X \Rightarrow \sim Y}$) – **indirect proof** – **proof by a contradiction** – if we know statement $\sim Y$, that is false by some definition or already disproved and is implicated by the negation of X, then X is true by virtue of the same fact, that true statement implicate only true statements, so the negation of X cannot be true, so it is false, if X is decidable. So, X is true, if X is decidable. And this is really a **proof by a contrariety**, because we assume $\sim X$ and we know that Y is true and we prove the contrariety between these two statements in the form $((\sim X) \Rightarrow \sim(Y))$. Since we know that Y is true and from the contrariety we know that $\sim X$ and Y cannot be both true, we conclude, that $\sim X$ is false or undecidable, so X is true, if X is decidable. We do not need Y, if we have self-contrariety in the form $(\sim X) \Rightarrow \sim(\sim X)$. Then $\sim X$ also cannot be true, because a true statement cannot implicate a false statement, so $\sim X$ is false or undecidable, so X is true, if X is decidable.

To simplify a proof, you can use operators $\rightarrow$, $\leftrightarrow$, so that you do not have to repeat all conditions of implication in every step of logical reasoning.

Of course, you are never sure, whether a statement is true or false, until you either prove it or disprove it. To disprove a statement is to prove its negation.

So, the basic method of proving any statement is to split it to a conjunction of statements, that are easier to be proved. These are special cases of the main problem or simply different problems.

Remember, that any limitations of a given language does not limit the possibility of the existence of a proof, because a proof does not need sentences, since either way it uses statements and sentences are only their expressions in a given language, that do not have to be a prefect tool, which can express everything. Every proof can be reduced to one step of implication – from a sufficient reason to a conclusion. So, do not think about a proof as something, that has finite number greater than 1 of steps, where each step is a finite size sentence. This is completely wrong. A proof can have even infinite number of steps (e.g. you can see countably infinite number of steps as steps done in a geometric sequence of pauses, the series of which converges to a finite real number, but it is not necessary) and even uncountable number of steps, that can be transformed to one step of implication. And in a given language, the expressions of statements used in that proof can be even impossible, in the same way as there are inexpressible numbers (a proof of it is further in this section). For example, the statements about inexpressible numbers, are for that reason inexpressible too. It does not matter. We all accept it in numbers, so why should we not accept it in statements? As I said, imperfect language does not limit the possibility of the existence of a proof, and it should go without saying. In the same way as there are statements that are inexpressible in any language, there are proofs that are inexpressible in any language. For example, the proofs, that use inexpressible statements, are for that reason inexpressible too. Fortunately, mind is not limited by a language and symbols in general, since it has more or less accurate reflections of statements and does not need words to operate on them, so it can understand more, but probably also every mind has its limitations, that will not allow it to understand everything. We should not expect and demand too much from such an imperfect tool as a language, that, to very high degree, allows communication between two minds. Maybe synthetic telepathy in the future will allow sending the reflections of statements directly between minds and language will not be necessary in the middle between them. Until then, you have to read books such as this one.

There exist inexpressible numbers for the same reason, for which there exist inexpressible statements. Any language can express at most countable number of statements and numbers and you can use at most countable number of languages, because you have to refer to them by finite symbols. So, the numbers of all statements and all numbers, that can be expressed any way (e.g. by expressing polynomials for algebraic numbers) in finite sentences, will be still countable, but the numbers of all statements and all numbers are uncountable. **Q.E.D.**

Remember, that some statements are undecidable, so down-top method can lead to a contradiction for the assumption, that the statement is true, as well as for the assumption, that it is false, which would be, in fact, a proof, that the statement is undecidable. For all undecidable statements $x \leftrightarrow \sim x$, which cannot be, of course, true for a decidable statement, so we conclude, that they are not decidable, which means that they are undecidable. A statement is decidable if and only if it is not a statement about the unpredictable future and its logical value does not depend on the logical value of any undecidable statement. For more details, see Section 5.23. If a statement, the assumption of the negation of which leads to a contradiction, is not undecidable, then it is decidable and true.

Sometimes, we can prove for parameterized statement $F(x)$, implications in the following form:

$$\bigwedge_{1 \le i \le n} \left(F\left(\coprod_{1 \le j \le n} x_j \right) \Rightarrow F\left(\coprod_{1 \le j < i} x_j, step_i(x_i), \coprod_{i < j \le n} x_j \right) \right)$$

Then, from any starting element $F(a)$, we can prove automatically (a **proof by induction**) all further statements, because, by the virtue of the above proved implications, first statement $F(a)$ proves the closest further statement in every dimension in given direction of parameters space and they can prove statements, that are closest further to them, and so on to infinity. The whole point is, that since you have proved implication for every single dimensional step $step_i$, you can assume, that you have proved all multidimensional steps, that can be constructed from these single dimensional steps, from starting element to infinity in parameters space. In other words, from starting statement $F(a)$, you proved this way, in given direction, all statements in the whole extent (of parameters space) to infinity.

In the simplest case, you have one natural parameter, that needs one dimension, in which you have to prove induction step in plus or minus direction to prove the statement for all values of parameter, respectively, greater or smaller than starting value.

More complex proofs can use more than one of these three techniques described above and even all of them multiple number of times.

In the following subsections, I will explain in details all three kinds of proofs on such a simple single dimensional example:

$$\boldsymbol{T = \left(T(n) = (2^n \geq n) \equiv true\, for\, n \geq 1\right)}$$

7.2 Example of a direct proof explained in details

Let $T(n)$ be the parameterized statement we want to prove for all $n \geq 1$ defined as:

$$T(n) = (2^n \geq n)$$

Here is a direct proof:

From mathematics, we already know, that $\binom{n}{i}$ for $0 < i < n$ is natural number greater than 1 and divisible by n for $n \geq 2$ and we will use it.

For $n = 1$, we can check, that statement T is fulfilled:

$$T(1) = \left(2^1 = 2 \geq 1\right) \equiv true$$

So, now we need to prove statement T only for $n \geq 2$:

$$T(n) = (2^n \geq n) = \left((1+1)^n \geq n\right) = \left(1^n + 1^n + \sum_{i=1}^{n-1} \binom{n}{i} 1^{n-1} 1^i \geq n\right)$$

$$= \left(2 + \sum_{i=1}^{n-1} \frac{n!}{i!(n-i)!} \geq n\right) = \left(2 + n \sum_{i=1}^{n-1} \frac{(n-1)!}{i!(n-i)!} \geq n\right)$$

$$= \left(\sum_{i=1}^{n-1} \frac{(n-1)!}{i!(n-i)!} \geq \frac{n-2}{n} = 1 - \frac{2}{n}\right)$$

So, for $n \geq 2$:

$$T(n) = \left(\sum_{i=1}^{n-1} \frac{(n-1)!}{i!(n-i)!} \geq \frac{n-2}{n} = 1 - \frac{2}{n}\right) \Leftarrow \left(\sum_{i=1}^{n-1} \frac{(n-1)!}{i!(n-i)!} \geq n-1 \geq 1 \geq 1 - \frac{2}{n}\right) \equiv true$$

Notice, that for $n \geq 2$ you can go from step $\left(\sum_{i=0}^{n} \frac{(n-1)!}{i!(n-i)!} \geq n-1 \geq 1 \geq 1 - \frac{2}{n}\right)$, that is trivially true or already proved to thesis $T(n)$, that we need to prove, by logical consequence, because logical equivalence is logical consequence true in both directions (forward and backward). And for $n = 1$, we have a trivial proof. In other words, already proved or trivially true statement implicates $T(n)$ for $n \geq 1$, so $T(n)$ must be true as an implication of a true statement, because a true statement has always only true implications.

In other words, we have proved directly from an already proved or trivial true statement, that $T(n)$ is true, as an implication of this already proved or trivial true statement:

$$\left(\boldsymbol{T(1)\ and} \left(\sum_{i=1}^{n-1} \frac{(\boldsymbol{n}-\boldsymbol{1})!}{\boldsymbol{i}!(\boldsymbol{n}-\boldsymbol{i})!} \geq \boldsymbol{n}-\boldsymbol{1} \geq \boldsymbol{1} \geq \boldsymbol{1} - \frac{\boldsymbol{2}}{\boldsymbol{n}}\right)\right)$$

$$\Rightarrow (T(1)\ and\ T(2)\ and\ T(3)\ and\ \ldots) = \bigwedge_{n=1}^{\infty} \boldsymbol{T(n)} = \boldsymbol{T}$$

Q.E.D.

7.3 Example of an indirect proof – proof by a contradiction – explained in details

Let $T(n)$ be the parameterized statement we want to prove for all $n \geq 1$ defined as:

$$T(n) = (2^n \geq n)$$

So, we want to prove statement T defined as:

$$T = \bigwedge_{n=1}^{\infty} T(n)$$

Here is an indirect (by a contradiction) proof:

From mathematics, we know, that exponentiation (e.g.: 2^n) is growing function and a^x is positive number for $a > 0$, and we will use that fact.

Assume that $\sim T$.

And we know, that:

$$\sim \bigwedge_{n=1}^{\infty} T(n) = \bigvee_{n=1}^{\infty} \sim T(n)$$

So, $T(n)$ is false for some $n \geq 1$.

$$\sim T(n) = \sim(2^n \geq n) = (2^n < n) = \left(2^{2^n} < 2^n < n\right) = \left(2^{2^{2^n}} < 2^{2^n} < 2^n < n\right) = \ldots$$

$$= \left(\overbrace{2^{\cdot^{\cdot^{\cdot^{2^n}}}}}^{k} < \overbrace{2^{\cdot^{\cdot^{\cdot^{2^n}}}}}^{k-1} < \overbrace{2^{\cdot^{\cdot^{\cdot^{2^n}}}}}^{2} = 2^{2^n} < \overbrace{2^{\cdot^{\cdot^{\cdot^{2^n}}}}}^{1} = 2^n < n \right)$$

And we know, that $\overbrace{2^{\cdot^{\cdot^{\cdot^{2^n}}}}}^{k} > 0$ for any natural k greater than 0, so we have arbitrarily large number of different natural numbers smaller than some natural number n, which is impossible.

So:

$$\sim \bigvee_{n=1}^{\infty} \sim T(n)$$

And we know, that:

$$\sim \bigvee_{n=1}^{\infty} \sim T(n) = \bigwedge_{n=1}^{\infty} T(n)$$

So this is true, that $T(n)$ is true for every $n \geq 1$ as a consequence of the opposite assumption, **so we have a contradiction**.

In other words, we have, that:

$$\sim T = \sim \bigwedge_{n=1}^{\infty} T(n) = \bigvee_{n=1}^{\infty} \sim T(n) \Rightarrow \sim \bigvee_{n=1}^{\infty} \sim T(n) = \bigwedge_{n=1}^{\infty} T(n) = T$$

So:

$$\sim T \Rightarrow T$$

So, T cannot be false, because then true statement $\sim T$ would implicate false statement T, which is impossible. So, we have proved indirectly, that T must be either true or undecidable. So, if we proved, that T is decidable, then T would be true. And T is decidable, because it is not a statement about the unpredictable future and its logical value does not depend on the logical value of any undecidable statement. So, finally, T is true.

Q.E.D.

7.4 Example of proof by induction explained in details

Let $T(n)$ be the parameterized statement we want to prove for all integers $n \geq 1$ defined as:

$$T(n) = (2^n \geq n)$$

Here is an induction proof, that $T(n)$ for $n \geq 1$:

First of all for $n = 1$, we can easily prove, that $T(1)$, that is equal to $(2^1 = 2 \geq 1)$, is true.

Now we need to prove induction step $T(n) \Rightarrow T(n+1)$:

So, let us assume, that $T(n)$ is true and prove $T(n+1)$ on the basis of that fact:

$$\boldsymbol{T(n+1)} = (2^{n+1} \geq n+1) = (2 \ast 2^n = 2^n + 2^n \geq n+1)$$
$$= (2 \ast 2^n = 2^n + 2^n \geq n + n + 1 - n) = (\boldsymbol{2^n + 2^n \geq n + n + 1 - n})$$

Now for $n \geq 1$, we have to use assumption, that $T(n)$ is true:

Remember, that from assumption of the theorem, we know, that $n \geq 1$, so $n - 1 \geq 0$:

$$\boldsymbol{T(n)} \Rightarrow (2^n + 2^n \geq n + n) \Rightarrow \left((n-1) + 2^n + 2^n \geq n + n\right) \Rightarrow (2^n + 2^n \geq n + n + 1 - n)$$
$$= \boldsymbol{T(n+1)}$$

So, we have proved, that for $n \geq 1$:

$$T(n) \Rightarrow T(n+1)$$

So, we have proved our implication for every step, starting from value of parameter n equal to 1, so we have proved, that:

$$T(1) \Rightarrow T(1+1) = T(2)$$

And we have also proved, that:

$$T(2) \Rightarrow T(2+1) = T(3)$$

And we have also proved, that:

$$T(3) \Rightarrow T(3+1) = T(4)$$

And so on to infinity, because in every place of that infinite sequence, we have step:

$$T(n) \Rightarrow T(n+1)$$

Which we have proved for $n \geq 1$.

In other words, we have proved, that:

$$T(1) \Rightarrow T(2) \Rightarrow T(3) \Rightarrow \ldots \Rightarrow T(n) \Rightarrow \ldots$$

And from logic, we know, that:

$$\left((A \Rightarrow B) \textit{ and } (B \Rightarrow C)\right) \Rightarrow (A \Rightarrow C)$$

So, we have proved:

$$T(1) \Rightarrow T(2)$$

$$T(1) \Rightarrow T(3)$$

$$T(1) \Rightarrow T(4)$$

And so on.

So:

$$\boldsymbol{T(1)} \Rightarrow \left(T(1) \textit{ and } T(2) \textit{ and } T(3) \textit{ and } T(4) \textit{ and so on}\right) = \bigwedge_{n=1}^{\infty} \boldsymbol{T(n)} = \boldsymbol{T}$$

So, trivial proof, that $T(1)$ is true, and nontrivial proof of induction step $T(n) \Rightarrow T(n+1)$ for $n \geq 1$ proves our statement $T(n)$ for every $n \geq 1$.

So, we proved statement $T(n)$ for every $n \geq 1$.

So, we proved statement T.

Q.E.D.

8 Subjective and objective logic

We can distinguish between subjective and objective declarences. For example, the sentence “Broccoli tastes good”, that was said by someone to express his opinion, is a subjective declarence. And the sentence “In my opinion Broccoli tastes good” and the sentence “I think, that Broccoli tastes good” and the sentence “Broccoli tastes good for me” have the same statement as the sentence “Broccoli tastes good”, so they mean the same. And, of course, only an objective declarence in given context has single statement, that is either true or false. A subjective declarence in given context can have different statement for each speaker, that declares it. The statement of a declarence is always objective. And the statement of an objective declarence has the same logical value for all speakers in the same context, whereas the statement of a subjective declarence does not have to have the same logical value for all speakers in the same context. A subjective declarence expresses some truth when said by a speaker in a context, if and only if its statement in case of the speaker and the context is true.

Subjective declarences are opinions, for example, “**My** dog is **cute.**” The interpretation of subjective declarences depends on the speaker, that expresses them. We have also contextual declarences like, for example, “This is a dog”, that can be objective non-strict declarences. The interpretation of contextual declarences depends on the context, in which they are expressed. A declarence can be both **subjective** and contextual like, for example, “This dog is **cute**”, “His dog is **cute.**” A subjective declarence and a contextual declarence both are non-strict declarences. The interpretation of non-strict declarences depends on the situation, in which they are expressed.

A declarence, that uses non-strict terms (e.g.: love, hate, beautiful, ugly, etc.) or non-strict references (e.g.: this, that, there, I, my, me, he, his, him, she, her, the car, the grand piano, etc.) and is not metaphorical, is a non-strict declarence. Non-strict terms include non-strict time quantifiers (e.g.: often, rarely, usually, etc.), but there are also strict time quantifiers (e.g.: always, all the time, never, ever, sometimes, etc.).

A declarence, that uses only non-strict references, is an objective non-strict declarence.

A declarence is a subjective declarence (an opinion) if and only if it uses non-strict terms.

Remember that the objective definitions of non-strict (subjective) terms do not exist.

An objective declarence that says something strict about the meaning of non-strict terms, is also an opinion.

Every speaker has its subjective dictionary – the set of all his subjective definitions of non-strict terms. Every subjective definition defines, what a non-strict term strictly means for a speaker, and must fit in some frame of allowed meaning of that term so that different speakers can communicate with and understand each other. All opinions of a speaker can be derived from his subjective definitions and from, what

https://doi.org/10.1515/9783111441382-008

he thinks is true. A speaker does not lie, when he declares, what he thinks is true, no matter whether it is true or not. The truth about a speaker's own meanings of non-strict terms and non-strict references, is the subjective point of view of this speaker, but it is not a subjective truth, because truth includes only objective statements. Remember, that an opinion does not express a subjective statement – it expresses objective statement for all speakers, because all statements are objective, though it is a subjective declarence of a speaker. In other words, a subjective declarence has always an objective statement.

Remember, that **situation modifiers** can change the speaker and the context of the statement of a declarence, e.g.:

$$(\boldsymbol{For\ people\ in\ India}\ cows\ are\ saint) = (Any\ human\ anytime\ in\ India)^{\text{"Cows are saint"}}$$

$$(\boldsymbol{For\ him}\ Mary\ is\ beautiful) = (He\ in\ any\ context)^{\text{"Mary is beautiful"}}$$

When you have mixed situations in sentences, then you have to divide such sentences into the logical expression of smaller parts, e.g.:

$$(Sophia\ is\ wonderful,\ but\ not\ \boldsymbol{for\ John})$$

$$= \left((Sophia\ is\ wonderful)\ but\ (John\ in\ any\ context)^{\text{"Sophia is not wonderful"}}\right)$$

For example, so-called fuzzy logic is nothing more than subjective logic, because all non-strict declarences, like for example: "It is quite hot weather", can mean something different for everyone and we can get to know their strict meaning only, when a speaker, that uses them, will explain, what they strictly mean for him – what are his strict definitions of non-strict terms used in them. And there is always the strict explanation of every non-strict declarence even if a speaker has not enough self-consciousness to explain, what he has on mind, if only it is not a pure caprice of the speaker, because then the speaker has not this declarence for no reason. So there are always some objective causes of any non-strict declarence, that is not a pure caprice, description of which is the strict explanation of this declarence. When a non-strict declarence is a pure caprice, then the definitions of non-strict terms used in the declarence does not fulfill any strict conditions, so such a declarence means strictly nothing.

So a non-strict declarence alone has not any statement. A non-strict declarence has a statement only, when it is a declarence of some speaker in given context and in case of every speaker and every context it can have different statement.

Remember, that a non-strict declarence is a meaningful sentence for a speaker in a context, if and only if all non-strict symbols used in it has their definitions for the speaker in the context.

Truth is the conjunction of all true statements, so there is nothing like the subjective part of truth, which means, that people can argue with each other only about (objective) truth, so they need to know each other's subjective dictionaries to communicate.

A non-strict declarence expresses some truth only if its statement is a part of truth.

A declarence is objective if and only if it does not use non-strict terms.

A subjective declarence can have different interpretation for every speaker in the same context.

A contextual declarence can have different interpretation in every context for the same speaker.

A non-strict declarence can have different interpretation in every situation.

A non-strict declarence expresses some truth for a speaker if and only if he thinks, that the statement of this declarence, that comes from his interpretation of this declarence, is true. Every non-strict declarence expresses some truth only if the statement of this declarence, that comes from a speaker's interpretation of this declarence, is true. A subjective declarence has not any statement, when interpreted in separation from any speaker. So it can express in case of any other speaker some untruth for this different speaker and just some untruth, because such a speaker can have different interpretation of the same subjective declarence. That is why the statements of non-strict declarences of different speakers do not have to be contrary to each other even if these declarences would be contrary to each other, if they were said by the same speaker (e.g.: "It is beautiful" vs. "It is not beautiful", "This is my dog" vs. "This is not my dog").

So the statement of a declarence is true, when it, whether the declarence is not strict (when the statement comes from a subjective interpretation of non-strict declarence) or the declarence is strict (when the statement is the meaning of a strict sentence), is a part of truth. And since a statement of a non-strict declarence can be different in every situation, the set of all realizations of such a statement can be also different in every situation.

Since behind any opinion is a statement, that is equal to the strict meaning of this opinion, the statement can implicate a fact, because we know, what that opinion strictly means.

For example:

$$^\wedge\text{“}\textit{It is cold}\text{”} \Rightarrow {}^\wedge\text{“}\textit{Temperature is below } 0 \textit{ celcius degree}\text{”}$$

For a speaker, that has the following subjective definition:

$$^\wedge\text{“}\textit{It is cold}\text{”} = {}^\wedge\text{“}\textit{Temperature is below } 0 \textit{ celcius degree}\text{”}$$

Another example:

$$^\wedge\text{“}\textit{All flowers in this garden are beautiful}\text{”} \Rightarrow {}^\wedge\text{“}\textit{All flowers in this garden are red}\text{”}$$

For a speaker, that has the following subjective definition:

$$^\wedge\text{“}\textit{The flowers are beautiful}\text{”} = {}^\wedge\text{“}\textit{The flowers are red}\text{”}$$

And yet another example:

$$^\wedge\text{“}He\ is\ intelligent\text{”} \Rightarrow {}^\wedge\text{“}He\ passed\ the\ exams\text{”}$$

For a speaker, that has the following subjective definition:

$$^\wedge\text{“}He\ is\ intelligent\text{”} = {}^\wedge\text{“}He\ passed\ the\ exams\text{”}$$

And you can, of course, combine opinions with objective declarences and the statements of such combinations can implicate the statements of opinions, e.g.:

$$^\wedge\text{“}Red\ color\ is\ the\ most\ beautiful\ color\ and\ I\ am\ dressed\ in\ blue\text{”}$$

$$\Rightarrow {}^\wedge\text{“}I\ am\ not\ dressed\ in\ the\ most\ beautiful\ color\text{”}$$

A fact is just a true statement – a part of truth. This term is used in this meaning in the whole book.

Someone's opinion is a subjective declarence, that expresses some truth for him due to his subjective interpretation, that is objectively true or false. A fact is, of course, true for everyone.

When someone has an opinion, that conflicts with some facts, then this is his mistake, because then this opinion has a false statement. So you should not have an opinion, that conflicts with some facts, unless you want to make such a mistake. So, whenever possible, put objective judgment above your earlier personal conviction, so that you simply do not make a mistake. For this reason we cannot say, that such an incorrect opinion or anything, that someone erroneously thinks is true, is subjective truth, because this is not any kind of truth anymore, since it is untruth. So, in general, we cannot say, that if someone is simply wrong, then he has his logical right to have this subjective truth due to the fact, that he is a subjective entity (a subject), that sees reality his own subjective way. He is always only either right or wrong – the statement of his subjective or objective declarence is either true (a part of truth) or false (not a part of truth). After English Wikipedia, "truth is the property of being in accord with fact or reality", so if someone, in what he thinks, is simply wrong about reality, then this cannot be any kind of truth. So we should not think, that someone has a subjective truth, if only this truth is in accord with his subjective perception of reality, because if his subjective perception of reality is not in accord with reality, then he is simply wrong about reality and did not invent his own truth, that has logical right to be called some kind of truth. Any truth is always a conjunction of true statements, so if someone is wrong about reality, then his "truth" is a conjunction of statements, that are not all true, so also the conjunction of them is not true, so it cannot be called some kind of truth, e.g. subjective truth.

Only individual interpretations (that comes from individual definitions) of non-strict words are subjective in opinions and such words are used only for convenience, so that you do not have to wordily explain complex matters, but to say about them something meaningful enough in a communication between minds. Because it is hard to strictly explain (define) non-strict notions, they allow to express briefly something more complex.

But keep in mind that if a speaker uses a subjective declarence without a speaker modifier, then the default speaker modifier is "for me," so another speaker can confirm (e.g. "Right", "I agree", "Yes, it's true") or deny (e.g.: "Wrong", "I disagree", "No, it's false") either the statement, that the declarence expresses some truth for the former speaker (which is equal to the statement of this declarence for this speaker), or what the declarence expresses for the latter speaker. In the second case, if the latter speaker confirms or deny, then it is possible that he does not really respectively agree or disagree with the former speaker, but to communicate effectively he should still agree more or less with the generally accepted constraints and boundaries of meanings of the non-strict terms, that are used in that declarence. So the latter speaker can confirm or deny either the piece of information expressed by the former speaker (less often) or that opinion (as his own) said by the former speaker (more often). To avoid ambiguity, an explicit speaker modifier should be used, e.g.: "Yes, it is beautiful for you.", "Yes, it is beautiful for me as well.".

Some terms are not strict only because of disagreements about their meaning, that is, they have several different definitions, but they should have one strict definition. Such a term should be either made strict (choose or invent one strict definition) or abandoned, if it cannot be made strict for some reason. For other definitions of it, at most, use different symbols.

Remember, that a citation in the form "*X* says/said '*Y*'" does not change the speaker to speaker *X* and the context to the context, in which he said "*Y*", but reported speech in the form "*X* says/said/think/etc., that *Y*" changes the definitions of non-strict terms (but not of non-strict references) to those of speaker *X*.

In formal language, the same symbol must not be used for a strict and a non-strict term at the same time.

Some declarences, whether they are strict or not strict, should not be interpreted literally but metaphorically. For example, the strict declarence "The Sun is smiling", that is always false, when interpreted literally, because the Sun, of course, is never smiling, because it cannot smile, since it does not have a face, can be interpreted metaphorically, for example, in such a context:

"Today is beautiful day. The Sun is smiling."

In this context this declarence can express some truth and means more or less, that the Sun is shining bright.

The same non-strict declarence "He is cold", that uses not strict notion "cold" can be interpreted metaphorically, for example, in such a context:

"He does not like her. He is cold."

In this context this declarence can express some truth and means more or less, that he has not positive feelings.

Metaphoric declarences are subjective and have not their literal meaning, so like any other subjective declarences require to be interpreted by a speaker to have objective meaning. But you should not confuse them with fuzzy (not strict) declarences, because we do not just interpret not strict notions in them, but analyze their context, whether it is a situation or a kind of a publication or the surrounding text. For example, when someone uses not strict notion “beautiful” to describe someone, we automatically know, that mostly he rates high his or her appearance or personality, because to some degree we know the subjective meaning of not strict notions, while we know nothing about the meaning of metaphoric declarences and we need the context of the declarence to clarify its meaning to us and then, being suggested by that context, we can only guess, what an author of a metaphoric declarence had on mind.

A term is contractual, if a group of people agree on the strict meaning of some non-strict term.

In some natural languages there can be ambiguous sentences, that do not necessarily use non-strict terms or non-strict references. If such an ambiguous sentence in some language in some context and said by some speaker is still ambiguous, then this sentence should not be allowed by this language at least when said by that speaker in that context, so it should not be ever used by that speaker in that context, because you cannot interpret it at all when it is said by that speaker in that context.

9 Definition – more advanced part

9.1 Conjunction and being a part – introduction

In this chapter, we will extend the relation of being a part to all things, including abstract ones.

We can always take into consideration many things together and treat them as one thing, the parts of which they are. Such a thing will be **the conjunction of these things**.

In other words, two or more things taken into consideration as one thing or treated or perceived as one thing are together this one thing that is equal to the conjunction of these things.

We will use the word "and" and the big and small symbol "$\wedge$" to express a conjunction of things.

So, for any X and Y:

$$TK(X) = (X\ is\ taken)$$

$$\big(TK(X) = TK(Y)\big) = (X = Y)$$

$$TK(X\ and\ Y) \Rightarrow \big(TK(X)\ and\ TK(Y)\big)$$

And for any concrete X and Y:

$$TK(X\ and\ Y) = \big(TK(X)\ and\ TK(Y)\big)$$

X equals Y if and only if X is a part of thing Y and Y is a part of thing X.

We have, that for any plain X and Y:

$$(intersection\ of\ X\ and\ Y) = (X \cap Y)$$

We have, that for any X and Y:

$$(conjunction\ of\ X\ and\ Y) = (union\ of\ X\ and\ Y) = (X \cup Y)$$

$$\begin{aligned}(X\ is\ a\ part\ of\ Y) &= (X\ is\ a\ subthing\ of\ Y) = (Y\ is\ a\ superthing\ of\ X)\\ &= (Y\ includes\ X) = (X \subset Y) = \big(TK(Y) \Rightarrow TK(X)\big)\\ &= \big((there\ is\ Y) \Rightarrow (there\ is\ X)\big) = \big((Y \in \Omega) \Rightarrow (X \in \Omega)\big)\end{aligned}$$

$$(X \cup Y) = (X\ and\ Y)$$

$$\big((X \subset Y)\ and\ (X \supset Y)\big) = (X = Y)$$

And:

$$(X\ isA\ Y) = (a\ X\ is\ a\ Y \backslash X\ are\ Y) = \big(X^{\uparrow} \subset Y^{\uparrow}\big)$$

https://doi.org/10.1515/9783111441382-009

For any things X and Y:

$$(\boldsymbol{X} \subset \boldsymbol{Y}) \Rightarrow ((\boldsymbol{X} \subset \boldsymbol{Y})\ \textbf{\textit{was meant to be true}}) = (\boldsymbol{X}\ \textbf{\textit{was meant to be a part of}}\ \boldsymbol{Y})$$

$$(\boldsymbol{X} \subset \boldsymbol{Y}) = \left(\begin{array}{l} \bigwedge\limits_{\boldsymbol{a} \in \boldsymbol{X}^{\uparrow}} \bigvee\limits_{\boldsymbol{b} \in \boldsymbol{Y}^{\uparrow}} ((\boldsymbol{a} \subset \boldsymbol{b})\ \textbf{\textit{was meant to be true}}) \\ \textbf{\textit{and}} \bigwedge\limits_{\boldsymbol{b} \in \boldsymbol{Y}^{\uparrow}} \bigvee\limits_{\boldsymbol{a} \in \boldsymbol{X}^{\uparrow}} ((\boldsymbol{a} \subset \boldsymbol{b})\ \textbf{\textit{was meant to be true}}) \end{array}\right)$$

For any concrete plain things a and b such that $(a\ and\ b) \subset \Lambda_A^{\mho}$:

$$(\boldsymbol{a} \subset \boldsymbol{b}) = (b = (a\ and\ b)) = \left(b = (\Lambda_A^{\mho} \cap (a\ and\ b))\right)$$

$$= \left(b = \left(\left(a\ and \sim_A^{\mho} a\right) \cap (a\ and\ b)\right)\right) = \left(\boldsymbol{b} = \left(\boldsymbol{a}\ \textbf{\textit{and}}\ (\boldsymbol{b} \cap \sim_A^{\mho} \boldsymbol{a})\right)\right)$$

So also:

$$(\boldsymbol{a} \subset \boldsymbol{b}) = (\sim_A^{\mho} b \subset \sim_A^{\mho} a) = \left(\sim_A^{\mho} a = \left(\sim_A^{\mho} b\ and\ (\sim_A^{\mho} a \cap b)\right)\right)$$

$$= \left(a = \sim_A^{\mho}\left(\sim_A^{\mho} b\ and\ (\sim_A^{\mho} a \cap b)\right) = \left(b \cap \left(a\ and \sim_A^{\mho} b\right)\right)\right)$$

$$= \left(\boldsymbol{a} = \left(\boldsymbol{b} \cap \left(\boldsymbol{a}\ \textbf{\textit{and}} \sim_A^{\mho} \boldsymbol{b}\right)\right)\right)$$

So for any things X and Y such that $(@X^{\to\infty}\ and\ @Y^{\to\infty}) \subset \Lambda_A^{\mho}$:

$$(\boldsymbol{X} \subset \boldsymbol{Y}) = \left(\bigwedge_{\boldsymbol{a} \in \boldsymbol{X}^{\uparrow}} \bigvee_{\boldsymbol{b} \in \boldsymbol{Y}^{\uparrow}} (\boldsymbol{a} \subset \boldsymbol{b})\ \textbf{\textit{and}} \bigwedge_{\boldsymbol{b} \in \boldsymbol{Y}^{\uparrow}} \bigvee_{\boldsymbol{a} \in \boldsymbol{X}^{\uparrow}} (\boldsymbol{a} \subset \boldsymbol{b})\right)$$

Where $X^{\to}$ is the expansion of X and $X^{\to\infty}$ is the infinite expansion (concretization) of X. For more detail, see Section 10.11.

And for any things X and Y such that $(@X^{\to}\ and\ @Y^{\to}) \subset \Lambda_A^{\mho}$:

$$(X \subset Y) = \left(\bigwedge_{a \in X^{\uparrow}} \bigvee_{b \in Y^{\uparrow}} (a \subset b)\ and \bigwedge_{b \in Y^{\uparrow}} \bigvee_{a \in X^{\uparrow}} (a \subset b)\right)$$

$$= \left(\begin{array}{l} \bigwedge\limits_{a \in X^{\uparrow}} \bigvee\limits_{b \in Y^{\uparrow}} \left(a = \left(b \cap \left(a\ and \sim_A^{\mho} b\right)\right)\right) \\ and \bigwedge\limits_{b \in Y^{\uparrow}} \bigvee\limits_{a \in X^{\uparrow}} \left(b = \left(a\ and\ (b \cap \sim_A^{\mho} a)\right)\right) \end{array}\right)$$

$$= \left(\left(X^{\uparrow} \subset \left(Y \cap (X\ and \sim_A^{\mho} Y)\right)^{\uparrow}\right) and \left(Y^{\uparrow} \subset \left(X\ and\ (Y \cap \sim_A^{\mho} X)\right)^{\uparrow}\right)\right)$$

$$= \left(\left(\boldsymbol{X\,isA} \left(\boldsymbol{Y} \cap (\boldsymbol{X\,and} \sim_{\boldsymbol{A}}^{\mho} \boldsymbol{Y}) \right) \right) \boldsymbol{and} \left(\boldsymbol{Y\,isA} \left(\boldsymbol{X\,and} \,(\boldsymbol{Y} \cap \sim_{\boldsymbol{A}}^{\mho} \boldsymbol{X}) \right) \right) \right)$$

$$= \left(\begin{array}{l} \bigwedge_{a \in X^{\uparrow}} \bigvee_{b \in Y^{\uparrow}} (a = (b \cap a)) \\ and \bigwedge_{b \in Y^{\uparrow}} \bigvee_{a \in X^{\uparrow}} (b = (a\,and\,b)) \end{array} \right)$$

$$= \left((\boldsymbol{X\,isA}\,(\boldsymbol{Y} \cap \boldsymbol{X}))\ \boldsymbol{and}\ (\boldsymbol{Y\,isA}\,(\boldsymbol{X\,and\,Y})) \right)$$

For any things X and Y such that $(X^{\to\infty}\ and\ Y^{\to\infty}) \subset Statement^{\Uparrow}$:

$$(\boldsymbol{X} \subset \boldsymbol{Y}) = \left(\bigwedge_{\boldsymbol{a} \in \boldsymbol{X}^{\uparrow}} \bigvee_{\boldsymbol{b} \in \boldsymbol{Y}^{\uparrow}} (\boldsymbol{a} \subset \boldsymbol{b})\ \boldsymbol{and} \bigwedge_{\boldsymbol{b} \in \boldsymbol{Y}^{\uparrow}} \bigvee_{\boldsymbol{a} \in \boldsymbol{X}^{\uparrow}} (\boldsymbol{a} \subset \boldsymbol{b}) \right)$$

For any things X and Y such that $(X^{\to} and\ Y^{\to}) \subset Statement^{\Uparrow}$:

$$(X \subset Y) = \left(\bigwedge_{a \in X^{\uparrow}} \bigvee_{b \in Y^{\uparrow}} (a \subset b)\ and \bigwedge_{b \in Y^{\uparrow}} \bigvee_{a \in X^{\uparrow}} (a \subset b) \right) = \left(\begin{array}{l} \bigwedge_{a \in X^{\uparrow}} \bigvee_{b \in Y^{\uparrow}} (a = (b\,or\,a)) \\ and \bigwedge_{b \in Y^{\uparrow}} \bigvee_{a \in X^{\uparrow}} (b = (a\ and\,b)) \end{array} \right)$$

$$= \left((\boldsymbol{X\,isA}\,(\boldsymbol{Y\,or\,X}))\ \boldsymbol{and}\ (\boldsymbol{Y\,isA}\,(\boldsymbol{X\,and\,Y})) \right)$$

For $(X \subset Y)$, of course, every representative of X must be a part of some representative of Y and every representative of Y must have some representative of X as a part of itself.

So the head of a dog is not a part of an animal, because when you take an animal then you can take other animal, that is not a dog, but, of course, the head of a dog is a part of a dog, because when you take a dog, then you take also its head. And the head of an animal is not a part of a dog, because it can be the head of a different animal and when you take a dog then you do not take the head of a different animal.

And " $\because\!\!-$ " is the symbol of general subtraction and for any things X, Y, Z, x and y:

$$((X \because\!\!- Y) = Z) = \left((x \subset Z) = \left((x \subset X)\ and\ ((y \subset x) \Rightarrow \sim(y \subset Y)) \right) \right)$$

And for any concrete plain things X and Y such, that $(X\,and\,Y) \subset \Lambda_A^{\mho}$:

$$(X \because\!\!- Y) = (X -_A^{\mho} Y)$$

An **atomic thing (*a*)** Y is a thing, that has itself as the only nonempty part of itself, so it cannot be $(X\,and\,(Y \because\!\!- X))$ for X being nonempty and being something else than this thing. Not every abstract thing has atomic parts other than itself, so every abstract thing, that is not a conjunction of nonempty things, is atomic.

So, the empty parts and their conjunctions are not atomic.

Things, that are neither atomic nor empty, are complex.

The empty parts (zeroes) are as follows:

$$\psi_{\varnothing} = \varnothing_{\psi} = \varnothing = (\mathit{The\ Empty\ Set})$$

$$\gamma_{\varnothing} = \varnothing_{\gamma} = \varnothing = (\mathit{The\ Empty\ Tuple})$$

$$\varsigma_{\varnothing} = \varnothing_{\varsigma} = (\mathit{The\ Empty\ Statement})$$

$$\xi_{\varnothing} = \varnothing_{\xi} = (\mathit{The\ Empty\ Attribute})$$

$$\chi_{\varnothing} = \varnothing_{\chi} = (\mathit{The\ Empty\ Shape})$$

$$\kappa_{\varnothing} = \varnothing_{\kappa} = (\mathit{The\ Empty\ Physcial\ Shape})$$

For any $K \in \{\psi, \gamma, \varsigma, \xi, \chi, \kappa\}$:

$$(X\ isA\ K) = (X^{\uparrow} \subset K^{\uparrow}) \Rightarrow (\varnothing_X = \varnothing_K)$$

And for any $X \in t^{\uparrow}$ for $t \in \{\psi, \gamma, \chi, \kappa\}$:

$$(X \cap \varnothing_t) = (X - X) = \varnothing_t$$

And for any $X \in t^{\uparrow}$ for $t \in \{\varsigma, \xi\}$:

$$(X\ or\ \varnothing_t) = (X\ or\ {\sim}X\ or\ X\ is\ undecidable) = \varnothing_t$$

And for any $X \in t^{\uparrow}$ for $t \in \{\psi, \gamma, \varsigma, \xi, \chi, \kappa\}$:

$$(X\ and\ \varnothing_t) = (X - \varnothing_t) = X$$

$$(\varnothing_t\ and\ \varnothing_t) = \varnothing_t$$

$$(\Lambda_t \breve{\cap} X) = X$$

$$(\Lambda_t\ and\ X) = \Lambda_t$$

$$(\Lambda_t - X) = {\sim}X$$

For any X:

$$(\Lambda\ and\ X) = \Lambda$$

$$\Lambda = \Upsilon = \mathit{totality}$$

Remember, that nothing is not a part of anything, because everything is a part of something. At first it may sound difficult, but it is true.

Simply, nothing is not a thing, so, for example, in the following condition of a quantifier, 'it' will not be considered:

$$\bigwedge_{x} R(x)$$

So, the above conjunction (a universal quantifier) will not contain $R(\Phi)$.

A **concrete thing** is a thing, that is not abstract, so the class of this thing has only one member.

The following operator ("$\pitchfork$") will be defined only for any atomic thing x and any naturally or artificially plain thing Y such that $(x \cup Y) \subset \Lambda_A^{\mho}$ and $x \in A$:

$$(x \pitchfork Y) = (x \subset Y) = (\textit{Atomic thing } x \textit{ is a part of plain thing } Y)$$

And we have:

$$\sim(x \pitchfork Y) = \left(x \pitchfork \sim_A^{\mho} Y\right)$$

For any things x and y:

$$(\mathbf{x} \subset \mathbf{y}) = \bigvee_{p \subset y} (p = x) = \sim \bigwedge_{\mathbf{p} \subset \mathbf{y}} (\mathbf{p} \neq \mathbf{x})$$

So for any things x and y:

$$(\mathbf{x} \subset \mathbf{y}) = \bigwedge_{q \subset x} (q \subset y) = \bigwedge_{q \subset x} \bigvee_{r \subset y} (q = r) = \bigwedge_{q \subset x} \sim \bigwedge_{r \subset y} (q \neq r) = \sim \bigvee_{\mathbf{q} \subset \mathbf{x}} \bigwedge_{\mathbf{r} \subset \mathbf{y}} (\mathbf{q} \neq \mathbf{r})$$

And for any plain x and y:

$$(\mathbf{x} \subset \mathbf{y}) = \bigwedge_{q \pitchfork x} (q \pitchfork y) = \bigwedge_{q \pitchfork x} \bigvee_{r \pitchfork y} (q = r) = \bigwedge_{q \pitchfork x} \sim \bigwedge_{r \pitchfork y} (q \neq r) = \sim \bigvee_{\mathbf{q} \pitchfork \mathbf{x}} \bigwedge_{\mathbf{r} \pitchfork \mathbf{y}} (\mathbf{q} \neq \mathbf{r})$$

And the relation $\subset$ is transitive relation, so for any things X, Y and Z:

$$((X \subset Y) \textit{ and } (Y \subset Z)) \Rightarrow (X \subset Z)$$

And we have:

$$\begin{aligned}(X \subset Y) &\Rightarrow (X \textit{ was meant to be a part of } Y) = ((Y \textit{ exists}) \Rightarrow (X \textit{ exists})) \\ &= ((Y \in \mathcal{B}) \Rightarrow (X \in \mathcal{B}))\end{aligned}$$

9.2 Conjunction as the fundament of logic and set theory

9.2.1 Introduction

This section is for readers, that are more inquiring and advanced in mathematics. If you do not feel as one of them, you can skip it and always come back here for deeper understanding of logic.

9.2.2 Every thing as the conjunction of all its parts

For any things X_1, X_2 and Y such that $(X_1^{\rightarrow}\ and\ X_2^{\rightarrow}\ and\ Y^{\rightarrow}) \subset \mathcal{B}$:

$$((X_1 \subset Y)\ and\ (X_2 \subset Y)) \Rightarrow \big((Y\ isA\ (X_1 \cup Y))\ and\ (Y\ isA\ (X_2 \cup Y))\big)$$
$$= \big(Y\ isA\ (X_1 \cup (X_2 \cup Y))\big) = \big(Y\ isA\ ((X_1 \cup X_2) \cup Y)\big)$$

Which can be easily generalized for all parts of Y.

From this, we have for any thing a such that $a^{\rightarrow} \subset \mathcal{B}$:

$$a\ isA \bigwedge_{a \supset b} b = \bigcup_{a \supset b} b$$

If X is a concrete thing, then:

$$(X\ and\ X) = X$$

And for any concrete plain things X and Y:

$$(X \subset Y) = \big(Y = (X \cup (Y - X)) = (X \cup Y) = (X\ and\ Y)\big) = (X = (X \cap Y))$$

And for any concrete statements X and Y:

$$(X \subset Y) = (Y = (X \cup Y) = (X\ and\ Y)) = (X = (X\ or\ Y))$$

So for concrete plain things Y_1, Y_2 and X:

$$((X \subset Y_1)\ and\ (X \subset Y_2)) = \big(X = (X \cap Y_1) = (X \cap Y_2) = ((X \cap Y_1) \cap Y_2) = (X \cap (Y_1 \cap Y_2))\big)$$

Which can be easily generalized for all things, that contain X.

From this, we have for any concrete plain thing a:

$$a = \bigcap_{b \supset a} b$$

So for concrete statements Y_1, Y_2 and X:

$$((X \subset Y_1)\ and\ (X \subset Y_2)) = \big(X = (X\ or\ Y_1) = (X\ or\ Y_2) = ((X\ or\ Y_1)\ or\ Y_2) = (X\ or\ (Y_1\ or\ Y_2))\big)$$

Which can be easily generalized for all things, that contain X.

From this, we have for any concrete statement a:

$$a = \bigvee_{b \supset a} b$$

So also for any concrete things X_1, X_2 and Y:

$$((X_1 \subset Y)\ and\ (X_2 \subset Y)) = \big(Y = (X_1 \cup Y) = (X_2 \cup Y) = (X_1 \cup (X_2 \cup Y)) = ((X_1 \cup X_2) \cup Y)\big)$$

Which can be easily generalized for all parts of Y.

From this, we have for concrete plain thing a:

$$a = \bigcup_{a \supset b} b = \bigcup_{b \pitchfork a} b = \bigwedge_{a \supset b} b = \bigwedge_{b \pitchfork a} b$$

And we have for any concrete statement a:

$$a = \bigwedge_{a \supset b} b$$

Summing up, every thing, the expansion of which (for the definition of expansion see Section 10.11) contains only concrete beings, is the conjunction/union of all its parts. And every concrete thing is equal to the conjunction/union of all its parts and every concrete plain thing is equal to the conjunction/union of all its atomic parts.

9.2.3 Elementary operations on things

An **atomic thing (*α*)** is a thing, that is not a conjunction of two different nonempty things, that do not have an intersection, so it is simply a thing, that is not a conjunction of two different nonempty things. We will use symbol **"*α*"** as the symbol of an atomic thing.

$$(a\ is\ Atomic) = (a\ is\ Atomic\ Thing) = (a\ is\ \alpha) \neq (\sim a\ is\ \alpha)$$

So a plain atomic thing cannot include any other thing.

Only every thing, consistently meant to be concrete X, is a member of the class of X.

For any X and Y:

$$(X \in Y^{\uparrow}) = (X\ is\ a\ member\ of\ the\ class\ of\ Y) = (X\ is\ a\ representative\ of\ the\ kind\ of\ Y)$$

$$(X = Y^{\uparrow}) = (X_{\downarrow} = Y)$$

$$(X@_Y) = \{p{:}(p \subset X)\ and\ (p \in Y^{\uparrow})\}$$

$$(@_Y X) = \left(\bigwedge_{e \in X\ and\ e \in Y^{\uparrow}} e\right)$$

$$(@_\omega X) = (@X)$$

$$(X@_\omega) = (X@)$$

$$(X \subset Y) = (X@ \subset Y@)$$

For any plain X and Y:

$$(X@ \subset Y@) = (X@_\alpha \subset Y@_\alpha)$$

For any thing X, that is not a concrete set:

$$@_K X @_L = @_K (X @_L)$$

$$@X@ = @(X@) = X$$

For any concrete set X:

$$@_K X @_L = (@_K X) @_L$$

$$@X@ = (@X)@ \supset X$$

9.2.4 Negation and complement

There are five kinds of a negation:

1. the artificial special negation (tilde "$\sim^{\mho}_{A}$"), which is the negation of a representative of the kind of T, which is any element of set $\mho$ (universe; where $T = \mho_{\downarrow}$) built from set A of atoms, where all these atoms are concrete things,
2. the natural special negation (tilde "$\sim$"), which is the special case of special negation (tilde "$\sim^{\mho}_{A}$") of a representative of the kind of T, where $\mho = T^{\uparrow}$ and $A = @\mho@_a$ and:
   a. T is equal to Set
   b. T is equal to Shape in Space S for any Space S
   c. T is equal to Physical Shape in Physical Space S for any Physical Space S
   d. T is equal to any other plain thing,
3. the natural special negation (tilde "$\sim$") for statements and attributes,
4. the artificial general negation ("not", vertical tilde "$\wr^{\mho}_{A}$"), based on artificial special negation,
5. the natural general negation ("not", vertical tilde "$\wr$") of anything.

$\sim^{\mho}_{A} X$ gives the complement of X, except for $T \in \{Statement, Attribute\}$ and the negation of X for $T \in \{Statement, Attribute\}$.

So, $\sim X$ gives the complement of X for X being, for example, a set or a shape or a physical shape.

Only every class, that has defined complement operation, has a plain inclusion operator and its every member is a **plain thing**, which means, that it is a conjunction of atomic parts.

For artificial special negation, the following condition must be fulfilled:

$$\Big((\{a,b\} \subset A) \; and \; (a \neq b)\Big) \Rightarrow ((a \cap b) = \varnothing^{\mho}_{A})$$

For any X such that $X^{\uparrow} \subset \mho$:

$$\sim^{\mho}_{A} X = \{\sim^{\mho}_{A} x : x \in X^{\uparrow}\}_{\downarrow}$$

And we have:

$$\Lambda_A^{\mho} = (@A \cap @\mho)$$

$\Lambda_A^{\mho}$ is the totality for the negation $\sim_A^{\mho}$. As you can see, it is the conjunction of all atoms (from set A), that are used in elements of universe $\mho$.

$$\varnothing_A^{\mho} = \varnothing_{(\Lambda_A^{\mho}@)_{\downarrow}}$$

$\varnothing_A^{\mho}$ is the empty element for type T.

$$\Lambda_A^{\mho} = \sim_A^{\mho}\varnothing_A^{\mho}$$

$$\sim_A^{\mho}\Lambda_A^{\mho} = \varnothing_A^{\mho}$$

For any X such that $X \subset \Lambda_A^{\mho}$:

$$\sim_A^{\mho}X = \sim_A^{\mho}(\varnothing_A^{\mho}\ and\ X) = (\Lambda_A^{\mho} -_A^{\mho} X)$$

So for any T except for $T \in \{Statement,\ Attribute\}$:

$$(X \cup \sim_A^{\mho}X) = \Lambda_A^{\mho}$$

For $T = (Statement)$:

$$(X\ and \sim X\ and\ X\ is\ decidable) = \Lambda_{\varsigma}$$

For $T = (Attribute)$:

$$(X\ and \sim X\ and\ X\ is\ decidable) = \Lambda_{\xi}$$

For any T except for $T \in \{Statement, Attribute\}$:

$$(X \cap \sim_A^{\mho}X) = (X -_A^{\mho} X) = \varnothing_A^{\mho}$$

For $T = (Statement)$:

$$(X\ or \sim X\ or\ X\ is\ undecidable) = \varnothing_{\varsigma}$$

For $T = (Attribute)$:

$$(X\ or \sim X\ or\ X\ is\ undecidable) = \varnothing_{\xi}$$

Again, for any T except for $T \in \{Statement, Attribute\}$:

$$\sim_A X = \sim_A^{@A@} X$$

$$\sim^{\mho} X = \sim_{@\mho@_a}^{\mho} X$$

For any X such that $X \in T^{\Uparrow}$:

$$\sim X = \sim^{T^{\Uparrow}} X$$

The natural general negation ("$\wr$") gives the opposite thing to given thing.

For any thing X:

$$\left(X^{\circlearrowleft} \cup (\wr X)^{\circlearrowleft}\right) = \Omega$$

$$\left(X^{\circlearrowleft} \cap (\wr X)^{\circlearrowleft}\right) = \varnothing$$

Important note: Although prefix unary operators (e.g. $\wr X$) have a higher priority than suffix unary operators, upper and lower index unary operators (e.g. $X^{\circlearrowleft}$) have even higher priority, so $\wr X^{\circlearrowleft} = \wr(X^{\circlearrowleft}) \neq (\wr X)^{\circlearrowleft}$.

A general negation does not give a complement, but we can say, that $\wr_A^{\mho} X$ gives a thing of complementary kind/class/type.

So for any X:

$$\wr_A^{\mho} X = Not_A^{\mho} X = \left(\sim_A^{\mho}(X^{\circlearrowleft})\right)_{\circlearrowleft}$$

$$\wr_A X = \wr_A^{@A@} X$$

$$\wr^{\mho} X = \wr_{@\mho@_\alpha}^{\mho} X$$

$$\wr X = \wr^{\Psi} X$$

So for any concrete set X, for which $X_{\circlearrowleft}$ is defined:

$$\sim_A^{\mho} X = \left(\wr_A^{\mho}(X_{\circlearrowleft})\right)^{\circlearrowleft}$$

And for any X:

$$\wr X = Not\, X = \left(\sim(X^{\circlearrowleft})\right)_{\circlearrowleft}$$

And:

$$\sim_A^{\mho} \sim_A^{\mho} X = X$$

$$\wr_A^{\mho} \wr_A^{\mho} X = X$$

For any concrete naturally or artificially plain X and Y such that $(X \cup Y) \subset \Lambda_A^{\mho}$:

$$(X -_A^{\mho} Y) = \sim_A^{\mho}(\sim_A^{\mho} X \; and \; Y) = (X \cap \sim_A^{\mho} Y)$$

So:

$$(X \mathbin{-^{\mho}_{A}} Y) = (\sim^{\mho}_{A} Y \mathbin{-^{\mho}_{A}} \sim^{\mho}_{A} X)$$

$$(X \cap Y) = (X \mathbin{-^{\mho}_{A}} \sim^{\mho}_{A} Y) = (Y \mathbin{-^{\mho}_{A}} \sim^{\mho}_{A} X)$$

And for any concrete statements X and Y:

$$(X - Y) = \sim(\sim X \; and \; Y) = (X \; or \sim Y)$$

So:

$$(X - Y) = (\sim Y - \sim X) = \sim(Y -_{or} X)$$

$$(X \; or \; Y) = (X - \sim Y) = (Y - \sim X)$$

So we have a relative complement (a subtraction) for any X and Y such, that $(@X^{\to\infty} \; and \; @Y^{\to\infty}) \subset \Lambda^{\mho}_{A}$:

$$(X \mathbin{-^{\mho}_{A}} Y) = \left\{ \begin{aligned} z: & \left(\left(z = (x \mathbin{-^{\mho}_{A}} y)\right) \; and \; (x \in X^{\uparrow}) \; and \; (y \in Y^{\uparrow}) \; and \; ((x \cap y) \neq \varnothing^{\mho}_{A})\right) \\ & or \left((z \in X^{\uparrow}) \; and \bigwedge_{y \in Y^{\uparrow}} ((z \cap y) = \varnothing^{\mho}_{A})\right) \end{aligned} \right\}$$

Where $X^{\to\infty}$ is the infinite expansion (concretization) of X. For more detail, see Section 10.11.

And we can easily verify, that for concrete naturally or artificially plain X and Y such that $(X \cup Y) \subset \Lambda^{\mho}_{A}$:

$$(X \; and \; Y) = (X \; and \; (Y \mathbin{-^{\mho}_{A}} X)) = ((X \mathbin{-^{\mho}_{A}} Y) \; and \; Y)$$

$$(X \cap Y) = (X \cap (Y \; and \sim^{\mho}_{A} X)) = ((X \; and \sim^{\mho}_{A} Y) \cap Y)$$

$$((X \mathbin{-^{\mho}_{A}} Y) \; and \; (Y \mathbin{-^{\mho}_{A}} X)) = (X \Delta Y)$$

Here is a proof of it:

$$(\boldsymbol{X} \; \boldsymbol{and} \; (\boldsymbol{Y} \mathbin{-^{\mho}_{A}} \boldsymbol{X})) = (X \; and \; (Y \cap \sim^{\mho}_{A} X)) = ((X \; and \; Y) \cap (X \; and \sim^{\mho}_{A} X))$$

$$= ((X \; and \; Y) \cap \Lambda^{\mho}_{A}) = (\boldsymbol{X} \; \boldsymbol{and} \; \boldsymbol{Y})$$

$$(\boldsymbol{X} \cap (\boldsymbol{Y} \; \boldsymbol{and} \sim^{\mho}_{A} \boldsymbol{X})) = ((X \cap Y) \; and \; (X \cap \sim^{\mho}_{A} X)) = ((X \cap Y) \; and \; \varnothing^{\mho}_{A}) = (\boldsymbol{X} \cap \boldsymbol{Y})$$

$$(\sim^{\mho}_{A} \boldsymbol{Y} \; \boldsymbol{and} \; (\boldsymbol{Y} \mathbin{-^{\mho}_{A}} \boldsymbol{X})) = (\sim^{\mho}_{A} Y \; and \; (Y \cap \sim^{\mho}_{A} X)) = ((\sim^{\mho}_{A} Y \; and \; Y) \cap (\sim^{\mho}_{A} Y \; and \sim^{\mho}_{A} X))$$

$$= (\Lambda^{\mho}_{A} \cap \sim^{\mho}_{A} (Y \cap X)) = \sim^{\mho}_{A} (\boldsymbol{X} \cap \boldsymbol{Y})$$

So:

$$\big((\boldsymbol{X} -^{\mho}_{A} \boldsymbol{Y})\ \boldsymbol{and}\ \boldsymbol{Y}\big) = \big(Y\ and\ (X -^{\mho}_{A} Y)\big) = (Y\ and\ X) = (\boldsymbol{X}\ \boldsymbol{and}\ \boldsymbol{Y})$$

$$\big((\boldsymbol{X}\ \boldsymbol{and}\ {\sim}^{\mho}_{A}\boldsymbol{Y}) \cap \boldsymbol{Y}\big) = \big(Y \cap (X\ and\ {\sim}^{\mho}_{A}Y)\big) = (Y \cap X) = (\boldsymbol{X} \cap \boldsymbol{Y})$$

$$\big((\boldsymbol{X} -^{\mho}_{A} \boldsymbol{Y})\ \boldsymbol{and}\ (\boldsymbol{Y} -^{\mho}_{A} \boldsymbol{X})\big) = \big((X \cap {\sim}^{\mho}_{A}Y)\ and\ (Y -^{\mho}_{A} X)\big)$$

$$= \Big(\big(X\ and\ (Y -^{\mho}_{A} X)\big) \cap \big({\sim}^{\mho}_{A}Y\ and\ (Y -^{\mho}_{A} X)\big)\Big) = \big((X\ and\ Y) \cap {\sim}^{\mho}_{A}(X \cap Y)\big)$$

$$= (\boldsymbol{X} \Delta \boldsymbol{Y})$$

The following analogous equations are not fulfilled for some concrete statements X and Y, because undecidable statements exist:

$$(X\ and\ Y) = \big(X\ and\ (Y - X)\big) = \big((X - Y)\ and\ Y\big)$$

$$(X\ or\ Y) = \big(X\ or\ (Y\ and\ {\sim}X)\big) = \big((X\ and\ {\sim}Y)\ or\ Y\big)$$

$$\big((X - Y)\ and\ (Y - X)\big) = {\sim}(X\ xor\ Y)$$

For any concrete statement X:

$${\sim}X = {\sim}\bigvee_{Y \Rightarrow X} Y = \bigwedge_{Y \Rightarrow X} {\sim}Y = \bigwedge_{{\sim}Y \Rightarrow X} Y = {\sim}\bigwedge_{X \Rightarrow Y} Y = \bigvee_{X \Rightarrow Y} {\sim}Y = \bigvee_{X \Rightarrow {\sim}Y} Y$$

9.2.5 Subtraction of statements

For any concrete statements a and b:

$$(a - b) = (a\ or\ {\sim}b)$$

Let us assume, that x is the conjunction of all nonempty parts of a, that do not implicate any nonempty part of b.

Then, in any situation (under the assumption, that undecidable statements do not exist):

$$a = (x\ and\ b)$$

$$(a - b) = \big((x\ and\ b)\ or\ {\sim}b\big) = \big((x\ or\ {\sim}b)\ and\ (b\ or\ {\sim}b)\big) \leftrightarrow \big((x\ or\ {\sim}b)\ and\ true\big)$$

$$\leftrightarrow (x\ or\ {\sim}b) = (x - b)$$

So:

$$(a - b) = (x\ or\ {\sim}b) = (x - b)$$

Since for any p such that $(b \Rightarrow p)$, we have:

$$\big((x\ or\ {\sim}b) \Rightarrow p\big) = \big((x \Rightarrow p)\ and\ ({\sim}b \Rightarrow p)\big) \Rightarrow \big(false\ and\ ({\sim}b \Rightarrow p)\big) \equiv false$$

In other words, $(a-b)$ does not implicate any part of b, like it should be in subtraction.

It is not fulfilled for undecidable statements, because for them, $(a-b)$ is also undecidable, so $(a-b)$ and b both implicate the common core of undecidable statements.

Q.E.D.

9.2.6 Logic for a plain things

Whole logic is also true for all plain things of any type t under the following assumptions for any things a and b such that $(a\, and\, b) \subset \Lambda_t$ and for some unary relation R:

$$a \neq (a\, is\, true) = R(a)$$

$$(a \subset b) = (R(b) \Rightarrow R(a))$$

$$\bigvee_{x \subset \Lambda_t} R(x)\, and \bigvee_{x \subset \Lambda_t} \sim R(x)$$

So, R is defined always as $R(x) = (x \subset X)$ for some X such that $X \subset \Lambda_t$ and $(\Lambda_t - X) \neq \varnothing_t$.

For example: $R(x) = (x\, is\, a\, shape\, from\, some\, bounded\, region\, B\, of\, space\, S)$

But remember, that while we have the rules:

$$(R(a)\, and\, R(b)) = R(a\, and\, b)$$

$$(R(a)\, or\, R(b)) \Rightarrow R(a \cap b)$$

We do not have the rule:

$$(R(a)\, or\, R(b)) \Leftarrow R(a \cap b)$$

9.2.7 Derivation of the main assumption of new logic

We can derive **the main assumption of new logic** the following way:

In general, for concrete things X, Y and Z:

$$(X \subset Y) = (Y \supset X)$$

$$(X = Y) = ((X \subset Y)\, and\, (X \supset Y))$$

$$(Y \subset (X\, and\, Y))$$

$$(Y \subset Y)$$

$$((X\, and\, Y) \subset Z) = ((X \subset Z)\, and\, (Y \subset Z))$$

So:

$$(\boldsymbol{Y} = (\boldsymbol{X}\ \boldsymbol{and}\ \boldsymbol{Y})) = \Big((Y \subset (X\ and\ Y))\ and\ (Y \supset (X\ and\ Y))\Big)$$
$$= \Big((Y \subset (X\ and\ Y))\ and\ (\boldsymbol{X} \subset \boldsymbol{Y})\ and\ (Y \subset Y)\Big)$$

Where "$(Y \subset (X\ and\ Y))$" and "$(Y \subset Y)$" are tautologies; therefore, we can eliminate them to get statement $(X \subset Y)$. For more details, see Section 5.20.

So:

$$(\boldsymbol{Y} = (\boldsymbol{X}\ \boldsymbol{and}\ \boldsymbol{Y})) = (\boldsymbol{X} \subset \boldsymbol{Y})$$

And for concrete statements X and Y:

$$(\boldsymbol{X} \Rightarrow \boldsymbol{Y}) = (\boldsymbol{X} \subset \boldsymbol{Y}) = (\boldsymbol{Y} = (\boldsymbol{X}\ \boldsymbol{and}\ \boldsymbol{Y}))$$

So, then also:

$$(\boldsymbol{X} = \boldsymbol{Y}) = ((X \subset Y)\ and\ (Y \subset X)) = ((\boldsymbol{X} \Leftarrow \boldsymbol{Y})\ \boldsymbol{and}\ (\boldsymbol{X} \Rightarrow \boldsymbol{Y}))$$

So:

$$(\boldsymbol{X} = \boldsymbol{Y}) = ((\boldsymbol{X} \Leftarrow \boldsymbol{Y})\ \boldsymbol{and}\ (\boldsymbol{X} \Rightarrow \boldsymbol{Y}))$$
$$(X \Rightarrow Y) \Rightarrow (X \rightarrow Y)$$
$$(X = Y) \Rightarrow (X \leftrightarrow Y)$$

We have also the following more intuitive derivation:

For concrete naturally or artificially plain X and Y such that $(X \cup Y) \subset \Lambda_A^{\mho}$:

$$\Big(\boldsymbol{X}\ \boldsymbol{and}\ (\boldsymbol{Y} \overset{\mho}{\underset{A}{-}} \boldsymbol{X})\Big) = \Big(X \cup (Y \cap \overset{\mho}{\underset{A}{\sim}} X)\Big) = \Big((X \cup Y) \cap (X \cup \overset{\mho}{\underset{A}{\sim}} X)\Big) = \Big((X \cup Y) \cap \Lambda_A^{\mho}\Big)$$
$$= (X \cup Y) = (\boldsymbol{X}\ \boldsymbol{and}\ \boldsymbol{Y})$$

So:

$$(\boldsymbol{X} \subset \boldsymbol{Y}) = \left(TK(Y) = \Big(TK(X)\ and\ TK\big(Y \overset{\mho}{\underset{A}{-}} X\big)\Big)\right) = \left(TK(Y) = TK\Big(X\ and\ \big(Y \overset{\mho}{\underset{A}{-}} X\big)\Big)\right)$$
$$= \left(\boldsymbol{Y} = \Big(\boldsymbol{X}\ \boldsymbol{and}\ \big(\boldsymbol{Y} \overset{\mho}{\underset{A}{-}} \boldsymbol{X}\big)\Big) = (\boldsymbol{Y} = (\boldsymbol{X}\ \boldsymbol{and}\ \boldsymbol{Y}))\right)$$

For concrete statements X and Y:

$$(\boldsymbol{X} \subset \boldsymbol{Y}) = \Big(TK(Y) = (TK(X)\ and\ TK(Y))\Big) = (TK(Y) = TK(X\ and\ Y)) = (\boldsymbol{Y} = (\boldsymbol{X}\ \boldsymbol{and}\ \boldsymbol{Y}))$$

9.2.8 Other laws of things

For any concrete naturally or artificially plain things X and Y such that $(Y \cup X) \subset \Lambda_A^{\mho}$:

$$(\boldsymbol{Y} \supset \boldsymbol{X}) = \bigwedge_{x \pitchfork X} (x \pitchfork Y) = \sim \bigvee_{x \pitchfork X} \sim (x \pitchfork Y) = \sim \bigvee_{x \pitchfork X} (x \pitchfork \sim_A^{\mho} Y) = \sim \bigvee_{y \pitchfork \sim_A^{\mho} Y} (y \pitchfork X)$$

$$= \bigwedge_{y \pitchfork \sim_A^{\mho} Y} \sim (y \pitchfork X) = \bigwedge_{y \pitchfork \sim_A^{\mho} Y} (y \pitchfork \sim_A^{\mho} X) = (\sim_A^{\mho} \boldsymbol{X} \supset \sim_A^{\mho} \boldsymbol{Y})$$

And for any concrete naturally or artificially plain things X and Y such that $(X \cup Y) \subset \Lambda_A^{\mho}$ and for any concrete atomic thing $\boldsymbol{a}$ such that $\boldsymbol{a} \in A$:

$$(\boldsymbol{X} \subset \boldsymbol{Y}) = ((a \pitchfork X) \Rightarrow (a \pitchfork Y)) = (\sim (a \pitchfork Y) \Rightarrow \sim (a \pitchfork X))$$

$$= \left((a \pitchfork \sim_A^{\mho} Y) \Rightarrow (a \pitchfork \sim_A^{\mho} X) \right) = (\sim_A^{\mho} \boldsymbol{Y} \subset \sim_A^{\mho} \boldsymbol{X})$$

So:

$$(\boldsymbol{X} \subset \boldsymbol{Y}) = (\sim_A^{\mho} \boldsymbol{Y} \subset \sim_A^{\mho} \boldsymbol{X})$$

Which is not true for "≀" in place of "~", because, for example:

$$((\textit{Head of a Dog}) \subset \textit{Dog})$$

And:

$$((\textit{Head of a Dog}) \subset \wr \textit{Dog})$$

So if in general, it was true, that $((X \subset Y) = (\wr Y \subset \wr X))$, then:

$$((\textit{Head of a Dog}) \subset \textit{Dog} \subset \wr (\textit{Head of a Dog}))$$

But:

$$\sim ((\textit{Head of a Dog}) \subset \wr (\textit{Head of a Dog}))$$

Q.E.D.

Which means also, that $(X \cup Y) = \sim_A^{\mho} (\sim_A^{\mho} X \cap \sim_A^{\mho} Y)$, which for concrete naturally or artificially plain things X and Y can be derived from the above equality, is not true for "≀" in place of "~".

Since a conjunction of concrete atomic things is a full analogy to a set, in case of which single element sets are concrete atomic parts of it, in general, for concrete naturally or artificially plain things X and Y such, that $(X \cup Y) \subset \Lambda_A^{\mho}$:

$$(X \cup Y) = \sim_A^{\mho}\left(\sim_A^{\mho} X \cap \sim_A^{\mho} Y\right)$$

Here is the following proof of it:

If we know, that for any concrete set X of concrete statements we have:

$$\bigwedge_{a \in X} a = \sim \bigvee_{a \in X} \sim a$$

Then for any concrete sets X and Y:

$$\begin{aligned}(\boldsymbol{X} \cap \boldsymbol{Y}) &= \{a{:}(a \in X)\ and\ (a \in Y)\} = \{a{:} \sim(\sim(a \in X)\ or \sim(a \in Y))\} \\ &= \sim\{a{:} \sim(a \in X)\ or \sim(a \in Y)\} = \sim(\{a{:} \sim(a \in X)\} \cup \{a{:} \sim(a \in Y)\}) \\ &= \sim(\sim\{a{:}(a \in X)\} \cup \sim\{a{:}\ (a \in Y)\}) = \sim(\sim\boldsymbol{X} \cup \sim\boldsymbol{Y})\end{aligned}$$

So for any concrete set A of concrete sets:

$$\begin{aligned}\bigcap_{X \in A} X &= \left\{a{:} \bigwedge_{X \in A} (a \in X)\right\} = \left\{a{:} \sim \bigvee_{X \in A} \sim(a \in X)\right\} = \sim\left\{a{:} \bigvee_{X \in A} \sim(a \in X)\right\} \\ &= \sim \bigcup_{X \in A} \{a{:} \sim(a \in X)\} = \sim \bigcup_{X \in A} \sim X\end{aligned}$$

And for any concrete set Y of plain things such, that $@Y \subset \Lambda_A^{\mho}$:

$$\begin{aligned}\bigcap_{X \in Y} X &= @\left\{a{:} \bigwedge_{X \in Y} (a \pitchfork X)\right\} = @\left\{a{:} \sim \bigvee_{X \in Y} \sim(a \pitchfork X)\right\} = @\sim_A\left\{a{:} \bigvee_{X \in Y} \sim(a \pitchfork X)\right\} \\ &= \sim_A^{\mho} @_{A_\downarrow} \bigcup_{X \in Y} \{a{:}\ \sim(a \pitchfork X)\} = \sim_A^{\mho} \bigcup_{X \in Y} @_{A_\downarrow} \{a{:}\ (a \pitchfork \sim_A^{\mho} X)\} = \sim_A^{\mho} \bigcup_{X \in Y} \sim_A^{\mho} X\end{aligned}$$

Because for any concrete set Y of atoms such, that $Y \subset A$:

$$@\sim_A Y = \sim_A^{\mho} @_{A_\downarrow} Y$$

And for any concrete plain things a and X such, that $a \in A$ (a is an appropriate atom) and $(a \cup X) \subset \Lambda_A^{\mho}$:

$$\sim(a \pitchfork X) = \left(a \pitchfork \sim_A^{\mho} X\right)$$

And for any concrete set Y of concrete sets and any K:

$$\widehat{@}_K \bigcup_{X \in Y} X = \bigcup_{X \in Y} \widehat{@}_K X$$

$$\widehat{@}_K \bigcap_{X \in Y} X \subset \bigcap_{X \in Y} \widehat{@}_K X$$

Q.E.D.

If we know, that for any concrete plain thing X:

$$\sim^{\mho}_{A}\bigcap_{a\pitchfork X} a=\bigcup_{a\pitchfork X}\sim^{\mho}_{A} a$$

Then we have another proof for any concrete plain things X and Y:

$$\sim^{\mho}_{A}(X\cup Y)=\sim^{\mho}_{A}\bigcup_{a\pitchfork (X\cup Y)} a=\bigcap_{a\pitchfork (X\cup Y)}\sim^{\mho}_{A} a=\left(\bigcap_{a\pitchfork X}\sim^{\mho}_{A} a\cap\bigcap_{a\pitchfork Y}\sim^{\mho}_{A} a\right)=\left(\sim^{\mho}_{A}\bigcup_{a\pitchfork X} a\cap\sim^{\mho}_{A}\bigcup_{a\pitchfork Y} a\right)$$
$$=(\sim^{\mho}_{A}X\cap\sim^{\mho}_{A}Y)$$

Q.E.D.

Here is another proof:

If we know, that for any concrete plain thing S and any concrete set X of concrete plain things:

$$\bigwedge_{x\in X}(x\supset S)=\left(\bigcap_{x\in X} x\supset S\right)$$

$$\bigwedge_{x\in X}(x\subset S)=\left(\bigwedge_{x\in X} x\subset S\right)$$

Then, for any concrete plain thing S and any concrete set X of concrete plain things such, that $(@X\cup S)\subset\Lambda^{\mho}_{A}$:

$$\left(\bigwedge_{\boldsymbol{x}\in\boldsymbol{X}}\boldsymbol{x}\subset\boldsymbol{S}\right)=\bigwedge_{x\in X}(x\subset S)=\bigwedge_{x\in X}(\sim^{\mho}_{A}x\supset\sim^{\mho}_{A}S)=\left(\bigcap_{x\in X}\sim^{\mho}_{A}x\supset\sim^{\mho}_{A}S\right)=\left(\sim^{\mho}_{\boldsymbol{A}}\bigcap_{\boldsymbol{x}\in\boldsymbol{X}}\sim^{\mho}_{\boldsymbol{A}}\boldsymbol{x}\subset\boldsymbol{S}\right)$$

So for $S=\sim^{\mho}_{A}\bigcap_{x\in X}\sim^{\mho}_{A}x$, we have:

$$\left(\bigwedge_{\boldsymbol{x}\in\boldsymbol{X}}\boldsymbol{x}\subset\sim^{\mho}_{\boldsymbol{A}}\bigcap_{\boldsymbol{x}\in\boldsymbol{X}}\sim^{\mho}_{\boldsymbol{A}}\boldsymbol{x}\right)=\left(\sim^{\mho}_{A}\bigcap_{x\in X}\sim^{\mho}_{A}x\subset\sim^{\mho}_{A}\bigcap_{x\in X}\sim^{\mho}_{A}x\right)\equiv \mathit{true}$$

And for $S=\bigwedge_{x\in X}x$, we have:

$$\left(\sim^{\mho}_{\boldsymbol{A}}\bigcap_{\boldsymbol{x}\in\boldsymbol{X}}\sim^{\mho}_{\boldsymbol{A}}\boldsymbol{x}\subset\bigwedge_{\boldsymbol{x}\in\boldsymbol{X}}\boldsymbol{x}\right)=\left(\bigwedge_{x\in X}x\subset\bigwedge_{x\in X}x\right)\equiv \mathit{true}$$

So, finally:

$$\bigcup_{\boldsymbol{x}\in\boldsymbol{X}}\boldsymbol{x}=\bigwedge_{\boldsymbol{x}\in\boldsymbol{X}}\boldsymbol{x}=\sim^{\mho}_{\boldsymbol{A}}\bigcap_{\boldsymbol{x}\in\boldsymbol{X}}\sim^{\mho}_{\boldsymbol{A}}\boldsymbol{x}$$

Q.E.D.

For concrete statements, instead we have:

$$(X \, and \, Y) = \sim(\sim X \, or \sim Y)$$

For a proof of it, see Section 5.4.

Here is another proof:

If we know, that for any concrete statement S and any concrete set X of concrete statements:

$$\bigwedge_{x \in X} (x \supset S) = \left(\bigvee_{x \in X} x \supset S \right)$$

$$\bigwedge_{x \in X} (x \subset S) = \left(\bigwedge_{x \in X} x \subset S \right)$$

Then, for any concrete statement S and any concrete set X of concrete statements:

$$\left(\bigwedge_{\boldsymbol{x} \in \boldsymbol{X}} \boldsymbol{x} \subset \boldsymbol{S} \right) = \bigwedge_{x \in X} (x \subset S) = \bigwedge_{x \in X} (\sim x \supset \sim S) = \left(\bigvee_{x \in X} \sim x \supset \sim S \right) = \left(\sim \bigvee_{\boldsymbol{x} \in \boldsymbol{X}} \sim \boldsymbol{x} \subset \boldsymbol{S} \right)$$

So for $S = \sim \bigvee_{x \in X} \sim x$, we have:

$$\left(\bigwedge_{\boldsymbol{x} \in \boldsymbol{X}} \boldsymbol{x} \subset \sim \bigvee_{\boldsymbol{x} \in \boldsymbol{X}} \sim \boldsymbol{x} \right) = \left(\sim \bigvee_{x \in X} \sim x \subset \sim \bigvee_{x \in X} \sim x \right) \equiv true$$

And for $S = \bigwedge_{x \in X} x$, we have:

$$\left(\sim \bigvee_{\boldsymbol{x} \in \boldsymbol{X}} \sim \boldsymbol{x} \subset \bigwedge_{\boldsymbol{x} \in \boldsymbol{X}} \boldsymbol{x} \right) = \left(\bigwedge_{x \in X} x \subset \bigwedge_{x \in X} x \right) \equiv true$$

So finally:

$$\bigcup_{\boldsymbol{x} \in \boldsymbol{X}} \boldsymbol{x} = \bigwedge_{\boldsymbol{x} \in \boldsymbol{X}} \boldsymbol{x} = \sim \bigvee_{\boldsymbol{x} \in \boldsymbol{X}} \sim \boldsymbol{x}$$

Q.E.D.

For any concrete naturally or artificially plain things X and Y such, that $(X \cup Y) \subset \Lambda_A^{\mho}$, we have also:

$$(X \subset Y) \Rightarrow \sim(X \subset \sim_A^{\mho} Y)$$

For any concrete statement a:

$$a = \bigwedge_{a \supset b} b = \bigvee_{b \supset a} b$$

For any concrete naturally or artificially plain thing a such, that $a \subset \Lambda_A^{\mho}$:

$$a = \bigcup_{a \supset b} b = \bigcap_{b \supset a} b = \bigcup_{b \pitchfork a} b = \sim_A^{\mho} \bigcap_{\sim_A^{\mho} b \pitchfork a} b = \bigcap_{\sim_A^{\mho} b \pitchfork \sim_A^{\mho} a} b = \sim_A^{\mho} \bigcup_{b \pitchfork \sim_A^{\mho} a} b$$

$$a = \bigwedge_{a \supset b} b = \bigwedge_{b \pitchfork a} b = \sim_A^{\mho} \bigwedge_{b \pitchfork \sim_A^{\mho} a} b$$

And for any concrete plain things X, Y and Z:

$$((X \cap Y) \cup Z) = ((X \cup Z) \cap (Y \cup Z))$$

$$((X \cup Y) \cap Z) = ((X \cap Z) \cup (Y \cap Z))$$

$$((X \cap Y) - Z) = ((X - Z) \cap (Y - Z))$$

$$((X \cup Y) - Z) = ((X - Z) \cup (Y - Z)) = ((X \, and \, Y) - Z) = ((X - Z) \, and \, (Y - Z))$$

So:

$$((X \, and \, Y) = Z) = ((Z - X) = (Y - X)) = ((Z - Y) = (X - Y))$$

For any concrete plain things X and Y:

$$((X \, and \, Y) = Z) = \left((Z - (X - Y)) = Y\right) = \left((Z - (Y - X)) = X\right)$$

For all things, we also have, that:

$$(X^{\uparrow} \subset Y^{\uparrow}) \Rightarrow \left((V^{\uparrow} \subset (X \, and \, W)^{\uparrow}) \Rightarrow (V^{\uparrow} \subset (Y \, and \, W)^{\uparrow})\right)$$

$$= ((X \, and \, W)^{\uparrow} \subset (Y \, and \, W)^{\uparrow}) = ((X \, and \, W) \, isA \backslash areA \, (Y \, and \, W))$$

And for any concrete plain things X, Y and Z, we have:

$$((X \, and \, Y) = Z) = \left(\left((X - (X \cap Y)) = (Z - Y)\right) \, and \, \left((Y - (X \cap Y)) = (Z - X)\right)\right)$$

And for any concrete statements X and Y, we have:

$$(X \, or \, Y) = \left(\begin{array}{c} (X \, and \sim Y) \, xor \, (\sim X \, and \, Y) \, xor \, (X \, and \, Y) \\ or \, (X \, is \, undecidable \, and \, Y) \, or \, (X \, and \, Y \, is \, undecidable) \end{array} \right) = \sim(\sim X \, and \sim Y)$$

And we know, that for any things X and Y:

$$(X \cup Y) = (X \, and \, Y)$$

So in case of any concrete plain things X and Y:

$$(X \cap Y) = \sim(\sim X \cup \sim Y) = \sim(\sim X \, and \sim Y)$$

And in case of any concrete statements X and Y:

$$(X\,or\,Y) = \sim(\sim X\,and \sim Y) = \sim(\sim X \cup \sim Y)$$

9.2.9 Some operations on abstract things

For any sets X and Y:

$$\left(X \underset{R}{\underbrace{\times}} Y\right) = \{R(x,y):(x,y) \in (X \times Y)\}$$

For any things X and Y:

$$\left(X \overset{R}{\overbrace{\times}} Y\right) = \left\{R(x,y):(x,y) \in \left(X^{\uparrow} \times Y^{\uparrow}\right)\right\}$$

$$\left(X \overset{[\![R]\!]}{\overbrace{\times}} Y\right) = \left\{R(x,y):(x,y) \in \left(X^{\Uparrow} \times Y^{\Uparrow}\right)\right\}$$

$$(X \cup Y)^{\uparrow} = (X\,and\,Y)^{\uparrow} = \left(X \overset{\cup}{\overbrace{\times}} Y\right) = \left(X \overset{and}{\overbrace{\times}} Y\right)$$

For any attributes or statements X and Y:

$$(X\,or\,Y)^{\uparrow} = \left(X \overset{or}{\overbrace{\times}} Y\right)$$

For any plain things X and Y:

$$(X \cap Y)^{\uparrow} = \left(X \overset{\cap}{\overbrace{\times}} Y\right)$$

For any things X and Y:

$$(X \cup Y)^{\Uparrow} = (X\,and\,Y)^{\Uparrow} = \left(X \overset{[\![\cup]\!]}{\overbrace{\times}} Y\right) = \left(X \overset{[\![and]\!]}{\overbrace{\times}} Y\right)$$

For any attributes or statements X and Y:

$$(X\,or\,Y)^{\Uparrow} = \left(X \overset{[\![or]\!]}{\overbrace{\times}} Y\right)$$

For any plain things X and Y:

$$(X \cap Y)^{\Uparrow} = \left(X \overset{[\![\cap]\!]}{\overbrace{\times}} Y\right)$$

9.2.10 Statements

For any concrete statement a:

$$(r \in [a]) = (r \text{ is a realization of } a)$$

$$= \left(((r \in \mathcal{B}) \Rightarrow a) \text{ and} \sim \bigvee_{q} \left((q \subset r) \text{ and } (q \neq r) \text{ and } ((q \in \mathcal{B}) \Rightarrow a)\right)\right)$$

$$a = \bigvee_{r \in [a]} (r \in \mathcal{B})$$

So:

$$\sim a = \bigvee_{r \in [\sim a]} (r \in \mathcal{B})$$

So:

$$a = \bigvee_{r \in [a]} (r \in \mathcal{B}) = \sim \bigvee_{r \in [\sim a]} (r \in \mathcal{B}) = \bigwedge_{r \in [\sim a]} (r \notin \mathcal{B})$$

$$\sim \bigvee_{r \in [a]} (r \in \mathcal{B}) = \bigvee_{r \in [\sim a]} (r \in \mathcal{B})$$

For any concrete statements a and b:

$$[a \text{ and } b] = \left([a] \underbrace{\times}_{and} [b]\right)$$

$$[a \text{ or } b] = ([a] \cup [b])$$

$$[a \text{ xor } b] = [(a \text{ and} \sim b) \text{ or } (\sim a \text{ and } b)] = \left(\left([a] \underbrace{\times}_{and} [\sim b]\right) \cup \left([\sim a] \underbrace{\times}_{and} [b]\right)\right)$$

And we have, that for every r:

$$[(r \in \mathcal{B})] = \{r\}$$

For any relation $R(x)$ for any tuple of concrete arguments (x):

$$[R(x)] = \left\{ r: \left(\begin{matrix} (R(x) \in primaryFacts(r)) \\ \text{and } \left((q \subset r) \text{ and } (R(x) \in primaryFacts(q)) \Rightarrow (q = r)\right) \end{matrix} \right) \right\}$$

For any relation $R(Any\ A)$ for any tuple of arguments A:

$$[R(Any\ A)] = [\sim(!R)(Some\ A)] = \left\{ @ \bigcup_{a \in A\Uparrow} [R(a)] \right\}$$

$$[(!R)(Any\,A)] = [\sim R(Some\,A)] = \left\{ @ \bigcup_{a \in A\Uparrow} [(!R)(a)] \right\}$$

For any relation $R(Some\,A)$ for any tuple of arguments A:

$$[R(Some\,A)] = [\sim(!R)(Any\,A)] = \bigcup_{a \in A\Uparrow} [R(a)]$$

$$[(!R)(Some\,A)] = [\sim R(Any\,A)] = \bigcup_{a \in A\Uparrow} [(!R)(a)]$$

For more details about function *primaryFacts*, see Section 6.2. For more details about relations, see Section 10.12.

For any concrete statements a and b:

$$(a \Rightarrow b) = \bigwedge_{x \in [a]} \bigvee_{y \in [b]} (x \supset y)$$

Here is the following proof of it:

Of course, universal and existential quantifiers are, respectively, generalized conjunction and generalized disjunction. And we have:

$$\left(\bigvee_{y \in \{a,b\}} (x \Rightarrow y) \right) = ((x \Rightarrow a)\; or\; (x \Rightarrow b)) \Rightarrow (x \Rightarrow (a\; or\; b)) = \left(x \Rightarrow \left(\bigvee_{y \in \{a,b\}} y \right) \right)$$

And:

$$\left(\bigwedge_{y \in \{a,b\}} (y \Rightarrow x) \right) = ((a \Rightarrow x)\; and\; (b \Rightarrow x)) = ((\sim x \Rightarrow \sim a)\; and\; (\sim x \Rightarrow \sim b))$$

$$= (\sim x \Rightarrow (\sim a\; and \sim b)) = ((a\; or\; b) \Rightarrow x) = \left(\left(\bigvee_{y \in \{a,b\}} y \right) \Rightarrow x \right)$$

That can be easily generalized to:

$$\left(\bigvee_{y \in A} (x \Rightarrow y) \right) \Rightarrow \left(x \Rightarrow \left(\bigvee_{y \in A} y \right) \right)$$

And:

$$\left(\bigwedge_{y \in A} (y \Rightarrow x) \right) = \left(\left(\bigvee_{y \in A} y \right) \Rightarrow x \right)$$

So:

$$\left(\bigwedge_{x\in[\boldsymbol{a}]}\bigvee_{\boldsymbol{y}\in[\boldsymbol{b}]}(\boldsymbol{x}\supset\boldsymbol{y})\right)\Rightarrow\left(\bigwedge_{x\in[a]}\bigvee_{y\in[b]}(y\ \textit{was meant to be a part of}\ x)\right)$$

$$=\left(\bigwedge_{x\in[a]}\bigvee_{y\in[b]}\big((x\in\mathcal{B})\Rightarrow(y\in\mathcal{B})\big)\right)\Rightarrow\left(\bigwedge_{x\in[a]}\left((x\in\mathcal{B})\Rightarrow\bigvee_{y\in[b]}(y\in\mathcal{B})\right)\right)$$

$$=\left(\bigvee_{x\in[a]}(x\in\mathcal{B})\Rightarrow\bigvee_{y\in[b]}(y\in\mathcal{B})\right)=(\boldsymbol{a}\Rightarrow\boldsymbol{b})$$

So:

$$(a\Rightarrow b)\Leftarrow\left(\bigwedge_{x\in[a]}\bigvee_{y\in[b]}(\boldsymbol{x}\supset\boldsymbol{y})\right)$$

Now the other way:

$$(a\Rightarrow b)=\big(a=(a\ \textit{and}\ b)\big)\Rightarrow$$

$$\left(\left(\bigwedge_{x\in[a]}\bigvee_{y\in[b]}(\boldsymbol{x}\supset\boldsymbol{y})\right)=\left(\bigwedge_{x\in[a\ \textit{and}\ b]}\bigvee_{y\in[b]}(\boldsymbol{x}\supset\boldsymbol{y})\right)=\left(\bigwedge_{x\in\left([a]\underset{\textit{and}}{\underbrace{\times}}[b]\right)}\bigvee_{y\in[b]}(\boldsymbol{x}\supset\boldsymbol{y})\right)\equiv \textit{true}\right)$$

So:

$$(a\Rightarrow b)\Rightarrow\left(\bigwedge_{x\in[a]}\bigvee_{y\in[b]}(\boldsymbol{x}\supset\boldsymbol{y})\right)$$

So finally:

$$(a\Rightarrow b)=\left(\bigwedge_{x\in[a]}\bigvee_{y\in[b]}(x\supset y)\right)$$

Q.E.D.

And we have for any concrete naturally or artificially plain things a and b such, that $(a\ \textit{and}\ b)\subset\Lambda_A^{\mho}$:

$$(\boldsymbol{a}\supset\boldsymbol{b})=\left(\bigcap_{\tilde{\mho}_A x\pitchfork\tilde{\mho}_A a}x\supset\bigcup_{y\pitchfork b}y\right)=\bigwedge_{\tilde{\mho}_A x\pitchfork\tilde{\mho}_A\boldsymbol{a}}\ \bigwedge_{\boldsymbol{y}\pitchfork\boldsymbol{b}}(\boldsymbol{x}\supset\boldsymbol{y})=\bigwedge_{\boldsymbol{x}\pitchfork\tilde{\mho}_A\boldsymbol{a}}\ \bigwedge_{\tilde{\mho}_A\boldsymbol{y}\pitchfork\boldsymbol{b}}(\boldsymbol{y}\supset\boldsymbol{x})$$

$$\bigvee_{\boldsymbol{x} \pitchfork \boldsymbol{a}} \bigvee_{\underset{A}{\overset{\mho}{\sim}} \boldsymbol{y} \pitchfork \underset{A}{\overset{\mho}{\sim}} \boldsymbol{b}} (\boldsymbol{x} \supset \boldsymbol{y}) \Rightarrow \left(\bigcup_{x \pitchfork a} x \supset \bigcap_{\underset{A}{\overset{\mho}{\sim}} y \pitchfork \underset{A}{\overset{\mho}{\sim}} b} y \right) = (\boldsymbol{a} \supset \boldsymbol{b})$$

And for any concrete statements a and b:

$$(\boldsymbol{a} \Rightarrow \boldsymbol{b}) = \left(\bigvee_{x \supset a} x \Rightarrow \bigwedge_{y \subset b} y \right) = \bigwedge_{\boldsymbol{x} \supset \boldsymbol{a}} \bigwedge_{\boldsymbol{y} \subset \boldsymbol{b}} (\boldsymbol{x} \Rightarrow \boldsymbol{y}) = \bigwedge_{\sim \boldsymbol{x} \supset \boldsymbol{a}} \bigwedge_{\sim \boldsymbol{y} \subset \boldsymbol{b}} (\boldsymbol{y} \Rightarrow \boldsymbol{x})$$

$$\bigvee_{\boldsymbol{x} \subset \boldsymbol{a}} \bigvee_{\boldsymbol{y} \supset \boldsymbol{b}} (\boldsymbol{x} \Rightarrow \boldsymbol{y}) \Rightarrow \left(\bigwedge_{x \subset a} x \Rightarrow \bigvee_{y \supset b} y \right) = (\boldsymbol{a} \Rightarrow \boldsymbol{b})$$

For any concrete statement a:

$$a = \bigwedge_{a \Rightarrow b} b = \bigvee_{b \Rightarrow a} b = \bigvee_{r \in [a]} (r \in \mathcal{B})$$

So:

$$\sim a = \sim \bigwedge_{a \Rightarrow b} b = \sim \bigvee_{b \Rightarrow a} b$$

$$\sim \boldsymbol{a} = \bigvee_{\boldsymbol{a} \Rightarrow \boldsymbol{b}} \sim \boldsymbol{b} = \bigwedge_{\boldsymbol{b} \Rightarrow \boldsymbol{a}} \sim \boldsymbol{b} = \bigvee_{\sim \boldsymbol{b} \Rightarrow \sim \boldsymbol{a}} \sim \boldsymbol{b}$$

9.2.11 Conjunctions and disjunctions of inclusions

And we have:

$$(a \supset b \supset c) = ((a \supset b) \; and \; (b \supset c))$$

And, in general, for any binary relations R, S:

$$(a R b S c) = ((a R b) \; and \; (b S c))$$

For all concrete things we have, that $\supset (a, b)$ is transitive relation:

$$((a \supset b) \; and \; (b \supset c)) \Rightarrow (a \supset c)$$

Here is the following proof of it:

$$((a \supset b) \; and \; (b \supset c)) = \big((a = (b \; and \; a)) \; and \; (b = (c \; and \; b)) \big)$$
$$\Rightarrow (a = (c \; and \; b \; and \; a) = (c \; and \; a)) = (a \supset c)$$

And left-distributive over "and" (axiom):

$$((\boldsymbol{a} \supset \mathbf{b})\ \boldsymbol{and}\ (\boldsymbol{a} \supset \mathbf{c})) = (\boldsymbol{a} \supset (\mathbf{b}\ \boldsymbol{and}\ \mathbf{c}))$$

Here is the following proof of it:

$$((a \supset b)\ and\ (a \supset c)) = \left((a = (b\ and\ a))\ and\ (a = (c\ and\ a))\right)$$
$$\Rightarrow \left(a = (b\ and\ c\ and\ a) = ((b\ and\ c)\ and\ a)\right) = (a \supset (b\ and\ c))$$

For the full proof, see Section 2.2.2.

And left-distributive over "or" in one direction:

$$((\boldsymbol{a} \supset \mathbf{b})\ \boldsymbol{or}\ (\boldsymbol{a} \supset \mathbf{c})) \Rightarrow (\boldsymbol{a} \supset (\mathbf{b}\ \boldsymbol{or}\ \mathbf{c}))$$

Here is the following proof of it:

$$((a \supset b)\ or\ (a \supset c)) = \left((a = (b\ and\ a))\ or\ (a = (c\ and\ a))\right)$$
$$\Rightarrow \left(a = ((b\ and\ a)\ or\ (c\ and\ a)) = ((b\ or\ c)\ and\ a)\right) = (a \supset (b\ or\ c))$$

So:

$$(\boldsymbol{a} \supset (\mathbf{b}\ \boldsymbol{and}\ \mathbf{c})) = ((a \supset b)\ and\ (a \supset c)) \Rightarrow ((a \supset b)\ or\ (a \supset c)) \Rightarrow (\boldsymbol{a} \supset (\mathbf{b}\ \boldsymbol{or}\ \mathbf{c}))$$

And for concrete plain things *a*, *b*, *c*:

$$((\boldsymbol{a} \supset \mathbf{c})\ \boldsymbol{and}\ (\mathbf{b} \supset \mathbf{c})) = ((\boldsymbol{a} \cap \mathbf{b}) \supset \mathbf{c})$$

Which is obvious, but can be also proved for concrete plain things *a*, *b*, *c* such, that $(a\ and\ b\ and\ c) \subset \Lambda_A^{\mho}$:

$$((a \supset c)\ and\ (b \supset c)) = \left((\sim_A^{\mho} c \supset \sim_A^{\mho} a)\ and\ (\sim_A^{\mho} c \supset \sim_A^{\mho} b)\right) = \left(\sim_A^{\mho} c \supset (\sim_A^{\mho} a\ and \sim_A^{\mho} b)\right)$$
$$= ((a \cap b) \supset c)$$

So, as a consequence, also for concrete statements:

$$((a \supset c)\ and\ (b \supset c)) = ((\sim c \supset \sim a)\ and\ (\sim c \supset \sim b)) = (\sim c \supset (\sim a\ and \sim b)) = ((a\ or\ b) \supset c)$$

$$((a \supset c)\ or\ (b \supset c)) = ((\sim c \supset \sim a)\ or\ (\sim c \supset \sim b)) \Rightarrow (\sim c \supset (\sim a\ or \sim b)) = ((a\ and\ b) \supset c)$$

So, for concrete statements:

$$((\boldsymbol{a} \supset \mathbf{c})\ \boldsymbol{and}\ (\mathbf{b} \supset \mathbf{c})) = ((\boldsymbol{a}\ \boldsymbol{or}\ \mathbf{b}) \supset \mathbf{c})$$

$$((\boldsymbol{a} \supset \mathbf{c})\ \boldsymbol{or}\ (\mathbf{b} \supset \mathbf{c})) \Rightarrow ((\boldsymbol{a}\ \boldsymbol{and}\ \mathbf{b}) \supset \mathbf{c})$$

$$((\boldsymbol{a}\ \boldsymbol{or}\ \mathbf{b}) \supset \mathbf{c}) = ((a \supset c)\ and\ (b \supset c)) \Rightarrow ((a \supset c)\ or\ (b \supset c)) \Rightarrow ((\boldsymbol{a}\ \boldsymbol{and}\ \mathbf{b}) \supset \mathbf{c})$$

So also for all concrete things with negation:

$$\sim\left(\boldsymbol{a} \supset (\boldsymbol{b}\, \boldsymbol{and}\, \boldsymbol{c})\right) = \left(\sim(\boldsymbol{a} \supset \boldsymbol{b})\, \boldsymbol{or} \sim(\boldsymbol{a} \supset \boldsymbol{c})\right)$$

$$\sim\left(\boldsymbol{a} \supset (\boldsymbol{b} \breve{\cap} \boldsymbol{c})\right) \Rightarrow \left(\sim(\boldsymbol{a} \supset \boldsymbol{b})\, \boldsymbol{and} \sim(\boldsymbol{a} \supset \boldsymbol{c})\right)$$

$$\sim\left((\boldsymbol{a} \breve{\cap} \boldsymbol{b}) \supset \boldsymbol{c}\right) = \left(\sim(\boldsymbol{a} \supset \boldsymbol{c})\, \boldsymbol{or} \sim(\boldsymbol{b} \supset \boldsymbol{c})\right)$$

$$\sim\left((\boldsymbol{a}\, \boldsymbol{and}\, \boldsymbol{b}) \supset \boldsymbol{c}\right) \Rightarrow \left(\sim(\boldsymbol{a} \supset \boldsymbol{c})\, \boldsymbol{and} \sim(\boldsymbol{b} \supset \boldsymbol{c})\right)$$

Of course, universal and existential quantifiers are respectively generalized conjunction and generalized disjunction.

So we have for any concrete S and any concrete set X of concrete things:

$$\bigwedge_{x \in X} (x \subset S) = \left(\bigwedge_{x \in X} x \subset S\right)$$

And for any concrete plain thing S and any concrete set X of concrete plain things:

$$\bigvee_{x \in X} (x \subset S) \Rightarrow \left(\bigcap_{x \in X} x \subset S\right)$$

And for any concrete statement S and any concrete set X of, respectively, concrete statements:

$$\bigvee_{x \in X} (x \subset S) \Rightarrow \left(\bigvee_{x \in X} x \subset S\right)$$

So for any concrete plain thing S and any concrete set X of concrete plain things:

$$\left(\bigwedge_{x \in X} x \subset S\right) = \bigwedge_{x \in X} (x \subset S) \Rightarrow \bigvee_{x \in X} (x \subset S) \Rightarrow \left(\bigcap_{x \in X} x \subset S\right)$$

And for any concrete statement S and any concrete set X of, respectively, concrete statements:

$$\left(\bigwedge_{x \in X} x \subset S\right) = \bigwedge_{x \in X} (x \subset S) \Rightarrow \bigvee_{x \in X} (x \subset S) \Rightarrow \left(\bigvee_{x \in X} x \subset S\right)$$

And we have for S being concrete plain thing and X being a set of concrete plain things:

$$\bigwedge_{x \in X} (x \supset S) = \left(\bigcap_{x \in X} x \supset S\right)$$

And for S being concrete statement and X being a set of, respectively, concrete statements:

$$\bigwedge_{x\in X}(x\supset S)=\left(\bigvee_{x\in X} x\supset S\right)$$

And for S being concrete thing and X being a set of concrete things:

$$\bigvee_{x\in X}(x\supset S)\Rightarrow\left(\bigwedge_{x\in X} x\supset S\right)$$

So for any concrete plain thing S and any concrete set X of concrete plain things:

$$\left(\bigcap_{x\in X} x\supset S\right)=\bigwedge_{x\in X}(x\supset S)\Rightarrow\bigvee_{x\in X}(x\supset S)\Rightarrow\left(\bigwedge_{x\in X} x\supset S\right)$$

So:

$$\left(\bigcap_{x\in X} x\supset S\right)\Rightarrow\left(\bigwedge_{x\in X} x\supset S\right)$$

And for any concrete statement S and any concrete set X of, respectively, concrete statements:

$$\left(\bigvee_{x\in X} x\supset S\right)=\bigwedge_{x\in X}(x\supset S)\Rightarrow\bigvee_{x\in X}(x\supset S)\Rightarrow\left(\bigwedge_{x\in X} x\supset S\right)$$

So:

$$\left(\bigvee_{x\in X} x\supset S\right)\Rightarrow\left(\bigwedge_{x\in X} x\supset S\right)$$

9.2.12 Additional rules

For any concrete naturally or artificially plain things a, b, c and d such, that $(a\cup b\cup c\cup d)\subset\Lambda_A^{\mho}$, we have quite obvious statements:

$$(a\supset b)\Rightarrow((a\,and\,c)\supset(b\,and\,c))\Rightarrow((a\,and\,c)\supset(b\,and\,c)\supset b)\Rightarrow((a\,and\,c)\supset b)$$

$$(a\supset b)\Rightarrow\left((a\mathop{-}\limits_{A}^{\mho}c)\supset(b\mathop{-}\limits_{A}^{\mho}c)\right)\Rightarrow\left(a\supset(a\mathop{-}\limits_{A}^{\mho}c)\supset(b\mathop{-}\limits_{A}^{\mho}c)\right)\Rightarrow\left(a\supset(b\mathop{-}\limits_{A}^{\mho}c)\right)$$

So also:

$$(a\supset b)\Rightarrow((a\,and\,c)\supset b)\Rightarrow\left((a\,and\,c)\supset(b\mathop{-}\limits_{A}^{\mho}d)\right)$$

So, for any concrete naturally or artificially plain things *a*, *b*, *x* and *y*:

$$\boldsymbol{(a \supset b) \Rightarrow ((a \cup x) \supset (b \cap y))}$$

The same for any concrete statements *a*, *b*, *c* and *d*:

$$(a \supset b) \Rightarrow ((a\, and\, c) \supset (b\, and\, c)) \Rightarrow ((a\, and\, c) \supset (b\, and\, c) \supset b) \Rightarrow ((a\, and\, c) \supset b)$$

$$(a \supset b) \Rightarrow ((a - c) \supset (b - c)) \Rightarrow (a \supset (a - c) \supset (b - c)) \Rightarrow (a \supset (b - c))$$

So also:

$$(a \supset b) \Rightarrow ((a\, and\, c) \supset b) \Rightarrow ((a\, and\, c) \supset (b - d))$$

And for any concrete statements *a*, *b*, *x* and *y*:

$$\boldsymbol{(a \Rightarrow b) \Rightarrow ((a\, and\, x) \Rightarrow (b\, or\, y))}$$

For any concrete naturally or artificially plain things *a*, *b* and *c* such, that $(a \cup b \cup c) \subset \Lambda_A^{\mho}$:

$$\boldsymbol{\left((a\, and\, b) \supset \sim_A^{\mho} c\right) = \left(a \supset \sim_A^{\mho}(c\, and\, b)\right) = \left((c\, and\, b) \supset \sim_A^{\mho} a\right)}$$

For any concrete statements *a*, *b* and *c* (under the assumption, that undecidable statements do not exist):

$$\boldsymbol{((a\, and\, b) \Rightarrow \sim c) = (a \Rightarrow \sim(c\, and\, b)) = ((c\, and\, b) \Rightarrow \sim a)}$$

Here is the following proof of it:

First of all:

$$(a \supset b) \Rightarrow ((a\, and\, c) \supset (b\, and\, c))$$

So, for concrete naturally or artificially plain things:

$$\left(\boldsymbol{(c\, and\, b) \supset \sim_A^{\mho} a}\right) = \left(\boldsymbol{a \supset \sim_A^{\mho}(c\, and\, b)}\right) = \left(a \supset (\sim_A^{\mho} c - \,_A^{\mho} b)\right)$$

$$\Rightarrow \left((a\, and\, b) \supset \left((\sim_A^{\mho} c - \,_A^{\mho} b)\, and\, b\right)\right) = \left((a\, and\, b) \supset (\sim_A^{\mho} c\, and\, b)\right)$$

$$= \left(\left((a\, and\, b) \supset \sim_A^{\mho} c\right) and\, ((a\, and\, b) \supset b)\right) = \left(\boldsymbol{(a\, and\, b) \supset \sim_A^{\mho} c}\right)$$

Where:

$$\left(\boldsymbol{(\sim_A^{\mho} c - \,_A^{\mho} b) \cup b}\right) = \left((\sim_A^{\mho} c \cap \sim_A^{\mho} b) \cup b\right) = \left((\sim_A^{\mho} c \cup b) \cap (\sim_A^{\mho} b \cup b)\right) = \left((\sim_A^{\mho} c \cup b) \cap \Lambda_A^{\mho}\right)$$

$$= \boldsymbol{(\sim_A^{\mho} c \cup b)}$$

So:

$$\boldsymbol{\left((c\, and\, b) \supset \sim_A^{\mho} a\right) = \left(a \supset \sim_A^{\mho}(c\, and\, b)\right) \Rightarrow \left((a\, and\, b) \supset \sim_A^{\mho} c\right)}$$

And for concrete statements:

$$\begin{aligned}((\boldsymbol{c}\ \textbf{and}\ \boldsymbol{b}) \Rightarrow \sim\boldsymbol{a}) &= (\boldsymbol{a} \Rightarrow \sim(\boldsymbol{c}\ \textbf{and}\ \boldsymbol{b})) = (a \Rightarrow (\sim c - b)) \\ &\Rightarrow \big((a \text{ and } b) \Rightarrow ((\sim c - b) \text{ and } b)\big) = ((a \text{ and } b) \Rightarrow (\sim c \text{ and } b)) \\ &= \big(((a \text{ and } b) \Rightarrow \sim c) \text{ and } ((a \text{ and } b) \Rightarrow b)\big) = ((\boldsymbol{a}\ \textbf{and}\ \boldsymbol{b}) \Rightarrow \sim\boldsymbol{c})\end{aligned}$$

Where:

$$\begin{aligned}((\sim\boldsymbol{c} - \boldsymbol{b})\ \textbf{and}\ \boldsymbol{b}) &= ((\sim c \text{ or } \sim b) \text{ and } b) = ((\sim c \text{ and } b) \text{ or } (\sim b \text{ and } b)) \\ &= ((\sim c \text{ and } b) \text{ or } \sim(b \text{ or } \sim b)) = (\sim\boldsymbol{c}\ \textbf{and}\ \boldsymbol{b})\end{aligned}$$

So:

$$((\boldsymbol{c}\ \textbf{and}\ \boldsymbol{b}) \Rightarrow \sim\boldsymbol{a}) = (\boldsymbol{a} \Rightarrow \sim(\boldsymbol{c}\ \textbf{and}\ \boldsymbol{b})) \Rightarrow ((\boldsymbol{a}\ \textbf{and}\ \boldsymbol{b}) \Rightarrow \sim\boldsymbol{c})$$

Secondly:

$$(a \supset b) \Rightarrow \big((a \,{-}^{\mho}_{A}\, c) \supset (b \,{-}^{\mho}_{A}\, c)\big)$$

So for any concrete naturally or artificially plain things:

$$\begin{aligned}\big((\boldsymbol{a}\ \textbf{and}\ \boldsymbol{b}) \supset {\sim}^{\mho}_{A} c\big) &= \Big(\big((a \text{ and } b) \,{-}^{\mho}_{A}\, b\big) \supset ({\sim}^{\mho}_{A} c \,{-}^{\mho}_{A}\, b)\Big) = \big(((a \cup b) \cap \sim b) \supset ({\sim}^{\mho}_{A} c \,{-}^{\mho}_{A}\, b)\big) \\ &= \big(((a \cap \sim b) \cup (b \cap \sim b)) \supset ({\sim}^{\mho}_{A} c \,{-}^{\mho}_{A}\, b)\big) \\ &= \Big(\big((a \,{-}^{\mho}_{A}\, b) \cup \varnothing^{\mho}_{A}\big) \supset ({\sim}^{\mho}_{A} c \,{-}^{\mho}_{A}\, b)\Big) = \big(a \supset (a \,{-}^{\mho}_{A}\, b) \supset ({\sim}^{\mho}_{A} c \,{-}^{\mho}_{A}\, b)\big) \\ &\Rightarrow \big(a \supset ({\sim}^{\mho}_{A} c \,{-}^{\mho}_{A}\, b)\big) = \big(a \supset ({\sim}^{\mho}_{A} c \text{ or } {\sim}^{\mho}_{A} b)\big) \\ &= \big(\boldsymbol{a} \supset {\sim}^{\mho}_{A}(\boldsymbol{c}\ \textbf{and}\ \boldsymbol{b})\big) = \big((\boldsymbol{c}\ \textbf{and}\ \boldsymbol{b}) \supset {\sim}^{\mho}_{A} \boldsymbol{a}\big)\end{aligned}$$

So:

$$((\boldsymbol{a}\ \textbf{and}\ \boldsymbol{b}) \supset {\sim}^{\mho}_{A} \boldsymbol{c}) \Rightarrow (a \supset {\sim}^{\mho}_{A}(\boldsymbol{c}\ \textbf{and}\ \boldsymbol{b})) = ((\boldsymbol{c}\ \textbf{and}\ \boldsymbol{b}) \supset {\sim}^{\mho}_{A} \boldsymbol{a})$$

And for any concrete statements:

$$\begin{aligned}((\boldsymbol{a}\ \textbf{and}\ \boldsymbol{b}) \Rightarrow \sim\boldsymbol{c}) &= ((a \text{ and } b - b) \Rightarrow (\sim c - b)) = \big(((a \text{ and } b) \text{ or } \sim b) \Rightarrow (\sim c - b)\big) \\ &= \big(((a \text{ and } b) \text{ or } \sim b) \Rightarrow (\sim c - b)\big) = \big(((a \text{ or } \sim b) \text{ and } (b \text{ or } \sim b)) \Rightarrow (\sim c - b)\big) \\ &= (a \Rightarrow (a - b) \Rightarrow (\sim c - b)) \Rightarrow (a \Rightarrow (\sim c - b)) = (\boldsymbol{a} \Rightarrow \sim(\boldsymbol{c}\ \textbf{and}\ \boldsymbol{b})) \\ &= ((\boldsymbol{c}\ \textbf{and}\ \boldsymbol{b}) \Rightarrow \sim\boldsymbol{a})\end{aligned}$$

So:

$$((\boldsymbol{a}\ \textbf{and}\ \boldsymbol{b}) \Rightarrow \sim\boldsymbol{c}) \Rightarrow (\boldsymbol{a} \Rightarrow \sim(\boldsymbol{c}\ \textbf{and}\ \boldsymbol{b})) = ((\boldsymbol{c}\ \textbf{and}\ \boldsymbol{b}) \Rightarrow \sim\boldsymbol{a})$$

So, finally:

For any concrete naturally or artificially plain things a, b and c such, that $(a \cup b \cup c) \subset \Lambda_A^{\mho}$:

$$\left((\boldsymbol{a}\ \boldsymbol{and}\ \boldsymbol{b}) \supset \sim_A^{\mho}\boldsymbol{c}\right) = \left(\boldsymbol{a} \supset \sim_A^{\mho}(\boldsymbol{c}\ \boldsymbol{and}\ \boldsymbol{b})\right) = \left((\boldsymbol{c}\ \boldsymbol{and}\ \boldsymbol{b}) \supset \sim_A^{\mho}\boldsymbol{a}\right)$$

For any concrete statements a, b and c (under the assumption, that undecidable statements do not exist):

$$((\boldsymbol{a}\ \boldsymbol{and}\ \boldsymbol{b}) \Rightarrow \sim\boldsymbol{c}) = (\boldsymbol{a} \Rightarrow \sim(\boldsymbol{c}\ \boldsymbol{and}\ \boldsymbol{b})) = ((\boldsymbol{c}\ \boldsymbol{and}\ \boldsymbol{b}) \Rightarrow \sim\boldsymbol{a})$$

Q.E.D.

9.3 The list of the rules of inclusion and inference

All the most important and useful rules are in the appropriate places in this book, where they are proved and used. Their complete list is below in this section.

In the below rules, t is the class for variables, that are used in a rule. You can assume, that the symbol of negation ("~") can be replaced by the symbol of any artificial special negation.

And, of course, every symbol of inclusion can be replaced by the symbol of implication.

Any of the following rules is correct for inclusion of any concrete things if and only if its derivation does not use negation, intersection and disjunction. Otherwise, such a rule is correct only for concrete plain things and concrete statements.

1. $\sim\sim\boldsymbol{p} = \boldsymbol{p} \subset \boldsymbol{p} \subset \boldsymbol{p} = \sim\sim\boldsymbol{p}$

$$\sim\sim\boldsymbol{p} = \boldsymbol{p} \supset \boldsymbol{p} \supset \boldsymbol{p} = \sim\sim\boldsymbol{p}$$

$$\sim\sim p = (\Lambda_t - (\Lambda_t - p)) = \left(\Lambda_t \breve{\cap} \sim (\Lambda_t \breve{\cap} \sim p)\right) = \left(\Lambda_t \breve{\cap} (\varnothing_t\ and\ p)\right) = (\Lambda_t \breve{\cap} p) = p$$

2. $(\boldsymbol{p} \Rightarrow \sim\boldsymbol{p}) \Rightarrow \sim\boldsymbol{p}$

$$(p \Rightarrow \sim p) = (p = (\sim p\ and\ p)) \Rightarrow (p = false) \Rightarrow \sim p$$

3. $\boldsymbol{p} \Rightarrow \boldsymbol{p}$

   Derived from the axiom "$(p\ and\ q) \Rightarrow p$": $p = (p\ and\ p) \Rightarrow p$
4. $(\boldsymbol{p} = \boldsymbol{q}) = ((\boldsymbol{p} \supset \boldsymbol{q})\ \boldsymbol{and}\ (\boldsymbol{q} \supset \boldsymbol{p}))$

   Axiom of new logic
5. $\boldsymbol{p} \subset (\boldsymbol{p}\ \boldsymbol{and}\ \boldsymbol{q})$

$$(\boldsymbol{p}\ \boldsymbol{and}\ \boldsymbol{q}) \supset \boldsymbol{p}$$

   Axiom of new logic

6. $((\boldsymbol{q}\,\textbf{and}\,\boldsymbol{r}) \subset \boldsymbol{p}) = ((\boldsymbol{q} \subset \boldsymbol{p})\,\textbf{and}\,(\boldsymbol{r} \subset \boldsymbol{p}))$

$$(\boldsymbol{p} \supset (\boldsymbol{q}\,\textbf{and}\,\boldsymbol{r})) = ((\boldsymbol{p} \supset \boldsymbol{q})\,\textbf{and}\,(\boldsymbol{p} \supset \boldsymbol{r}))$$

Axiom of new logic

7. $\left(\boldsymbol{p} \subset (\boldsymbol{q} \breve{\cap} \boldsymbol{r})\right) = ((\boldsymbol{p} \subset \boldsymbol{q})\,\textit{and}\,(\boldsymbol{p} \subset \boldsymbol{r}))$

$$\left((\boldsymbol{q} \breve{\cap} \boldsymbol{r}) \supset \boldsymbol{p}\right) = ((\boldsymbol{q} \supset \boldsymbol{p})\,\textbf{and}\,(\boldsymbol{r} \supset \boldsymbol{p}))$$

Axiom of new logic

8. $(\boldsymbol{q} \subset \boldsymbol{p}) = (\sim\boldsymbol{p} \subset \sim\boldsymbol{q})$

$$(\boldsymbol{p} \supset \boldsymbol{q}) = (\sim\boldsymbol{q} \supset \sim\boldsymbol{p})$$

For a proof of it, see Section 5.4.

9. $\left(\boldsymbol{p}\,\textbf{and}\,(\boldsymbol{q} \breve{\cap} \boldsymbol{p})\right) = \boldsymbol{p}$

$$\left(\left(p\,\textit{and}\,(q \breve{\cap} p)\right) = p\right) = \left(p \supset (q \breve{\cap} p)\right) = ((\sim q\,\textit{and} \sim p) \supset \sim p) = \varnothing_{\varsigma}$$

10. $\left(\boldsymbol{p} \breve{\cap} (\boldsymbol{q}\,\textbf{and}\,\boldsymbol{p})\right) = \boldsymbol{p}$

$$\left(\left(p \breve{\cap} (q\,\textit{and}\,p)\right) = p\right) = ((q\,\textit{and}\,p) \supset p) = \varnothing_{\varsigma}$$

11. $((\boldsymbol{r} \subset \boldsymbol{q})\,\textbf{and}\,(\boldsymbol{q} \subset \boldsymbol{p})) \Rightarrow (\boldsymbol{r} \subset \boldsymbol{p})$

$$((\boldsymbol{p} \supset \boldsymbol{q})\,\textbf{and}\,(\boldsymbol{q} \supset \boldsymbol{r})) \Rightarrow (\boldsymbol{p} \supset \boldsymbol{r})$$

For a proof of it, see Section 5.37.

12. $((\boldsymbol{q} \subset \boldsymbol{p})\,\textbf{or}\,(\boldsymbol{r} \subset \boldsymbol{p})) \Rightarrow \left((\boldsymbol{q} \breve{\cap} \boldsymbol{r}) \subset \boldsymbol{p}\right)$

$$((\boldsymbol{p} \supset \boldsymbol{q})\,\textbf{or}\,(\boldsymbol{p} \supset \boldsymbol{r})) \Rightarrow \left(\boldsymbol{p} \supset (\boldsymbol{q} \breve{\cap} \boldsymbol{r})\right)$$

For a proof of it, see Section 5.37.

13. $((\boldsymbol{p} \subset \boldsymbol{q})\,\textbf{or}\,(\boldsymbol{p} \subset \boldsymbol{r})) \Rightarrow (\boldsymbol{p} \subset (\boldsymbol{q}\,\textbf{and}\,\boldsymbol{r}))$

$$((\boldsymbol{q} \supset \boldsymbol{p})\,\textbf{or}\,(\boldsymbol{r} \supset \boldsymbol{p})) \Rightarrow ((\boldsymbol{q}\,\textbf{and}\,\boldsymbol{r}) \supset \boldsymbol{p})$$

For a proof of it, see Section 5.37.

14. $((\boldsymbol{p} \Rightarrow \boldsymbol{q})\,\textbf{and}\,(\boldsymbol{p} \Rightarrow \boldsymbol{r})\,\textbf{and}\,\boldsymbol{p}) \Rightarrow (\boldsymbol{q}\,\textbf{and}\,\boldsymbol{r})$
Trivial

15. $((\boldsymbol{q} \Rightarrow \boldsymbol{p})\,\textbf{and}\,(\boldsymbol{r} \Rightarrow \boldsymbol{p})\,\textbf{and}\,(\boldsymbol{q}\,\textbf{or}\,\boldsymbol{r})) \Rightarrow \boldsymbol{p}$

$$\left((q = (p\,\textit{and}\,q))\,\textit{and}\,(r = (p\,\textit{and}\,r))\,\textit{and}\,(q\,\textit{or}\,r)\right) \Rightarrow ((p\,\textit{and}\,q)\,\textit{or}\,(p\,\textit{and}\,r)) \Rightarrow p$$

16. $((\boldsymbol{p} \Rightarrow \boldsymbol{q})\,\textbf{and}\,(\boldsymbol{r} \Rightarrow \boldsymbol{s})\,\textbf{and}\,(\boldsymbol{p}\,\textbf{or}\,\boldsymbol{r})) \Rightarrow (\boldsymbol{q}\,\textbf{or}\,\boldsymbol{s})$

$$\left((p = (q\,\textit{and}\,p))\,\textit{and}\,(r = (s\,\textit{and}\,r))\,\textit{and}\,(p\,\textit{or}\,r)\right) \Rightarrow ((q\,\textit{and}\,p)\,\textit{or}\,(s\,\textit{and}\,r)) \Rightarrow (q\,\textit{or}\,s)$$

17. $((q \subset p) \Rightarrow (r \subset p)) = (r \subset q)$

$$((p \supset q) \Rightarrow (p \supset r)) = (q \supset r)$$

For a proof of it, see Section 6.3.1.

18. $((p \subset q) \Rightarrow (p \subset r)) = (q \subset r)$

$$((q \supset p) \Rightarrow (r \supset p)) = (r \supset q)$$

For a proof of it, see Section 6.3.1.

19. $((q \subset p) \text{ and } (s \subset r)) \Rightarrow ((r \subset q) \Rightarrow (s \subset p))$

$$((p \supset q) \text{ and } (r \supset s)) \Rightarrow ((q \supset r) \Rightarrow (p \supset s))$$

For a proof of it see Section 6.3.1.

20. $((q \subset p) \text{ and } (s \subset r)) \Rightarrow ((p \text{ and } r) \Rightarrow (q \text{ and } s) \Rightarrow (q \text{ or } s))$

$$((p \supset q) \text{ and } (r \supset s)) \Rightarrow ((p \text{ and } r) \supset (q \text{ and } s) \supset (q \breve{\cap} s))$$

$$((p \supset q) \text{ and } (r \supset s)) = \Big((p = (q \text{ and } p)) \text{ and } (r = (s \text{ and } r))\Big)$$

$$\Rightarrow \Big((p \text{ and } r) = ((q \text{ and } p) \text{ and } (s \text{ and } r))\Big)$$

$$= \Big((p \text{ and } r) = ((q \text{ and } s) \text{ and } (p \text{ and } r))\Big)$$

$$= ((p \text{ and } r) \supset (q \text{ and } s))$$

21. $(q \subset p) \Rightarrow ((q \text{ and } r) \subset (p \text{ and } r))$

$$(p \supset q) \Rightarrow ((p \text{ and } r) \supset (q \text{ and } r))$$

$$(q \subset p) = (p = (q \text{ and } p)) \Rightarrow ((p \text{ and } r) = (q \text{ and } p \text{ and } r))$$

$$= \Big((p \text{ and } r) = ((q \text{ and } r) \text{ and } (p \text{ and } r))\Big) = ((p \text{ and } r) \supset (q \text{ and } r))$$

22. $(\sim r \subset (p \text{ and } q)) = (\sim(q \text{ and } r) \subset p) = (\sim p \subset (q \text{ and } r))$

$$((p \text{ and } q) \supset \sim r) = (p \supset \sim(q \text{ and } r)) = ((q \text{ and } r) \supset \sim p)$$

[not verified for statements and attributes]

For a proof of it, see Section 9.2.12.

23. $(q \subset p) \Rightarrow (q \subset (p \text{ and } r))$

$$(p \supset q) \Rightarrow ((p \text{ and } r) \supset q)$$

For a proof of it, see Section 9.2.12.

24. $(q \subset p) \Rightarrow \Big((q \breve{\cap} r) \subset p\Big)$

$$(p \supset q) \Rightarrow \Big(p \supset (q \breve{\cap} r)\Big)$$

For a proof of it see Section 9.2.12.

25. $(\boldsymbol{q} \subset \boldsymbol{p}) \Rightarrow \left((\boldsymbol{q} \breve{\cap} \boldsymbol{s}) \subset (\boldsymbol{p}\, \textbf{and}\, \boldsymbol{r})\right)$

$$(\boldsymbol{p} \supset \boldsymbol{q}) \Rightarrow \left((\boldsymbol{p}\, \textbf{and}\, \boldsymbol{r}) \supset (\boldsymbol{q} \breve{\cap} \boldsymbol{s})\right)$$

For a proof of it, see Section 9.2.12.

26. $((\boldsymbol{p} \Rightarrow \boldsymbol{q}) \Rightarrow \boldsymbol{p}) \Rightarrow ((\boldsymbol{p} \Rightarrow \boldsymbol{q}) \Rightarrow \boldsymbol{q})$
For a proof of it see Section 6.3.2.1.

27. $(\boldsymbol{p} \Rightarrow (\boldsymbol{p} \Rightarrow \boldsymbol{q})) \Rightarrow (\boldsymbol{p} \Rightarrow \boldsymbol{q})$
For a proof of it, see Section 6.3.2.1.

28. $((\boldsymbol{p} \Rightarrow \boldsymbol{q})\ \textbf{and}\ (\boldsymbol{p} \Rightarrow \sim\boldsymbol{q})) \Rightarrow \sim\boldsymbol{p}$

$$((p \Rightarrow q)\ and\ (p \Rightarrow \sim q)) = (p \Rightarrow (q\ and \sim q)) \Rightarrow (p \Rightarrow false) \Rightarrow \sim p$$

29. $((\boldsymbol{q} \Rightarrow \boldsymbol{p})\ \textbf{and}\ (\sim\boldsymbol{q} \Rightarrow \boldsymbol{p})) \Rightarrow \boldsymbol{p}$
$((\sim p \supset q)\ and\ (\sim p \supset \sim q)) \Rightarrow \sim\sim p = p$

30. $\left((\boldsymbol{p} \Rightarrow (\boldsymbol{q} \Rightarrow \boldsymbol{r}))\right) \Rightarrow \left(((\boldsymbol{p} \Rightarrow \boldsymbol{q}) \rightarrow (\boldsymbol{p} \Rightarrow \boldsymbol{r}))\right)$

$$p \Rightarrow (q \Rightarrow r)$$
$$p \Rightarrow q$$
$$\boldsymbol{p} \Rightarrow (q\, and\, (q \Rightarrow r)) \Rightarrow \boldsymbol{r}$$

31. $(p \Rightarrow (q \Rightarrow r)) \Rightarrow ((p\ and\ q) \Rightarrow r)$
$(\boldsymbol{p} \Rightarrow (\boldsymbol{q} \Rightarrow \boldsymbol{r})) = \left(p = (p\ and\ (q \Rightarrow r))\right)$

$$\Rightarrow \left((p\ and\ q) = (p\ and\ (q \Rightarrow r)\ and\ q) \Rightarrow (q\ and\ (q \Rightarrow r)) \Rightarrow r\right)$$
$$\Rightarrow ((\boldsymbol{p}\ \textbf{and}\ \boldsymbol{q}) \Rightarrow \boldsymbol{r})$$

For the other proof of it see Section 5.37.

32. $(\sim(p \Rightarrow q) \Rightarrow r) \Rightarrow (p \Rightarrow (q\ or\ r))$

$$(\sim(\boldsymbol{p} \Rightarrow \boldsymbol{q}) \Rightarrow \boldsymbol{r}) = (\sim r \Rightarrow (\sim q \Rightarrow \sim p)) \Rightarrow ((\sim r\ and \sim q) \Rightarrow \sim p) = (\boldsymbol{p} \Rightarrow (\boldsymbol{q}\ \textbf{or}\ \boldsymbol{r}))$$

For the other proof of it see Section 5.37.

33. $((p\ and\ q) \Rightarrow r) \Rightarrow (p \Rightarrow (q \rightarrow r))$

Trivial.

34. $(p \Rightarrow (q\ or\ r)) \Rightarrow (\sim(p \rightarrow q) \Rightarrow r)$

$$(\boldsymbol{p} \Rightarrow (\boldsymbol{q}\ \textbf{or}\ \boldsymbol{r})) = ((\sim r\ and \sim q) \Rightarrow \sim p) \Rightarrow (\sim r \Rightarrow (\sim q \rightarrow \sim p)) = (\sim(\boldsymbol{p} \rightarrow \boldsymbol{q}) \Rightarrow \boldsymbol{r})$$

Q.E.D.

10 Theory of things – operations on things

10.1 Introduction

Here you will find *Theory of Things*, that will be an introduction to more complex topics of logic. You do not need to know whole of this section to understand other sections except for the operation $X^{\uparrow}$, that returns the set of all representatives of the kind of X, and its inversion $X_{\downarrow}$.

Before reading this section read at least Section 5.1.

For any thing X:

$$\Phi \neq X = (thing\, X) \neq (the\, kind\, of\, X) = (the\, type\, of\, X) = (the\, class\, of\, X) \neq \Phi$$

$$(X\, is\, the\, thing\, of\, the\, class\, Y) = (Y\, is\, the\, class\, of\, X)$$

$$(X\, is\, the\, thing\, of\, the\, class\, of\, X)$$

$$(the\, type \backslash kind \backslash class\, for\, X) = (the\, type \backslash kind \backslash class\, of\, X)$$

For example: $(the\, kind\, for\, Element\, of\, Set\, S) = (the\, kind\, of\, Element\, of\, Set\, S)$

$$\begin{aligned}(a\, member\, of\, the\, class\, of\, X) &= (a\, representative\, of\, the\, kind\, of\, X)\\ &= (an\, example\, of\, the\, type\, of\, X)\\ &= (a\, thing\, that\, was\, consistently\, meant\, to\, be\, a\, concrete\, X)\end{aligned}$$

The set of all members of the class of X contains only every thing, that was consistently meant to be concrete X. A thing is consistently meant to be concrete X if and only if it was meant to be concrete X and it was not meant to be not concrete X. For example, a square circle was not consistently meant to be a circle, though it was meant to be a circle, because it was also meant to be a square, which means that it was meant to be not a circle.

The kind/type/class of X can be interpreted as the set of all members of the class of X (the representation of the kind of X), but does not have to be interpreted this way and then it is an independent being, that **has** its representation and members. To shorten sentences the class of X will mean always the current class of X, which is a concrete being. To get the name of abstract class of X, the class of which has the impossible, past, present, future and potential classes of X as members, simply use the phrase "full class of X"

Remember, that the class of the class of X is always a concrete thing, even if the class of X is an abstract class, because it has the class of X as the only member. But the class of full class of X is an abstract class, even if the class of X is a concrete class, because full class of X is an abstract thing, because it is eternal, so there are always the impossible, past, present, future and potential classes of X. Remember, that every abstract X is also eternal the same as its class.

https://doi.org/10.1515/9783111441382-010

Every class makes its thing, and every thing has its class.
For any things X and Y:

$$(X\ isA\backslash areA\ Y) = (X^{\uparrow} \subset Y^{\uparrow}) = (attrp(Y) \subset attrp(X))$$

$$(X\ isBp\backslash areBp\ Y) = (X \in Y^{\uparrow})$$

$$(X\ isp\backslash arep\ concrete\ Y) = (X\ isBp\backslash areBp\ Y) = (X \in Y^{\uparrow})$$

$$= \left(\begin{array}{c} (attrp(Y) \subset attr(X)) \\ and \quad \bigwedge_{\substack{attr(X) \subset p \subset Attribute^{\Uparrow}, \\ attr(X) \neq p}} (\sim @attrp(Y) \subset @p) \end{array} \right)$$

$$(X\ isp\backslash arep\ abstract\ Y) = ((X\ isA\backslash areA\ Y)\ and \sim(X\ isBp\backslash areBp\ Y))$$

$$= ((X^{\uparrow} \subset Y^{\uparrow})\ and \sim(X \in Y^{\uparrow}))$$

Functions *attrp* and *attr* are defined in Section 6.2.

For more details about relations *isA*, *isBp*/*areBp*, see Section 10.8.

And we have:

$$(x\ has\ attribute\ a) \Rightarrow (x\ was\ meant\ to\ have\ attribute\ a)$$

The class of a concrete thing has this thing as the only member. The class of an abstract thing has more members than one.

Examples:
Hunting Dog is an abstract thing and an abstract Dog.
My dog is a concrete thing and a concrete Dog.
Dog is an abstract thing and a concrete Species.

A concrete thing cannot be abstract X for any X, but concrete X can be an abstract thing (e.g., Dog is a concrete Species, but an abstract thing).

Every X is concrete X, so it is a member of the class of X, which means that it is a representative of the kind of X.

Every logically possible X at least was consistently meant to be concrete X, so it is also a member of the class of and a representative of the kind of X.

For example, Warm-blooded Species is not a concrete Species, because it is an abstract Species, since it is a subclass of Species.

Concrete X can have many members, but then the class of at least one of them is not a subclass of X, because otherwise it would be abstract X.

Every x was consistently meant to be concrete X if and only if:

$$(x\ isp\backslash arep\ X)\ and \left(\begin{array}{c} \sim \bigvee_{y \in \Omega, y \neq x} (y\ isp\backslash arep\ x) \\ or \bigvee_{y \in \Omega} ((y\ isp\backslash arep\ x)\ and \sim(y\ isp\backslash arep\ X)) \end{array} \right)$$

For example, the Dog is a Species, but My Dog, that is a member of the class of Dog, is not a Species, and the Rottweiler, that as a member of classification Race of Dog is a subclass of (the class of) Dog, is not a Species.

And we have, that for any x:

$$(x\ isp\backslash arep\ concrete\ thing(s)) = (x \in \omega^{\uparrow})$$

$$(x\ isp\backslash arep\ concrete\ being(s)) = ((x \in \omega^{\uparrow})\ and\ (x \in \mathcal{B}))$$

$$(x\ isp\backslash arep\ abstract\ thing(s)) = \sim(x \in \omega^{\uparrow})$$

$$(x\ isp\backslash arep\ abstract\ being(s)) = (\sim(x \in \omega^{\uparrow})\ and\ (x \in \mathcal{B}))$$

And we have that for any X and Y:

$$(X\ isA\backslash areA\ Y) = ((a)\ X\ is\backslash are\ (a)\ Y) = (X^{\uparrow} \subset Y^{\uparrow})$$

$$(X\ isp\backslash arep\ [concrete\ or\ abstract]\ Y) = ((X \in Y^{\uparrow})\ or\ (X^{\uparrow} \subset Y^{\uparrow}))$$

$$(X\ isp\backslash arep\ concrete\ Y) = (X\ isBp\backslash areBp\ Y) = (X \in Y^{\uparrow})$$

$$(X\ isp\backslash arep\ abstract\ Y) = ((X^{\uparrow} \subset Y^{\uparrow})\ and \sim(X \in Y^{\uparrow}))$$

$$(X\ isB\backslash areB\ Y) = ((the)\ X\ is\backslash are\ a\ Y) = (X \in Y^{\Uparrow})$$

And also:

$$(X\ is\backslash are\ concrete\ Y) = ((X\ isp\backslash arep\ concrete\ Y)\ and\ R(X, Y))$$

$$(X\ is\backslash are\ abstract\ Y) = (X\ isp\backslash arep\ abstract\ Y)$$

Where:

$$R(X, Y) = (X\ is\ a\ being\ or\ it\ is\ logically\ impossible,\ but\ it\ is\ not\ a\ double\ nobeing)$$

And we have:

$$(X\ was\ consistently\ meant\ to\ be\ concrete\ Y) = (X \in Y^{\uparrow})$$

For more details about relations $isp/arep, isA/areA, isBp/areBp, isB/areB$, see Section 10.8.

Nothing (Φ) has not the kind, that has its representation, because everything has the kind, that has its representation. The empty set is the representation of the kind of nothing, so nothing has the kind, that has the empty set as its representation. And it is not a contradiction. To understand it better, see Sections 5.1.2 and 5.54.

Nothing has not its class and nothing has not any subclass, because nothing is not a thing.

In the formal language we distinguish between concrete *X*, abstract *X* and the class of *X*, e.g., it is true, that:

$$\textit{The class of every abstract } X \textit{ is a subclass of } [\textit{the class of}]\ X$$

$$\textit{Every } [\textit{concrete}]\ X \textit{ is a member of the class of } X$$

So in the formal language use the phrases "concrete *X*" and "abstract *X*" to distinguish members of the class of *X* from the things of subclasses of *X*, and the phrases "the kind of *X*", "the type of *X*" and "the class of *X*" to distinguish the class of *X* from the thing *X* itself. For example, the class of *X* has members and its representation, while for *X* not being a class thing *X* can have some other attributes, but is not a class, so has not members, representation, etc.

So:

$$(X\ isp\backslash arep\ concrete\ Y) = (the\ X\ is\ a\ member\ of\ the\ class\ of\ Y)$$

$$(X\ is\backslash isp\backslash arep\ abstract\ Y) = (the\ X\ is\ the\ thing\ of\ an\ abstract\ subclass\ of\ [the\ class\ of]\ Y)$$

Important: To allow omitting of the phrase "the class of", when saying "subclass of [the class of] *X*", we will never say about the subclass of *X* for *X* being a class.

Remember, that in the practice of (informal) language we skip for convenience the following determiners:

"[Every member of the class of] (a) Dog is [a member of the class of] (an) Animal"

"[All members of the class of] Dogs are [members of the class of] Animals"

which is equivalent to:

"[Every representative of the kind of] (a) Dog is [a representative of the kind of] (an) Animal"

"[All representatives of the kind of] Dogs are [representatives of the kind of] Animals"

which is equivalent to:

"[Every example of type of] (a) Dog is [an example of type of] (an) Animal"

"[All examples of type of] Dogs are [examples of type of] Animals"

Remember also, that being an Animal is not an attribute of Dog, though it is an attribute of every Dog, so here only fulfilling, that every member of its class is an Animal, is an attribute of Dog. But being a species is an attribute of Dog.

For example, a circle has its radius, because every circle has its radius, but (the) circle has not its radius, because if it had its radius, then we could ask, how long is this radius? But in case of (the) circle, which is not a concrete circle, there would be no answer. So having its radius is not an attribute of (the) circle, though it is an attribute of every circle. On the other hand, for example, being a kind of a shape is an attribute of (the) circle.

Important note: For simplicity I decided to use informal language in the text (not in expressions) of this book whenever it does not mislead reader.

10.2 Physical and nonphysical things

A thing is physical if and only if it cannot be abstracted from physical totality. So every thing intended to be physical, that is a nobeing, is still physical, because all its definitions sill reference physical totality, for example, in the form "It was meant to be *X*", where *X* references physical totality, so it cannot be abstracted from physical totality. So all past, future and potential physical beings are physical nobeings, that are parts of physical totality.

The same every abstract thing, that references physical totality, cannot be abstracted from physical totality, so it is also a physical thing. For example, "height of a tower in France" is an abstract physical thing, where France is the reference to physical totality, but "height of a tower" is an abstract nonphysical thing, because it has no reference to physical totality.

Of course, not every physical being is material, but every material being is physical. And only concrete being can be material, so every abstract being is not material.

Some things are physical and some are nonphysical. Every part of physical totality is a physical thing and every physical thing is a part of physical totality. Every part of physical reality is a physical being, even time and space, and every physical being is a part of physical reality. A thing is physically present if and only if it is a part of physical totality. A being exists physically if and only if it is a part of physical reality. A thing is physical if and only if its every definition references physical totality. The definitions of physical things always reference physical totality, so they cannot be ever abstracted from physical totality. Nonphysical things can be defined without any reference to physical totality, so they can be abstracted from physical totality, if we do not care about their physical attributes.

Time and space exist physically always and everywhere. All physical beings exist physically in time and their definitions must reference physical totality (e.g., the age of United Nations, the height of the Eiffel Tower, the color of this apple, the number of bottles in this crate, etc.) and they can be often experienced sensually and always intellectually. A nonphysical thing cannot exist physically in space, so it can be experienced only intellectually, and its definition, that does not reference physical totality, is inalterable in physical time and if it is a being, then it is eternal in physical time and it is always an universal and abstract take of some aspect of physical totality. So the only necessary and sufficient condition for all such things to be is possibility of existence of physical reality. The existence of nonphysical things is only a result of the existence in reality of the potential of possible physical reality to be understood, so they are accessible only to a mind and not outer senses.

10.3 Past, present, future and potential things

Past, present and future nonphysical nobeings, are those nonphysical double nobeings, that respectively were present, are present or in the predictable future will be present nonphysically (as a part of nonphysical totality) at some time everywhere in space. Past, present and future physical nobeings are those physical double nobeings, that were present, are present or in the predictable future will be present physically (as a part of physical totality) at some time and their definitions reference physical totality.

Past, present and future nonphysical beings, are those nonphysical nobeings, that respectively existed, exist or in the predictable future will exist nonphysically (as a part of nonphysical reality) at some time everywhere in space. Past, present and future physical beings, are those physical nobeings, that existed, exist or in the predictable future will exist physically (as a part of physical reality) at some time and their definitions reference physical totality (e.g., my car, Mount Everest, the age of someone, the length of something material, European Union, etc.).

Potential nonphysical nobeings, are those nonphysical double nobeings, that theoretically can be present nobeings nonphysically (so they could be present nobeings nonphysically at some moment in time everywhere in space, e.g., tomorrow's square circle, that John is going to say about tomorrow at time *t*), but they are not yet past or present or future nonphysical nobeings. Potential physical nobeings are those physical double nobeings, that theoretically can be present nobeings physically (so they could be present nobeings physically at some moment in time and their definitions reference physical totality, e.g., the place of this time, that we are going to think about in the future at time *t*), but they are not yet past or present or future physical nobeings. Of course, all potential nobeings can become and some of them become future or present at some moment in time and they become past after they are present, while the rest of them become impossible things.

Potential nonphysical beings, are those nonphysical nobeings, that theoretically are possible nonphysically (so they could exist nonphysically at some moment in time everywhere in space, e.g., a circle, that Mary is going to have on mind at time *t*), but they are not yet past or present or future nonphysical beings. Potential physical beings are those physical nobeings, that theoretically are possible physically (so they could exist physically at some moment in time and their definitions reference physical totality, e.g., my tomorrow's dog, that is going to run in the park tomorrow at time *t*), but they are not yet past or present or future physical beings. Of course, all potential beings can become and some of them become future or present at some moment in time and they become past after they are present, while the rest of them become impossible things.

This is this way, because only impossible things cannot exist.

Abstract things in general are eternal.

10.4 Attributes of things

A disjunction of contrary attributes does not implicate the conjunction of these contrary attributes, so it does not implicate, that the attempt to define a being by this disjunction must fail.

In certain moments of the presence of the same thing it becomes either impossible or future, present and eventually past being, while any stripped attempt to define a being does not ever change.

For any subset of attributes of given thing x, fulfilling of the conjunction of the facts obtained from these attributes is also an attribute of this thing. And, of course, any implication, that is a fact about the thing, of the fact obtained from any attribute of the thing, can be transformed to an attribute of the thing.

A set of attributes of given thing x, from which all other attributes of this thing can be inferred and neither of the attributes from this set infer from the conjunction of the facts obtained from the rest of these attributes, is one of the definition cores for this thing, that defines it. The smallest set of attributes, that defines this thing, is among these cores.

10.5 Equality, sameness and equality of identities

Any past, present, future and potential physical beings in different places or orientations in space or at different moments in time are not equal (negation of relation $(=)(a,b)$) in the strict sense, because the sets of their attributes are different. But the same things can be perceived at different moments, because, for example:

$$X = \left(A\ wheel\ invention\ is\ (current\ year) - Y\ years\ old\right)$$
$$= (A\ wheel\ invention\ took \backslash takes \backslash will\ take\ place\ at\ year\ Y)$$

$$Z = \left(A\ circle\ was\ discovered\ (current\ year) - V\ years\ ago\right)$$
$$= (A\ circle\ is\ known\ from\ year\ V)$$

where, as you can see, statement X does not change in time. So things, as the invention of a wheel above, can get older and older, but all their new physical, seemingly relative, intended attributes stay always the same. Of course, as you can see in case of statement Z, the same and even more concerns a circle, that is eternal, so is not getting older and older in the same manner – all its new physical attributes stay always the same, though they are seemingly relative, and all its abstract attributes stay always the same.

Another consequence of the fact, that the same things can be perceived at different moments, is that things are the same thing (relation $(||)(a,b)$) in the strict sense, if and only if the difference in their definitions comes only from their longer presence in time, so their attempts of definition are equal. So present, future and potential beings, stay the same things, when they become respectively past, present and either future beings or impossible things. So you can, for example, say, that you are thinking about the same thing someone else was thinking yesterday, but you cannot say, that these things are equal:

$$((\textit{the stripped attempt to define } A) = (\textit{the stripped attempt to define } B)) = (A \| B)$$

$$((\textit{the result of the attempt to define } A) = (\textit{the result of the attempt to define } B)) = (A = B)$$

The equality of identities of things, that were not meant to be physical beings, is an objective (strict) term, because such things can be abstracted from physical totality. For example, the number one used 100 years ago has the same identity as the number one used today.

The equality of identities of physical things is always a subjective (not strict) term and usually it means, that certain thing has not changed its physical form or content too much and it continuously stays in the same physical bounds, that can change its physical form over time. So things, that are the same thing for someone at different moments in time, do not have to be strictly the same thing and then anyone else can always point out any difference in their stripped attempts to define them to argue, that for him these are two different things. That is, why I will use the terms "the same" and "the same thing" only in the strict sense and otherwise I will use the term "equal identities" or I will use these terms in quotation marks and thereby say, that some thing is "the same" or "the same thing" for someone. If you want to use a symbol, that refers for you to "the same thing", then see Section 5.1.3.

Remember, that for any observer Z:

$$(\textit{X and Y are the same thing}) \Rightarrow (\textit{X and Y are the same thing for observer Z})$$

10.6 Representatives that are past, future or potential

If there are representatives of some kind that are impossible, past, future or potential, then a representative of this kind can be respectively impossible, past, future or potential. For example, if you define a Planet (what every planet is, e.g.: "A Planet is X"), then you will have many impossible, past, future and potential planets, so not every representative will be a planet, but all of them will have an attempt to define them as a planet. A Planet is material by its definition, so impossible, past, future and potential Planets are not planets, though the attempts to define them assume that they are planets. Although a Planet is defined as material (so a Planet is material), the Planet itself is still an abstract thing, so it is not material, but every planet is material. And if you ask, for example, how many planets are in the Solar System, then you have on mind only material planets, so then you really ask, how many planets exist in the Solar System, and not how many impossible, past, future, potential and material planets are in the Solar System.

Remember that, for example, an abstract thing unicorn is also a being, representation of the class of which is not empty, even if any unicorn does not exist, did not ever exist and will not ever exist in reality, because the attempt to define unicorn does not fail (and it is not an attempt to define a nobeing), because it is not self-contrary and it is not contrary to (truth of) reality, so all impossible, past, future and all potential unicorns, that could exist in the future, are members of the class of unicorn and elements of its representation.

10.7 Singular and plural forms

For non-concrete thing X in singular form:

$$\&X = (plural\, form\, of\, X)$$

For thing X in plural form, that is not a concrete set of things:

$$X\& = (singular\, form\, of\, X)$$

In general:

$$(X\, in\, plural\, form) = (plural\, form\, of\, X)$$

$$(X\, in\, singular\, form) = (singular\, form\, of\, X)$$

Plural form of singular form X allows to identify this singular form.

Every representative of a plural form is a concrete set of at least two representatives of its singular form. Otherwise there would not be any difference in the definition of a representative of a plural form and the definition of a set. Also a conjunctive plural form is an abstract or concrete set, that has at least two elements.

Every kind is equal to the common kind for all elements of the set of its all representatives.

For non-concrete thing Y:

$$(Y\, is\, X) = (\&Y\, are\, \&X)$$

and in general:

$$(\&X = Y) = (X = Y\&)$$

Relation "*isp*" and the following two operations will be extended and explained in further sections:

$$X^{\uparrow\#S} = \left\{ x \in S\colon (x\, isp\, X)\, and \sim \bigvee_{y \in (\Omega - \{x\})} ((y\, isp\, x)\, and\, (y\, isp\, K)) \right\}$$

$$= (Set,\ that\ contains\ every\ thing,\ that\ was\ consistently\ meant\ to\ be\ concrete\ X\ from\ set\ S)$$

$$X_{\downarrow} = (Kind,\ that\ is\ common\ for\ all\ elements\ of\ set\ X\ and\ no\ other\ thing)$$

In an informal language we have that:

$$(all\,[concrete]\,\&X) = X^{\Uparrow}$$

In the formal language we cannot skip the word "concrete", so we have only, that:

Every concrete *X* is a representative of the kind of *X*.

All concrete &*X* are representatives of the kind of *X*.

And that is why we cannot use the phrase "all &*X*" in the formal language at all to not confuse it with its use in an informal language. So if you want to say something about all things, that are *X*, you must write "all concrete or abstract &*X*".

Using concrete sets for plural forms, representations and expansions of things, that will be explained further, is necessary, because their elements can be conjunctions, so to avoid ambiguities they must not be just conjunctions, but precludes treating a concrete set as a normal concrete thing. So special function, that transforms a concrete set into an *nset*, that from the set differs only by the fact, that it is treated by operators of getting expansion as a normal concrete thing, and vice versa, is necessary:

$$\_(x) = \begin{cases} nset\,from\,x\,for\,x\,being\,a\,set \\ set\,from\,x\,for\,x\,being\,a\,nset \\ x,\ else \end{cases}$$

You may need also the functions, that will transform not only a given set but also recursively its elements, if they are *sets* or *nsets*.

The following function will transform recursively all *nsets* into *sets*:

$$\_\mathrm{deep}\_(S) = \left(((S\,is\,a\,set)\,or\,(S\,is\,an\,nset))\ ?\left\{\coprod_{S} \_\mathrm{deep}\_(e)\,by\,e\right\} : S\right)$$

The following function will transform recursively all sets into *nsets*:

$$\_\mathrm{ndeep}\_(S) = \left(((S\,is\,a\,set)\,or\,(S\,is\,an\,nset))\ ?\,\_\left(\left\{\coprod_{S} \_\mathrm{ndeep}\_(e)\,by\,e\right\}\right) : S\right)$$

Another possible solution to this problem is to use a special form of a set for operators of getting expansion, e.g., for normal set *X*:

$$\bigwedge_{e\in X} \langle e\rangle$$

Since an abstract conjunctive plural form is an abstract set, we have to use square brackets to distinguish abstract elements of such a set, e.g.:

$$[Cat]\ and\ [Dog]$$

$$[My\,Animal]\ and\ [Your\,Plant]$$

In an abstract set we still can use concrete elements, e.g.:

$$[My\ Dog]\ and\ [Your\ Cat]\ and\ \{Dumbo\}$$

Examples of representations of plural forms:

$$([cat]\ and\ [dog])^{\uparrow}$$

$$= \left\{ \begin{array}{l} \{firstCat\}\ and\ \{firstDog\},\ \{firstCat\}\ and\ \{secondDog\},\ \ldots, \\ \{firstCat\}\ and\ \{lastDog\}, \\ \{secondCat\}\ and\ \{firstDog\},\ \{secondCat\}\ and\ \{secondDog\},\ \ldots, \\ \{secondCat\}\ and\ \{lastDog\} \\ \ldots \\ \{lastCat\}\ and\ \{firstDog\},\ \{lastCat\}\ and\ \{secondDog\},\ \ldots, \\ \{lastCat\}\ and\ \{lastDog\}, \end{array} \right\}$$

$$([my\ animal]\ and\ [your\ plant])^{\uparrow}$$

$$= \left\{ \begin{array}{l} \{my\ dog\}\ and\ \{your\ rose\}, \{my\ dog\}\ and\ \{your\ cactus\}, \\ \{my\ cat\}\ and\ \{your\ rose\}, \{my\ cat\}\ and\ \{your\ cactus\} \end{array} \right\}$$

An abstract plural form is an abstract set, that has representatives, that are concrete sets of things. In the example above this is an abstract set of beings, that has the abstract element, that is my animal, and the abstract element, that is your plant, and, as you can see above, it has four representatives, that are concrete sets of beings. For another example, in informal language *two dogs and a cat* (in formal language it will be *two dogs and* $[cat]$) is an abstract set of beings, that has representatives, that are concrete sets, each one of which has one element, that is some dog, and another element, that is some another dog, and one more element, that is some cat, where

$$(two\ dogs) = ([dog]\ and\ [otherDog]) = \left(\bigwedge_{x \in dog^{\uparrow}} \bigwedge_{y \in dog^{\uparrow}, x \neq y} \{\{x\}\ and\ \{y\}\} \right)_{\downarrow}$$

If you want to use in plural form some abstract thing itself, then just use curly brackets, that are used for concrete elements (concrete element of a set can be anything, so it does not have to be a concrete thing – do not confuse these two things) of a set, for example:

$$(\{Cat\}\ and\ \{Dog\})^{\uparrow} = \{\{Cat\}\ and\ \{Dog\}\}$$

Here is the rule of creating plural form from singular form X, that is not a concrete thing:

$$\&_{\#S}X = \left(\bigwedge_{Y \subset X^{\uparrow \#S}} \{Y\} - \bigcup_{y \in X^{\uparrow \#S}} \{\{y\}\} - \{\varnothing\} \right)_{\downarrow}$$

$$\&X\& = (\&X)\& = X$$

For example:

$$\&\left(\{my\ dog, your\ cat\}_{\downarrow}\right) = \{\{my\ dog\}\ and\ \{your\ cat\}\}_{\downarrow}$$

$$\&\left(\{my\ dog,\ your\ cat,\ his\ horse\}_{\downarrow}\right)$$

$$= \left\{ \begin{matrix} \{my\ dog\}\ and\ \{your\ cat\},\ \{my\ dog\}\ and\ \{his\ horse\},\ \{your\ cat\}\ and\ \{his\ horse\}, \\ \{my\ dog\}\ and\ \{your\ cat\}\ and\ \{his\ horse\} \end{matrix} \right\}_{\downarrow}$$

As you can see, every singular form has its plural form, but not every plural form has its singular form. For example, plural form $(\{my\ dog\}\ and\ \{your\ cat\}\ and\ \{his\ horse\})$ has not its singular form.

A group is a set, that has at least two elements.

All representatives of a non-conjunctive plural form are groups, that contain at least two representatives of the appropriate singular form.

You can name every non-conjunctive plural form X a "group of X" to make it a singular form and then you can create the plural form of it again – one degree higher, e.g.:

$$Dog \longrightarrow Dogs = Group\ of\ Dogs \longrightarrow Groups\ of\ Dogs = Group\ of\ Groups\ of\ Dogs \longrightarrow \ldots$$

Examples of plural form:

$$[Dog]\ and\ [Cat]$$

$$\{Dog\}\ and\ \{Cat\}$$

$$[my\ dog]\ and\ [your\ cat]\ and\ \{Dumbo\}$$

$$Horses$$

Examples of singular form:

$$\{my\ dog,\ your\ cat,\ his\ horse\}_{\downarrow}$$

$$\{my\ dog,\ Cat\}_{\downarrow}$$

$$\{Dog,\ Cat\}_{\downarrow}$$

$$(Dog^{\uparrow}\ and\ Cat^{\uparrow})_{\downarrow}$$

$$Horse$$

And:

$$\&X = (element\ of\ the\ set\ of\ all\ groups\ of\ representatives\ of\ X)$$

Although brackets are inexistent in a natural language, they are necessary in logic, because otherwise we would have ambiguous symbols, e.g.:

$$(the\ dog\ and\ the\ cat) = (the\ head\ of\ the\ dog\ and\ the\ rest\ of\ the\ dog\ and\ the\ cat)$$

And:

$$the\ dog\ and\ the\ cat\ are\ animals$$

But this is not true, that:

$$the\ head\ of\ the\ dog\ and\ the\ rest\ of\ the\ dog\ and\ the\ cat\ are\ animals$$

So it should be written in formal language this way:

$$\{the\ head\ of\ the\ dog\ and\ the\ rest\ of\ the\ dog\}\ and\ \{the\ cat\}\ are\ animals$$

Another example:

$$(dog^{\uparrow}\ and\ cat^{\uparrow})_{\downarrow} = (dog^{\uparrow}\ and\ \{FirstCat\}\ and\ \{SecondCat\}\ldots)_{\downarrow}$$

But $(dog^{\uparrow}\ and\ cat^{\uparrow})_{\downarrow} \neq (\{dog\}\ and\ \{cat\})_{\downarrow}$

Remember also, that e.g.:

$$dog\ is\ dog\ without\ the\ head\ and\ head\ of\ a\ dog$$

But:

$$\sim(dog\ without\ the\ head\ and\ head\ of\ a\ dog\ is\ dog)$$

Some examples of singular and plural forms:

$$\&Dog = Dogs$$

$$Dogs\& = Dog$$

$$\&(Breed\ of\ Dog) = (Breeds\ of\ Dog)$$

$$(Breeds\ of\ Dog)\& = (Breed\ of\ Dog)$$

$$\&Animal = Animals$$

$$Animals\& = Animal$$

$$\&(Breed\ of\ Horse) = (Breeds\ of\ Horse)$$

$$(Breeds\ of\ Horse)\& = (Breed\ of\ Horse)$$

10.8 Relations "is" and "are"

If your language distinguishes between relations $(a\,X\,is\,a\,Y\backslash X\,are\,Y)$ and $((the)\;X\,is\backslash are\,a\,Y)$, then use the following relations for any things X and Y:

$$(Any\;X\;is\backslash are\;Some\;Y) = (X^{\Uparrow} \subset Y^{\Uparrow}) \Leftarrow (a\;X\;is\;a\;Y\backslash X\;are\;Y) = (X\;isA\backslash areA\;Y)$$
$$= (X^{\uparrow} \subset Y^{\uparrow})$$

$$((the)\;X\,is\backslash are\,Some\,Y) = ((the)\;X\,is\backslash are\,a\,Y) = (X\,isB\backslash areB\,Y)$$
$$= (X \in Y^{\Uparrow}) \Rightarrow (X\,isBp/areBp\,Y) = (\$X\,isA\backslash areA\,Y)$$
$$= \left((\$X)^{\uparrow} \subset Y^{\uparrow}\right) = (X \in Y^{\uparrow})$$

$$((the)\,X\,is\backslash are\,a\,Y) = ((X\,isBp/areBp\,Y)\;and\,R(X,Y))$$

where

$$R(X,Y) = (X\;is\;a\;being\;or\;it\;is\;logically\;impossible,\;but\;it\;is\;not\;a\;double\;nobeing)$$

If your language does not distinguish these relations, like, for example, Polish language, then this language is incorrect, but you can use the following relation for any things X and Y:

$$(X\;isp\backslash arep\,[concrete\,or\,abstract]\;Y) = ((X\;isBp\backslash areBp\;Y)\;or\;(X\;isA\backslash areA\;Y))$$
$$= \left((X \in Y^{\uparrow})\;or\;(X^{\uparrow} \subset Y^{\uparrow})\right)$$

Then, for example, the statements of the following sentences will be true: Dog *isp* Animal, Dog *isp* Species. In English language, that distinguishes relations *isA*, *areA*, *isB*, *areB*, you should write: a Dog is an Animal, the Dog is a Species. Remember, that for this reason being an Animal is not an attribute of Dog, though it is an attribute of every Dog, so here only fulfilling, that every member of its class is an Animal, is an attribute of Dog. But being a species is an attribute of Dog.

10.9 Representation of a class

$A\#S$ is set A filtered by set S, which means, that:

$$(e \in A\#S) = (e \in (A \cap S))$$

$$(x \in S) = \left(\bigvee_{e \in S} (e = x)\right)$$

X was meant to be concrete K if and only if X has all attributes common for all members of the class of K alternatively fortified with some other attributes.

X was consistently meant to be concrete K if and only if X has all attributes common for all members of the class of K alternatively fortified with some other attributes, that are not contrary to the attributes common for all members of the class of K. So X was consistently meant to be concrete K if and only if X was meant to be concrete K and X was not meant to be not concrete K.

X was consistently meant to be a concrete K if and only if it is a member of the class of K.

For example, a square circle was not consistently meant to be a circle, because it was also meant to be not a circle, since it was meant to be a square, that is not a circle.

For any present thing K (e.g., My Dog, Horse, Circle, Point x, etc. and not My Past Dog, Future Horse, Potential Circle, Impossible Point x, etc.):

$$K^{\circlearrowleft\#S} = \left\{x: (x \in S)\ \mathit{and}\ \left((x\ \mathit{isA} \backslash \mathit{areA}\ K)\ \mathit{or}\ (x\ \mathit{isBp} \backslash \mathit{areBp}\ K)\right)\right\}$$
$$= \{x: (x \in S)\ \mathit{and}\ (x\ \mathit{isp} \backslash \mathit{arep}\ K)\}$$
$$= \begin{pmatrix} \text{the set that contains every thing,} \\ \text{that was consistently meant to be} \\ [\text{concrete or abstract}]\ K\text{, from set } S \end{pmatrix}$$

$$K^{\circlearrowleft\circlearrowright\#S} = \left\{x: (x \in S)\ \mathit{and}\ \left((x\ \mathit{isA} \backslash \mathit{areA}\ K)\ \mathit{or}\ (x\ \mathit{isB} \backslash \mathit{areB}\ K)\right)\right\}$$
$$= \{x: (x \in S)\ \mathit{and}\ (x\ \mathit{is} \backslash \mathit{are}\ [\mathit{concrete\ or\ abstract}]\ K)\}$$
$$= (\text{set that contains any thing(s), that is}\backslash\text{are [concrete or abstract] } K\text{, from set } S)$$

$$K^{\uparrow\#S} = \{x: (x \in S)\ \mathit{and}\ (x\ \mathit{isBp} \backslash \mathit{areBp}\ K)\}$$
$$= \left\{ \begin{matrix} x: \\ \left(\begin{matrix} (x \in S)\ \mathit{and}\ (x\ \mathit{isp} \backslash \mathit{arep}\ K) \\ \mathit{and} \left(\begin{matrix} \sim \bigvee_{y \in \Omega, y \neq x} (y\ \mathit{isp} \backslash \mathit{arep}\ x) \\ \mathit{or} \bigvee_{y \in \Omega} \left((y\ \mathit{isp} \backslash \mathit{arep}\ x)\ \mathit{and} \sim (y\ \mathit{isp} \backslash \mathit{arep}\ K)\right) \end{matrix} \right) \end{matrix} \right) \end{matrix} \right\}$$
$$= \begin{pmatrix} \text{the set that contains every thing,} \\ \text{that was consistently meant to be concrete } K \text{ from set } S \end{pmatrix}$$
$$= (\text{the representation of the kind of } K \text{ in set } S)$$

$$K^{\Uparrow\#S} = \{x\colon (x \in S)\ and\ (x\ isB\backslash areB\ K)\}$$

$$= \left\{ \begin{array}{c} x\colon \\ \left(\begin{array}{c} (x \in S)\ and\ (x\ is\backslash are\ K) \\ and \left(\begin{array}{c} \sim \bigvee\limits_{y \in \Omega, y \neq x} (y\ isp\backslash arep\ x) \\ or \bigvee\limits_{y \in \Omega} ((y\ isp\backslash arep\ x)\ and \sim (y\ isp\backslash arep\ K)) \end{array} \right) \end{array} \right) \end{array} \right\}$$

$$= (set\ that\ contains\ any\ thing(s),\ \ that\ is\backslash are\ concrete\ K,\ from\ set\ S)$$

$$K^{\circlearrowleft} = K^{\circlearrowleft\#\Omega} = \left(\begin{array}{c} the\ set\ that\ contains\ every\ thing, \\ that\ was\ consistently\ meant\ to\ be\ [concrete\ or\ abstract]\ K \end{array} \right)$$

$$K^{\uparrow} = K^{\uparrow\#\Omega} = \left(\begin{array}{c} the\ set\ that\ contains\ every\ thing, \\ that\ was\ consistently\ meant\ to\ be\ concrete\ K \end{array} \right)$$

$$= (the\ representation\ of\ the\ kind\ of\ K)$$

The class of Past/Future/Potential/Impossible *Y* has only members, that were consistently meant to be *Y*, but now are past/future/potential/impossible.

$$K^{\uparrow\#S} \subset K^{\circlearrowleft\#S}$$

$$K^{\Uparrow\#S} \subset K^{\circlearrowleft\circlearrowright\#S}$$

For any things X and Y and any sets S and T:

$$\left(X^{\uparrow\#S} \subset Y^{\uparrow\#S}\right) = \left(X^{\circlearrowleft\#S} \subset Y^{\circlearrowleft\#S}\right)$$

For any thing x:

$$\left(x \in X^{\uparrow\#S}\right) \Rightarrow \left(x \in X^{\circlearrowleft\#S}\right)$$

$$\left(X \in Y^{\circlearrowleft\#S}\right) = \left(\left(X \in Y^{\uparrow\#S}\right)\ or\ \left(X^{\uparrow\#S} \subset Y^{\uparrow\#S}\right)\right)$$

$$\left(X^{\uparrow\#S} \subset Y^{\uparrow\#S}\right) \Rightarrow \left(X^{\Uparrow\#S} \subset Y^{\Uparrow\#S}\right)$$

And:

$$(T \subset S) \Rightarrow \left(\left(X^{\uparrow} \subset Y^{\uparrow}\right) \Rightarrow \left(X^{\uparrow\#S} \subset Y^{\uparrow\#S}\right) \Rightarrow \left(X^{\uparrow\#T} \subset Y^{\uparrow\#T}\right)\right)$$

So:

$$\left(X^{\uparrow} \subset Y^{\uparrow}\right) \Rightarrow \left(X^{\uparrow\#\mathcal{B}} \subset Y^{\uparrow\#\mathcal{B}}\right)$$

The **representation** of kind K is the set of all representatives of kind K.

Only every **piece of X** is representative of the uncountable kind of X. So representation of the uncountable kind of X is the set of all pieces of X.

For any things X and Y:

$$(X = Y) = \left(X^{\uparrow} = Y^{\uparrow}\right) = \left(X^{\circlearrowleft} = Y^{\circlearrowleft}\right)$$

10.10 Not strict classes and representations

The not strict noun phrases, that are meant to name some things, are a large part of the not strict (subjective) terms. And many of them are obviously subjective as, for example, a Nice Picture. But there are many such terms, that seem like strict ones, but in practice they are not strict at all. For example, a Dog, a Cat, a Flower. We all know, what is a Dog or a Flower, and it seems like it is the same for all of us, but unfortunately we also all do not know, how to answer the simple questions, for example, from what moment it is a Dog or a Flower and even more difficult to answer is, how long it is a Dog or a Flower, in other words, when it stops to be a Dog or a Flower. That is, why a Dog and a Flower will be no strict terms, until we will decide to define strictly, for example, from what point to what point it is a Dog or a Flower and not something else. So it should not surprise, that by now representations of such terms as a Dog, a Cat, a Flower and so on are not strict, so they are subjective, so they are different for everyone. But notice, that, for example, relation "A Dog is an Animal" is strictly true, because in the definition of a Dog we all agreed, that as long as something is a Dog so long it is also an Animal.

10.11 Expansion of a class

$$K^{\to C\#S} = \left(K^{\to C} \cap S\right)$$

$K^{\to C}$ is nonstandard expansion of K.

To find $K^{\to C}$, when K is not a concrete set, you have to find the way from K to C in the form $K \to X_1 \to \cdots \to X_n \to C$, where $A \to B$ means, that operation $A^{\to B}$ is trivial, and execute appropriate sequence of trivial operations $Y_{i+1} = \left(Y_i^{\to X_i}\right)_{\downarrow}$, where $Y_1 = K$, to finally execute trivial operation $Y_n^{\to C}$. If there is not such a way, then $K^{\to C} = \varnothing$. If there is more than one such way, then all of them should give the same result.

Assuming, that $X = \mathit{Classification}\, B \;\mathit{of}\; D$, operation $A^{\to X}$ is trivial, when for every element C of $A^{\uparrow}$ there exists classification B of C and $C^{\uparrow} \subset D^{\uparrow}$

So we have the following generating rule:

$$\left(\left(C \in A^{\uparrow}\right) \text{ and } \left(C^{\uparrow} \subset D^{\uparrow}\right)\right) \Rightarrow \left((\text{Classification } B \text{ of } C)^{\uparrow} \subset A^{\to \text{Classification } B \text{ of } D}\right)$$

e.g.:

$(\text{Species of Animal})^{\to \text{Breed of Animal}}$ contain every Breed of Animal, because every Species of Animal can be expanded to Breeds of Animal.

If every element of $A^{\uparrow}$ is X, then $A^{\to X} = A^{\uparrow}$, so it is also trivial, e.g.:

$\text{Animal}^{\to \text{Individual}} = \text{Animal}^{\uparrow}$ contains every Animal.

For example, $\text{Category}^{\to \text{Subsubcategory}}$ would have the way $\text{Category} \to \text{Subcategory} \to \text{Subsubcategory}$:

$$X^{\to X} = X^{\uparrow}$$

$(\text{Breed of Animal})^{\to \text{Breed of Animal}} = (\text{Breed of Animal})^{\uparrow}$ and contains every Breed of Animal.

Important note: we assume, that the sentence "(the) Rottweilers are Breed(s) of Dog" is incorrect. The only correct form in this case is "(the) Rottweiler is a Breed of Dog".

$X^{\to}$ is standard expansion of X.

For X that is not a concrete set:

$$\boldsymbol{X}^{\to} = \boldsymbol{X}^{\uparrow}$$

We have the following generating rules of expansion for concrete set Y:

$$(\boldsymbol{X} \in \boldsymbol{Y}) \Rightarrow \left(\boldsymbol{X}^{\uparrow} \subset \boldsymbol{Y}^{\to}\right)$$

$$(\boldsymbol{X} \in \boldsymbol{Y}) \Rightarrow \left(\boldsymbol{X}^{\to \boldsymbol{C}} \subset \boldsymbol{Y}^{\to \boldsymbol{C}}\right)$$

And the following generating rule of quasi-inversion of expansion for any concrete set S:

$$(\boldsymbol{X} \subset \boldsymbol{S}) \Rightarrow \left(\boldsymbol{X}_{\downarrow} \in \boldsymbol{S}^{\leftarrow}\right)$$

So:

$$\left(X \subset Y^{\to[n]}\right) \Rightarrow \left(X^{\to} \subset Y^{\to[n+1]}\right)$$

$$\left(X^{\to^{k}}\right)^{\to} = X^{\to^{k}\to} = X^{\to^{k+1}}$$

$$\left(X^{\to^{k}}\right)^{\to^{n}} = X^{\to^{k}\to^{n}} = X^{\to^{k+n}}$$

$$X^{\to^{k}} = X^{\to[k]}$$

$X^{\rightarrow[n]\#S}$ and $X^{\rightarrow C\#S}$ are always concrete sets for any X:

$$K^{\rightarrow[\infty]} = K^{\rightarrow[degree(K)]} = (concretization\ of\ K) = (concretized\ K)$$

$degree(X)$ – degree of abstraction of X – the smallest natural number n for which $X^{\rightarrow[n+1]} = X^{\rightarrow[n]}$ or $X^{\rightarrow} = \{X\}$ for $n = 0$. This is the number of abstract layers (abstraction depth), that lead to only concrete things, where $X^{\rightarrow[n]}$ is a set of concrete things, e.g.: His Dog Ace has degree 0, Rottweiler and Dog have degree 1, Breed and Species have degree 2 and Methods of Classification of Living Beings have degree 3 and in general Methods of Classification of Concrete Beings have degree 3:

$$Methods\ of\ Classification\ \{abstract\ deg.3\} \rightarrow Clasification\ \{abstract\ deg.2\}$$
$$\rightarrow Class\ \{abstract\ deg.1\} \rightarrow Concrete\ Thing\ \{abstract\ deg.0 - concrete\}$$

$(Breed\ of\ Animal)^{\rightarrow Individual} = (Breed\ of\ Animal)^{\rightarrow[2]}$ contains every Individual of Animal, because every Breed of Animal can be expanded to Individuals.

10.12 Relations

10.12.1 Definition of "Some/Any" and "The" relations

We have the function domain generating rule for tuple of arguments A:

$$(A_1, \ldots, A_n)^{\Uparrow\#S} = A_1^{\Uparrow\#S} \times \ldots \times A_n^{\Uparrow\#S}$$
$$(A_1, \ldots, A_n)^{\Uparrow\#(S_1, \ldots, S_n)} = A_1^{\Uparrow\#S_1} \times \ldots \times A_n^{\Uparrow\#S_n}$$

We have three possible function argument declarations:

$$Some\ A_{)(} = \coprod_A Some\ A_i\ by\ A_i$$
$$Any\ A_{)(} = \coprod_A Any\ A_i\ by\ A_i$$
$$The\ A_{)(} = \coprod_A The\ A_i\ by\ A_i$$

where we have unpacking operation, that returns the interior expression of an expression in specified brackets, e.g.:

$$(X)_{)(} = X$$
$$\{X\}_{\}\{} = X$$

And exchange of brackets operation, e.g.:

$$X_{\{\}()} = \{X_{)(}\}$$

$$X_{\{\}()} = \{X_{\}\{}\}$$

e.g.:

$$R\left(Some\ (A_1, \ldots, A_n)_{)(}\right) = R(Some\ A_1, \ldots, Some\ A_n)$$

$$R\left(Any\ (A_1, \ldots, A_n)_{)(}\right) = R(Any\ A_1, \ldots, Any\ A_n)$$

$$R\left(The\ (A_1, \ldots, A_n)_{)(}\right) = R(The\ A_1, \ldots, The\ A_n)$$

For function $R(The\, A_{)(})$ we have for any tuple a that fits into tuple A:

$$\left((a \in A^{\Uparrow}) = (R(a_{)(}) \in B)\right) = (R{:}A^{\Uparrow} \to B) = (R \div A \to B)$$

where $The\, A_{)(}$ symbolizes tied (not untied) arguments.

For function $R(Some\backslash Any\, A_{)(})$ we have for any tuple a that fits into tuple A:

$$\left((a^{\Uparrow} \subset A^{\Uparrow}) = (R(a_{)(}) \in B)\right) = (R :: A \to B)$$

A function is of type $R(Some\, A_{)(}, Any\, B_{)(}, Not\, Some\, C_{)(}, Not\, Any\, D_{)(})$ if and only if in its definition it has one of quantifiers *Some, Any, Not Some, Not Any* before every use of any argument from respectively *A, B, C, D*.

The declaration of a function of this type is in the following form:

$$relationName\begin{pmatrix} quantifier_1\ argument_1(name_1[argumentAlias_1]), \\ \ldots, \\ quantifier_n\ argument_n(name_n[argumentAlias_n]), \end{pmatrix}$$

where names of arguments and argument aliases are optional.

The definition of a function of this type is in the following form:

$$declaration \to body$$

We assume, that any function, that returns a statement, is some relation and any relation is some function, that returns a statement. So a relation is a function and not vice versa.

So single argument function, that returns a statement, is also a relation.

A relation cannot be a subset of the Cartesian product of some sets, because for arguments out of domain it is nonmeaningful (its statement is undefined, e.g., *Beauty* $\in$ *Horse*) and for some arguments it can be also undecidable (e.g.: $R(Some\ Thing\ (x)) = \left(R(Some\ Thing\ (x))\ is\, false\right)$ is undecidable). If a relation was a subset of the Cartesian product, then it would not have information about the domain, for which it is meaningful. Of course, the most important reason, why a relation cannot be a subset of the Cartesian product of some sets, is, that it is a function, that returns a statement.

A full relation of this type has an argument for every use of quantifier *Some*, *Any*, *Not Some*, *Not Any*.

For a full relation of this type we have:

$$\sim R(Some\,A_{)(}, Any\,B_{)(}, Not\,Some\,C_{)(}, Not\,Any\,D_{)(})$$
$$= (!\,R)(Any\,A_{)(}, Some\,B_{)(}, Not\,Any\,C_{)(}, Not\,Some\,D_{)(})$$

So the negation of such a relation has interchanged quantifiers and the negated relation.

A normalized relation of this type is a relation of this type, that does not have the negation before any quantifier *Some* and *Any*, so it has no quantifiers *Not Some* and *Not Any*. For example, it cannot begin with the negation of a statement.

Every full normalized relation $R(Some\backslash Any\,A_{)(})$ of this type has associated basic relation $\overline{R}(A_{)(})$ such that:

$$\overline{R}(A_{)(}) = R(The\,A_{)(})$$

"The" relation is a basic relation, that is explained further in this section. For more details about "The" relation see Section 12.6.

You can declare or define relation $\overline{R}$ and if you declare this relation, then you can also declare or define concrete variant (with concrete quantifiers) of relation R, but if you define this relation, then you cannot define any variant of relation R, because then any variant of relation R is already defined.

So normalized relation can be, for example, in the following form:

$$R(Some\,A,\ Any\,B,\ Any\,C) = \bigvee_{a \in A^{\Uparrow}} \bigwedge_{b \in B^{\Uparrow}} \bigwedge_{c \in C^{\Uparrow}} \overline{R}(a, b, c)$$

But not in the following form:

$$S(Some\,A,\ Not\,Any\,B,\ Any\,C) = \bigvee_{a \in A^{\Uparrow}} \sim \bigwedge_{b \in B^{\Uparrow}} \bigwedge_{c \in C^{\Uparrow}} \overline{S}(a, b, c)$$

As you can see, you can easily normalize a relation of this type. For example, you can turn the above non-normalized relation $S(Some\,A,\ Not\,Any\,B,\ Any\,C)$ to normalized relation $T(Some\,A,\ Some\,B,\ Some\,C)$, where $T = (!\,S)$:

$$S(Some\,A,\ Not\,Any\,B,\ Any\,C) = (!\,S)(Some\,A,\ Some\,B,\ Some\,C)$$
$$= T(Some\,A,\ Some\,B,\ Some\,C) = \bigvee_{a \in A^{\uparrow}} \bigvee_{b \in B^{\uparrow}} \bigvee_{c \in C^{\uparrow}} \overline{!\,S}(a, b, c)$$
$$= \bigvee_{a \in A^{\uparrow}} \bigvee_{b \in B^{\uparrow}} \bigvee_{c \in C^{\uparrow}} \sim\overline{S}(a, b, c)$$

For a full normalized relation of this type we have:

$$\sim R(Some\,A_{)(},Any\,B_{)(}) = (!\,R)(Any\,A_{)(},Some\,B_{)(})$$

Here are examples of the interchange of quantifiers in the negations of full normalized relations of such type:

For:

$$R(Some\,Animal\,(X),\ Any\,Animal\,(Y)) \rightarrow \big((Some\,Animal\,(X))\ eats\ (Any\,Animal\,(Y))\big)$$

We have for example:

$$\begin{aligned}\sim R(Cat,\ Mouse) = \sim R(Some\,Cat,\ Any\,Mouse) &= \sim((Some\,Cat)\ eats\ (Any\,Mouse))\\ &= (Neither\ of\ Cats\ eat\ Some\,Mouse)\\ &= ((Any\,Cat)\ does\ not\ eat\ (Some\,Mouse))\end{aligned}$$

$$\begin{aligned}\sim R(My\,Cat,\ Your\,Mouse) &= \sim((My\,Cat)\ eats\ (Your\,Mouse))\\ &= ((My\,Cat)\ does\ not\ eat\ (Your\,Mouse))\end{aligned}$$

For $R(Some\,Animal) \rightarrow (Any\,Cat\ eats\ (Some\,Animal))$:

$$\begin{aligned}\sim R(Mouse) = \sim R(Some\,Mouse) &= \sim(Any\,Cat\ eats\ (Some\,Mouse))\\ &= (Some\,Cat\ does\ not\ eat\ (Any\,Mouse))\end{aligned}$$

Of course, you can also pass a representative of a class as such argument and use it without quantifiers *Some* and *Any*, e.g.:

For:

$$R(Some\,Animal\,(X),\ Some\,Animal\,(Y)) \rightarrow \big((Some\,Animal\,(X))\ eats\ (Some\,Animal\,(Y))\big)$$

We have, for example:

$$R(Cat,\ Your\,Mouse) = R(Some\,Cat,\ Your\,Mouse) = ((Some\,Cat)\ eats\ (Your\,Mouse))$$

Where:

$$\begin{aligned}R(Cat,\ Your\,Mouse) &= R(Some\,Cat,\ Some\,Your\,Mouse)\\ &= ((Some\,Cat)\ eats\ (Some\,Your\,Mouse))\\ &= ((Some\,Cat)\ eats\ (Your\,Mouse))\end{aligned}$$

Because inside $R(Some\backslash Any\,K)$ we have:

$$(X \in K^{\uparrow}) \Rightarrow (Some\,X = Any\,X = X)$$

If a class has the class of a thing as a subclass and this thing as a representative at the same time, then you have to use keyword *repr* before the name of the thing, e.g.:

$$X = \{Dog^{\Uparrow} \cup \{Dog\}\}_{\downarrow}$$

Then for:

$$R(Some\,X) \to ((Some\,X)\ do\ something)$$

We have:

$$R(Dog) = ((Some\,Dog)\ do\ something) \neq R(repr\,Dog) = ((Dog)\ do\ something)$$

You can distinguish two or more arguments of the same kind using different names or indexes in the sequence of arguments of the same kind, e.g.:

$$R(Some\,Place,\ Some\,Animal\,(X),\ Some\,Animal\,(Y))$$

$$\to \big((Some\,Animal\,(X))\ eats\ (Some\,Animal\,(Y))\ at\ (Some\,Place)\big)$$

$$R(Some\,Place,\ Some\,Animal,\ Some\,Animal)$$

$$\to \big((Some\,Animal\,(1))\ eats\ (Some\,Animal\,(2))\ at\ (Some\,Place)\big)$$

If function is defined as $R(Some\backslash Any\,A_{)(})$ then it can be used as $R(The\,A_{)(})$, because for any functions $R(Some\backslash Any\,A_{)(})$ and $S(The\,A_{)(})$, that have their results from any set B:

$$((R :: A \to B)\ and\ (S \div A \to B)) \Rightarrow (S \subset R)$$

And we have for any tuple a that fits into tuple A:

$$(a^{\Uparrow} \subset A^{\Uparrow}) = (a\ is\ an\ instantiation\ of\ arguments\ of\ R) = \big(R(a_{)(})\ is\ an\ instantiation\ of\ R\big)$$

$$((a^{\Uparrow} \subset A^{\Uparrow})\ and\ (|a^{\Uparrow}| > 1)) = (a\ is\ an\ abstract\ instatiantion\ of\ arguments\ of\ R)$$

$$= \big(R(a_{)(})\ is\ an\ abstract\ instantiation\ of\ R\big)$$

$$(a \in A^{\Uparrow}) = (a\ is\ a\ concrete\ instantiation\ of\ arguments\ of\ R)$$

$$= \big(R(a_{)(})\ is\ a\ concrete\ instantiation\ of\ R\big)$$

where $Some\backslash Any\,A_{)(}$ symbolizes untied (not tied) arguments.

And we can have function $R(The\,A_{)(})$, that returns a statement and that we call a basic relation:

$$(R{:}\ A^{\Uparrow} \to Statement^{\Uparrow}) \Rightarrow (R\ is\ a\ basic\ relation)$$

$$\big((R\ is\ a\ basic\ relation)\ and\ (R(a) \equiv true)\big)$$

$$= (R(a)\ is\ a\ positive\ instantiation\ of\ a\ basic\ relation)$$

$$\big((R\ is\ a\ basic\ relation)\ and\ (R(a) \equiv false)\big)$$
$$= (R(a)\ is\ a\ negative\ instantiation\ of\ a\ basic\ relation)$$

And we assume, that any function R once declared as $R(Some\backslash Any\, A_{)(})$ or $R(The\, A_{)(})$ can be used without modifiers *Some\Any* and *The*, so we can call it simply $R(A_{)(})$. This is done to shorten formulas. But remember, that then you cannot shorten other variants (with different quantifiers) of relation R.

For any relation $R(The\, A_{)(})$:

$$(!R) = \{(p, \sim r) : (p, r) \in R\}$$
$$!!R = !(!R) = R$$
$$(!R)(a) = \sim R(a) = \big((a) \in \langle \sim R(A_{)(})\rangle\big) = \sim\big((a) \in \langle R(A_{)(})\rangle\big)$$

where:

$$\langle R(A_{)(})\rangle = \{a : (a \in A^{\Uparrow})\ and\ R(a_{)(})\}$$

So relation $R(a)$ must be decidable for any tuple (a) that fits into tuple A. For binary relation $R(The\, A,\ The\, B)$:

$$R(a, b) = (a\, R\, b) = ((a, b) \in \langle R(A, B)\rangle)$$
$$(!R)(a, b) = (a\, !R\, b) = \sim(a\, R\, b) = \sim R(a, b) = ((a, b) \in \langle \sim R(A, B)\rangle) = \sim((a, b) \in \langle R(A, B)\rangle)$$

For any relation $R(The\, A_{)(})$ and X being a list of tuples of arguments, each of which has the same size and contain values for the next successive argument of the relation:

$$\overset{f}{\frown}{R}\{X\} = \bigwedge_{\{X\}} R\left(\overset{f}{\frown}{x}_{)(}\right)\ by\ x$$
$$\left(!\overset{f}{\frown}{R}\right)\{X\} = \sim\overset{f}{\frown}{R}\{X\} = \sim\bigwedge_{\{X\}} R\left(\overset{f}{\frown}{x}_{)(}\right)\ by\ x = \bigvee_{\{X\}} \sim R\left(\overset{f}{\frown}{x}_{)(}\right)\ by\ x$$
$$= \bigvee_{\{X\}} (!R)\left(\overset{f}{\frown}{x}_{)(}\right)\ by\ x$$

And for full normalized relation R we have:

$$R(Some\, X) = R(Some^{\Uparrow}\, X) = \bigvee_{e \in X^{\Uparrow}} \overline{R}(e)$$
$$R(Any\, X) = R(Any^{\Uparrow}\, X) = \bigwedge_{e \in X^{\Uparrow}} \overline{R}(e)$$

$$R(Some^{\uparrow} X) = \bigvee_{e \in X^{\uparrow}} \overline{R}(e)$$

$$R(Any^{\uparrow} X) = \bigwedge_{e \in X^{\uparrow}} \overline{R}(e)$$

10.12.2 Examples of normalized relations of this type

For example, we can have the following logical function (relation) definition:

(Some Dog runs to Some Place) = ((Some Dog) runs to (Some Place)) ← runTo(Some Dog, Some Place)

which can have, for example, the following instantiations:

((Some Rottweiler) runs to (Some Park)) = runTo(Some Rottweiler, Some Park)

((My Dog) runs to (The Park)) = runTo(My Dog, The Park) = (My Dog runs to The Park)

where we know My Dog and The Park as well.

And there is inclusion of the sets of all materializations:

[((Some Rottweiler) runs to (Some Park))] ⊂ [((Some Dog) runs to (Some Place))]

[(My Dog runs to The Park)] ⊂ [((Some Dog) runs to (Some Place))]

And *[(My Dog runs to The Park)]* has exactly two elements and they belong to *[Some Dog runs to Some Place].*

Another example, which involves plural form:

(Some Dogs are playing in Some Place) = ((Some Dogs) are playing in (Some Place)) ← arePlayingIn(Some Dogs, Some Place)

That can have instantiations:

(({My Dog} and {His Dog}) are playing in (The Garden)) = (({My Dog} and {His Dog}) are playing in (The Garden)) = arePlayingIn({My Dog} and {His Dog}, The Garden)

[({My Dog} and {His Dog} are playing in The Garden)] ⊂ [(Some Dogs are playing in Some Place)]

((Some Hunting Dogs) are playing in (Some Forest)) = ((Some Hunting Dogs) are playing in (Some Forest)) = arePlayingIn(Some Hunting Dogs, Some Forest)

[(Some Hunting Dogs are playing in Some Forest)] ⊂ [(Some Dogs are playing in Some Place)]

We can also have named arguments to be able to reference them many times:

(Some Dog (Y) runs to Some Place (X)) = ((Some Dog (Y)) runs to (Some Place (X))) ← runTo(Some Dog (Y), Some Place (X))

(Some Dog (Y) will stay in Some Place (X) for an hour) = ((Some Dog (Y)) will stay in (Some Place (X)) for an hour) ← willStayInForAnHour(Some Dog (Y), Some Place (X))

Then instantiation:

((Some Rottweiler (Y)) runs to (Some Park (X)) and (The Rottweiler (Y)) will stay in (The Park (X)) for an hour) = (runTo(Some Rottweiler (Y), Some Park (X)) and willStayInForAnHour(The Rottweiler (Y), The Park (X)))

means, that some Rottweiler will run to the same park, in which it will stay for an hour.

But instantiation:

((Some Rottweiler (Y)) runs to (Some Park (X)) and (The Rottweiler (Y)) will stay in (Some Park (Z)) for an hour) = (runTo(Some Rottweiler (Y), Some Park (X)) and willStayInForAnHour(The Rottweiler (Y), Some Park (Z)))

means, that some Rottweiler will run to some park and will stay in some possibly other park for an hour.

And:

((My Dog) runs to (The Park) and (My Dog) will stay in (The Park) for an hour) = (runTo(My Dog, The Park) and willStayInForAnHour(My Dog, The Park)) = (My Dog runs to The Park and will stay in The Park for an hour)

means, that my dog runs to the park and will stay in the same park for an hour.

Here we know My Dog and the Park as well.

As you can see, if an argument of a relation is the symbol of a representative of the kind, that is the kind of the appropriate declared argument of this relation, then you do not have to give him another name in parenthesis, because you can use directly his symbol, that is his natural name.

And there is inclusion of the sets of all materializations:

[((Some Rottweiler (Y)) runs to (Some Park (X)) and (The Rottweiler (Y)) will stay in (The Park (X)) for an hour)] ⊂ [((Some Dog (Y)) runs to (Some Place (X)) and (The Dog (Y)) will stay in (The Place (X)) for an hour)] ⊂ [((Some Dog) runs to (Some Place) and (Some Dog) will stay in (Some Place) for an hour)]

[((My Dog) runs to (The Park) and (My Dog) will stay in (The Park) for an hour)] ⊂ [((Some Dog (Y)) runs to (Some Place (X)) and (The Dog (Y)) will stay in (The Place (X)) for an hour)] ⊂ [((Some Dog) runs to (Some Place) and (Some Dog) will stay in (Some Place) for an hour)]

Additionally if my dog is Rottweiler, then, for example:

[((My Dog) runs to (The Park) and (My Dog) will stay in (The Park) for an hour)] ⊂ [((Some Rottweiler (Y)) runs to (Some Park (X)) and (The Rottweiler (Y)) will stay in (The Park (X)) for an hour)]

For more information about relations see Chapter 12.

10.12.3 More complex examples of normalized relations of this type

More complex examples of normalized relations of type $R(Some \backslash Any\, A_{)(})$:

$$\boldsymbol{R(Some\,Man\,(x[X]), Any\,Women\,(y[Y]), Some\,Man\,(z[Z]))}$$

$$\rightarrow \Big((Some\,Man\,(x))\ is\ loved\ by\ (Any\,Women\,(y))\ and\ (Some\,Man\,(z))\ hates\ (x)\Big)$$

$$= \bigvee_{x \in X^{\Uparrow}} \left(\bigwedge_{y \in Y^{\Uparrow}} ((x)\ is\ loved\ by\ (y))\ and \bigvee_{z \in Z^{\Uparrow}} ((z)\ hates\ (x)) \right)$$

$$= \bigvee_{x \in X^{\Uparrow}} \left(\bigwedge_{y \in Y^{\Uparrow}} \bigvee_{z \in Z^{\Uparrow}} ((x)\ is\ loved\ by\ (y))\ and \bigwedge_{y \in Y^{\Uparrow}} \bigvee_{z \in Z^{\Uparrow}} ((z)\ hates\ (x)) \right)$$

$$= \bigvee_{x \in X^{\Uparrow}} \bigwedge_{y \in Y^{\Uparrow}} \left(\bigvee_{z \in Z^{\Uparrow}} ((x)\ is\ loved\ by\ (y))\ and \bigvee_{z \in Z^{\Uparrow}} ((z)\ hates\ (x)) \right)$$

$$= \bigvee_{x \in X^{\uparrow}} \bigwedge_{y \in Y^{\uparrow}} \bigvee_{z \in Z^{\uparrow}} \Big(((x)\ is\ loved\ by\ (y))\ and\ ((z)\ hates\ (x)) \Big) = \bigvee_{x \in X^{\Uparrow}} \bigwedge_{y \in Y^{\Uparrow}} \bigvee_{z \in Z^{\Uparrow}} \boldsymbol{\overline{R}(x, y, z)}$$

$$\boldsymbol{S(Some\,Man\,(x[X]), Any\,Women\,(y[Y]), Some\,Man\,(z[Z]))}$$

$$\rightarrow \begin{pmatrix} (Some\,Man\,(x))\ is\ loved\ by\ (Any\,Women\,(y)) \\ and\ (Some\,Other\,Man\,(z))\ hates\ (x) \end{pmatrix}$$

$$= \begin{pmatrix} (Some\,Man\,(x))\ is\ loved\ by\ (Any\,Women\,(y)) \\ and\ (Some\,Man\,(z))\ hates\ (x) \\ and\ (z)\ is\ not\ equal\ to\ (x) \end{pmatrix}$$

$$= \bigvee_{x \in X^{\Uparrow}} \bigwedge_{y \in Y^{\Uparrow}} \bigvee_{z \in Z^{\Uparrow}} (\overline{R}(x, y, z)\ and \sim (z = x))$$

$$\boldsymbol{T(Some\,Guy\,(g[G]), Any\,Movie\,(m[M]), Some\,Special\,Effects\,(e[E]))}$$

$$\rightarrow \Big((Some\,Guy\,(g))\ likes\ (Any\,Movie\,(m)),\ that\ has\ (Some\,Special\,Effects\,(e))\Big)$$

$$= \bigvee_{g \in G^{\Uparrow}} \bigwedge_{m \in M^{\Uparrow}} \Big(\big((m)\ has\ (Some\,Special\,Effects\,(e))\big) \Rightarrow ((g)\ likes\ (m)) \Big)$$

$$= \bigvee_{g \in G^{\Uparrow}} \bigwedge_{m \in M^{\Uparrow}} \left(\bigvee_{e \in E^{\Uparrow}} ((m)\ has\ (e)) \Rightarrow ((g)\ likes\ (m)) \right)$$

$$= \bigvee_{g \in G^{\Uparrow}} \bigwedge_{m \in M^{\Uparrow}} \bigwedge_{e \in E^{\Uparrow}} \Big(((m)\ has\ (e)) \Rightarrow ((g)\ likes\ (m)) \Big)$$

$$= \bigvee_{g \in G^{\Uparrow}} \bigwedge_{m \in M^{\Uparrow}} \bigwedge_{e \in E^{\Uparrow}} \Big(\sim \boldsymbol{((m)\ has\ (e))\ or\ ((g)\ likes\ (m))} \Big)$$

Or equivalently:

$$X_{M,E} = \left\{x\colon \{1,\{x\}\} \in e \in \langle((The\,M)\ has\ (The\,E))\rangle\right\}_{\downarrow}$$

$$\begin{aligned}
&\boldsymbol{T}(\textbf{\textit{Some Guy}}\ (\boldsymbol{g}[G]), \textbf{\textit{Any Movie}}\ (\boldsymbol{m}[M]), \textbf{\textit{Some Special Effects}}\ (\boldsymbol{e}[E]))\\
&= \boldsymbol{T}_1(\textbf{\textit{Some Guy}}\ (\boldsymbol{g}[G]), \textbf{\textit{Any}}\ \boldsymbol{X}_{M,E}\ (\boldsymbol{m}_1))\\
&= \left((Some\ Guy\ (g))\ likes\ (Any\ X_{M,E}\ (m_1))\right) = \bigvee_{g \in G^{\Uparrow}} \bigwedge_{m_1 \in X^{\Uparrow}_{M,E}} \overline{T}_1(\boldsymbol{g}, \boldsymbol{m}_1)
\end{aligned}$$

where operator $\langle R(The\,A)\rangle$ returns the set of all tuples (a) of possible arguments that fits into tuple A, for which $R(a)$ is true. For more details see Section 12.6. And operator $X_{\downarrow}$ is the inversion of operator $X^{\uparrow}$:

$$\begin{aligned}
&\boldsymbol{U}(\textbf{\textit{Any Man}}(\boldsymbol{m}[M]), \textbf{\textit{Some Woman}}(\boldsymbol{w}[W]))\\
&= \left((Any\ Man\ (m))\ loves\ (Some\ Woman\ (w)),\ that\ loves\ him\right)\\
&= \bigwedge_{m \in M^{\Uparrow}} \bigvee_{w \in W^{\Uparrow}} \left(((\boldsymbol{m})\ \textbf{loves}\ (\boldsymbol{w}))\ \textbf{\textit{and}}\ ((\boldsymbol{w})\ \textbf{loves}\ (\boldsymbol{m}))\right)
\end{aligned}$$

10.12.4 More complex examples of normalization of relations of this type

More complex examples of normalization of relations of type $R(Some\backslash Any\,A_{)(})$:

$$\begin{aligned}
&\boldsymbol{R}(\textbf{\textit{Some Man}}\ (x[X]), \textbf{\textit{Not Any Women}}\ (y[Y]), \textbf{\textit{Not Some Man}}\ (z[Z]))\\
&\to \begin{pmatrix} \left((Some\ Man\ (x))\ is\ loved\ by\ (Not\ Any\ Women\ (y))\right)\\ and\ \left((Not\ Some\ Man\ (z))\ hates\ him\right) \end{pmatrix}\\
&= \bigvee_{x \in X^{\Uparrow}} \left(\sim \bigwedge_{y \in Y^{\Uparrow}} ((x) is\ loved\ by\ (y))\ and \sim \bigvee_{z \in Z^{\Uparrow}} ((z)\ hates\ (x))\right)\\
&= \bigvee_{x \in X^{\Uparrow}} \left(\bigvee_{y \in Y^{\Uparrow}} \sim((x)\ is\ loved\ by\ (y))\ and \bigwedge_{z \in Z^{\Uparrow}} \sim((z)\ hates\ (x))\right)\\
&= \bigvee_{x \in X^{\Uparrow}} \bigvee_{y \in Y^{\Uparrow}} \bigwedge_{z \in Z^{\Uparrow}} \left(((\boldsymbol{x})\ \textbf{is not loved by}\ (\boldsymbol{y}))\ \textbf{\textit{and}}\ ((\boldsymbol{z})\ \textbf{does not hate}\ (\boldsymbol{x}))\right)\\
&= \begin{pmatrix} (\textbf{\textit{Some Man}}\ (\boldsymbol{x}))\ \textbf{is not loved by}\ (\textbf{\textit{Some Woman}}\ (\boldsymbol{y}))\\ \textbf{\textit{and}}\ (\textbf{\textit{Any Man}}\ (\boldsymbol{z}))\ \textbf{does not hate him} \end{pmatrix}
\end{aligned}$$

$$\boldsymbol{T}(\textbf{\textit{Not Any Guy}}(\boldsymbol{g}[\boldsymbol{G}]), \textbf{\textit{Any Movie}}(\boldsymbol{m}[\boldsymbol{M}]), \textbf{\textit{Some Special Effects}}(\boldsymbol{e}[\boldsymbol{E}]))$$

$$\to \left((\textit{Not Any Guy}(g)) \textit{ likes } (\textit{AnyMovie}(m)), \textit{ that has } (\textit{Some Special Effects}(e))\right)$$

$$= \sim \bigwedge_{g \in G^{\Uparrow}} \bigwedge_{m \in M^{\Uparrow}} \left(\left((m) \textit{ has } (\textit{Some Special Effects}(e))\right) \Rightarrow ((g) \textit{ likes } (m))\right)$$

$$= \sim \bigwedge_{g \in G^{\Uparrow}} \bigwedge_{m \in M^{\Uparrow}} \left(\bigvee_{e \in E^{\Uparrow}} \left(((m) \textit{ has } (e))\right) \Rightarrow ((g) \textit{ likes } (m))\right)$$

$$= \sim \bigwedge_{g \in G^{\Uparrow}} \bigwedge_{m \in M^{\Uparrow}} \bigwedge_{e \in E^{\Uparrow}} \left(((m) \textit{ has } (e)) \Rightarrow ((g) \textit{ likes } (m))\right)$$

$$= \bigvee_{g \in G^{\Uparrow}} \bigvee_{m \in M^{\Uparrow}} \bigvee_{e \in E^{\Uparrow}} \sim\left(((m) \textit{ has } (e)) \Rightarrow ((g) \textit{ likes } (m))\right)$$

$$= \bigvee_{g \in G^{\Uparrow}} \bigvee_{m \in M^{\Uparrow}} \bigvee_{e \in E^{\Uparrow}} \sim\left(\sim((m) \textit{ has } (e)) \textit{ or } ((g) \textit{ likes } (m))\right)$$

$$= \bigvee_{g \in G^{\Uparrow}} \bigvee_{m \in M^{\Uparrow}} \bigvee_{e \in E^{\Uparrow}} \left(((m) \textit{ has } (e)) \textit{ and } ((g) \textit{ does not like } (m))\right)$$

$$= \begin{pmatrix} (\textbf{\textit{Some Guy}}(\boldsymbol{g})) \textbf{\textit{ does not like }} (\textbf{\textit{Some Movie}}(\boldsymbol{m})), \\ \textbf{\textit{that has }} (\textbf{\textit{Some Special Effects}}(\boldsymbol{e})) \end{pmatrix}$$

Or equivalently:

$$X_{M,E} = \left\{x: \{1, \{x\}\} \in e \in \left\langle\left(((\textit{The } M) \textit{ has } (\textit{The } E))\right)\right\rangle\right\}_{\downarrow}$$

$$\boldsymbol{T}\,(\textbf{\textit{Not Any Guy}}(\boldsymbol{g}[\boldsymbol{G}]), \textbf{\textit{Any Movie}}(\boldsymbol{m}[\boldsymbol{M}]), \textbf{\textit{Some Special Effects}}(\boldsymbol{e}[\boldsymbol{E}]))$$

$$= \boldsymbol{T}_1\,(\textbf{\textit{Not Any Guy}}(\boldsymbol{g}[\boldsymbol{G}]), \textbf{\textit{Any}}\, \boldsymbol{X}_{\boldsymbol{M,E}}(\boldsymbol{m}_1))$$

$$= \left((\textit{Not Any Guy}(g)) \textit{ likes } (\textit{Any } X_{M,E}(m_1))\right) = \sim \bigwedge_{g \in G^{\Uparrow}} \bigwedge_{m_1 \in X_{M,E}^{\Uparrow}} \overline{T}_1(g, m_1)$$

$$= \bigvee_{\boldsymbol{g} \in \boldsymbol{G}^{\Uparrow}} \bigvee_{\boldsymbol{m}_1 \in \boldsymbol{X}_{\boldsymbol{M,E}}^{\Uparrow}} \sim\overline{\boldsymbol{T}}_1(\boldsymbol{g}, \boldsymbol{m}_1)$$

where operator $\langle R(\textit{The } A)\rangle$ returns the set of all tuples (a) of possible arguments that fits into tuple A, for which $R(a)$ is true. For more details see Section 12.6. And operator $X_{\downarrow}$ is the inversion of operator $X^{\uparrow}$:

$$U\big(\textbf{\textit{Not Any Man}}\,(\textbf{\textit{m}}[\textbf{\textit{M}}]),\ \textbf{\textit{Some Woman}}\,(\textbf{\textit{w}}[\textbf{\textit{W}}])\big)$$

$$= \Big(\big(\textit{Not Any Man}\,(m)\big)\ \textit{loves}\ \big(\textit{Some Woman}\,(w)\big),\ \textit{that loves him}\Big)$$

$$= \sim \bigwedge_{m \in M^{\Uparrow}} \bigvee_{w \in W^{\Uparrow}} \Big(\big((m)\ \textit{loves}\ (w)\big)\ \textit{and}\ \big((w)\ \textit{loves}\ (m)\big)\Big)$$

$$= \bigvee_{m \in M^{\Uparrow}} \bigwedge_{w \in W^{\Uparrow}} \sim\Big(\big((m)\ \textit{loves}\ (w)\big)\ \textit{and}\ \big((w)\ \textit{loves}\ (m)\big)\Big)$$

$$= \bigvee_{m \in M^{\Uparrow}} \bigwedge_{w \in W^{\Uparrow}} \Big(\big((m)\ \textit{does not love}\ (w)\big)\ \textit{or}\ \big((w)\ \textit{does not love}\ (m)\big)\Big)$$

$$= \Big(\big(\textbf{\textit{Some Man}}\,(\textbf{\textit{m}})\big)\ \textbf{\textit{does not love}}\ \big(\textbf{\textit{Any Women}}\,(\textbf{\textit{w}})\big),\ \textbf{\textit{that loves him}}\Big)$$

10.12.5 Quantifier "Some"/"Any" and conversion of every other quantifier to it

If normalized relation R does not use quantifier *Any*, then:

$$R_{\mho}\big(\textit{Some}\,A_{)(}\big) = \bigvee_{A^{\Uparrow\#\mho}} R\big(a_{)(}\big)\ \textit{by}\ a$$

If normalized relation R does not use quantifier *Some*, then:

$$R_{\mho}\big(\textit{Any}\,A_{)(}\big) = \bigwedge_{A^{\Uparrow\#\mho}} R\,\big(a_{)(}\big)\ \textit{by}\ a$$

We can also decompose mixed *Some**Any* normalized relations, but we must start such decomposition from the subject group of a relation and continue from less dependent to more dependent objects (usually, if not always, objects from the left to the right of the sentence), e.g.:

$$\big(\textit{Some Technician}\ \{1\}\ \textit{tries to connect Some Cable}\ (2)\ \textit{to Any Compter}\ (3)\big)$$

$$\neq \big(\textit{Some Technician}\ \{1\}\ \textit{tries to connect Any Computer}\ (2)\ \textit{to Some Cable}\ (3)\big)$$

And we have for full normalized relation $R_{\mho}\big(\textit{Some}\,A_{)(}, \textit{Any}\,B_{)(}\big)$ and $N_{\mho}\big(\textit{Any}\,A_{)(},\ \textit{Some}\ B_{)(}\big)\ \Big(= \sim R_{\mho}\big(\textit{Some}\,A_{)(},\ \textit{Any}\,B_{)(}\big) = (!R_{\mho})\big(\textit{Any}\,A_{)(},\ \textit{Some}\,B_{)(}\big)\Big)$:

$$\left(\left(A \overbrace{\supset_{\mho}}^{attrp} C\right)\ \textit{and}\ \left(B \overbrace{\supset_{\mho}}^{attrp} D\right)\right)$$

$$= \Big(\big(\mathrm{A}^{\uparrow\#\mho} \subset \mathrm{C}^{\uparrow\#\mho}\big)\ \textit{and}\ \big(\mathrm{B}^{\uparrow\#\mho} \supset \mathrm{D}^{\uparrow\#\mho}\big)\Big) \Rightarrow \Big(R_{\mho}\big(A_{)(}, B_{)(}\big) \Rightarrow R_{\mho}\big(C_{)(}, D_{)(}\big)\Big)$$

$$= \Big(N_{\mho}\big(C_{)(}, D_{)(}\big) \Rightarrow N_{\mho}\big(A_{)(}, B_{)(}\big)\Big)$$

If you want to use argument X without quantifier then you can use $\$X$ in quantifiers *Some\Any*, e.g.:

$$(Dog\ is\ a\ species) = ((\mathbf{Some}\ \$\mathbf{Dog})\ \mathbf{is}\ (\mathbf{Some\ Species})) =$$
$$= ((\mathbf{Any}\ \$\mathbf{Dog})\ \mathbf{is}\ (\mathbf{Some\ Species}))$$

We can turn every quantifier to *Some\Any*, e.g.:

$$(There\ is\ a\ Dog,\ that\ likes\ Some\ Bones) = ((\mathbf{Some\ Dog})\ \mathbf{likes}\ (\mathbf{Some\ Bones}))$$
$$= \sim((\mathbf{Any\ Dog})\ \mathbf{does\ not\ likes}\ (\mathbf{Any\ Bones}))$$

Any dog, that likes some bones, would be a successful materialization of the above statement:

$$(There\ is\ a\ Dog,\ that\ does\ not\ like\ Some\ Bones)$$
$$= ((\mathbf{Some\ Dog})\ \mathbf{does\ not\ like}\ (\mathbf{Some\ Bones})) = \sim((\mathbf{Any\ Dog})\ \mathbf{likes}\ (\mathbf{Any\ Bones}))$$

Any dog, that does not like some bones, would be a successful materialization of the above statement.

$$(Some\ Dog\ likes\ Any\ Bones) = ((\mathbf{Some\ Dog})\ \mathbf{likes}\ (\mathbf{Any\ Bones}))$$
$$= \sim((\mathbf{Any\ Dog})\ \mathbf{does\ not\ like}\ (\mathbf{Some\ Bones}))$$

Any dog, that likes any bones, would be a successful materialization of the above statement.

$$(Some\ Dog\ does\ not\ like\ Any\ Bones) = ((\mathbf{Some\ Dog})\ \mathbf{does\ not\ like}\ (\mathbf{Any\ Bones}))$$
$$= \sim((\mathbf{Any\ Dog})\ \mathbf{likes}\ (\mathbf{Some\ Bones}))$$

Any dog, that does not like any bones, would be a successful materialization of the above statement.

$$(Every\ Dog\ likes\ Some\ Bones) = ((\mathbf{Any\ Dog})\ \mathbf{likes}\ (\mathbf{Some\ Bones}))$$
$$= \sim(There\ is\ a\ Dog,\ that\ does\ not\ like\ Any\ Bones)$$
$$= \sim((\mathbf{Some\ Dog})\ \mathbf{does\ not\ like}\ (\mathbf{Any\ Bones}))$$

If all dogs like some bones, then the conjunction of all dogs would be the successful materialization of the above statement.

$$(Neither\ of\ Dogs\ likes\ Some\ Bones) = \sim(There\ is\ a\ Dog,\ that\ likes\ Some\ Bones)$$
$$= \sim((\mathbf{Some\ Dog})\ \mathbf{likes}\ (\mathbf{Some\ Bones}))$$
$$= ((\mathbf{Any\ Dog})\ \mathbf{does\ not\ like}\ (\mathbf{Any\ Bones}))$$

If all dogs do not like any bones, then the conjunction of all dogs would be the successful materialization of the above statement.

The same we can do with arbitrarily complex examples:

$$(\textit{Neither of Dogs wants All Bones to taste The Same})$$

$$= \sim(\textit{Some Dog wants it to be true, that} \sim(\textit{Some Bone does not taste The Same}))$$

$$\begin{pmatrix} \textit{Some Dog will chase Every Cat, that is Near To It,} \\ \textit{until Neither of Cats will remain There} \end{pmatrix}$$

$$= \begin{pmatrix} \sim(\textit{Any Dog will not chase Some Cat, that is Near To It}) \\ \textit{until} \sim(\textit{Some Cat will remain There}) \end{pmatrix}$$

10.13 Negation of a thing

We have natural general negation ("$\wr$" – vertical tilde) of any thing X:

$$\wr \boldsymbol{X} = \boldsymbol{Not\ X}$$

$$\wr\wr \boldsymbol{X} = \boldsymbol{X}$$

$$(\wr \boldsymbol{X})^{\uparrow} \subset (\wr \boldsymbol{X})^{\mho} = \sim(\boldsymbol{X}^{\mho}) = \sim\boldsymbol{X}^{\mho} = \Omega - \boldsymbol{X}^{\mho}$$

$$\boldsymbol{\Phi} = \wr\omega = \boldsymbol{\Phi}$$

So:

$$(\wr\omega)^{\uparrow} \subset (\wr\omega)^{\mho} = \sim(\omega^{\mho}) = \sim\Omega = \varnothing$$

So:

$$(\wr\boldsymbol{\omega})^{\uparrow} = \varnothing$$

For any things X and Y:

$$(\boldsymbol{X}^{\uparrow} = \boldsymbol{Y}^{\uparrow}) = (\boldsymbol{X} = \boldsymbol{Y}) = (\wr\boldsymbol{X} = \wr\boldsymbol{Y}) = \left((\wr\boldsymbol{X})^{\uparrow} = (\wr\boldsymbol{Y})^{\uparrow}\right)$$

And we have the following equalities for ω (Thing):

$$\omega^{\uparrow} = \Omega_{\rightarrow \textit{Concrete Thing}}$$

$$\omega^{\mho} = \Omega$$

$$\sim\omega^{\uparrow} = \Omega - \omega^{\uparrow} = \Omega - \Omega_{\rightarrow Concrete\ Thing} = \Omega_{\rightarrow Abstract\ Thing}$$

$$\sim\omega^{\circlearrowleft} = \sim\Omega = \varnothing$$

The rest of this section is for very advanced readers.

Here we will prove some of the above equalities.

If a sentence, that use the symbol "Φ", is a declarence, then it uses the symbol only as a quantifier. So we have the following equalities, that are not what you probably may think:

$$\begin{aligned}&(\Phi^{\uparrow} = \varnothing) = (\textit{the representation of the kind of nothing is equal to the empty set})\\&= (\varnothing = \Phi^{\uparrow}) = (\textit{the empty set is equal to the representation of the kind of nothing})\\&= \sim \bigvee_X (X^{\uparrow} = \varnothing) = \bigwedge_X (X^{\uparrow} \neq \varnothing) = \varnothing_{\varsigma} \equiv true\end{aligned}$$

The above equality is a tautology, because "$X^{\uparrow} \neq \varnothing$" is a tautology.

So:

$$true \equiv (\Phi^{\uparrow} = \varnothing) = \varnothing_{\varsigma} = (\varnothing = \Phi^{\uparrow}) \equiv true$$

And we have:

$$\begin{aligned}(\Phi^{\circlearrowleft} = \varnothing) &= (\textit{the exteneded representation of nothing is equal to the empty set})\\&= (\varnothing = \Phi^{\circlearrowleft})\\&= (\textit{the empty set is equal to the exteneded representation of nothing})\\&= \sim \bigvee_X (X^{\circlearrowleft} = \varnothing) = \bigwedge_X (X^{\circlearrowleft} \neq \varnothing) = \varnothing_{\varsigma} \equiv true\end{aligned}$$

The above equality is a tautology, because "$X^{\circlearrowleft} \neq \varnothing$" is a tautology.

So:

$$true \equiv (\Phi^{\circlearrowleft} = \varnothing) = \varnothing_{\varsigma} = (\varnothing = \Phi^{\circlearrowleft}) \equiv true$$

Analogously we can prove, that:

$$\sim\Phi^{\uparrow} = \sim\varnothing = \Omega = \sim\Phi^{\uparrow}$$

$$\sim\Phi^{\circlearrowleft} = \sim\varnothing = \Omega = \sim\Phi^{\circlearrowleft}$$

And we have for every X (remember that we cannot put the symbol "Φ" in place of any variable):

$$(X \neq \Phi) = (X \text{ is different from nothing}) = \sim \bigvee_{Y} (X \neq Y) = \bigwedge_{Y} (X = Y) = \sim\varnothing_{\varsigma}$$
$$\equiv \mathit{false}$$

But also:

$$(X = \Phi) = (X \text{ is equal to nothing}) = \sim \bigvee_{Y} (X = Y) = \bigwedge_{Y} (X \neq Y) = \sim\varnothing_{\varsigma} \equiv \mathit{false}$$

$$\sim(X = \Phi) = \sim(X \text{ is equal to nothing}) = \sim\sim \bigvee_{Y} (X = Y) = \bigvee_{Y} (X = Y) = \varnothing_{\varsigma} \equiv \mathit{true}$$

So for every X:

$$(X = \Phi) = (X \neq \Phi)$$
$$\sim(X = \Phi) \neq (X \neq \Phi)$$

And we have:

$$(\Phi = \Phi) = (\text{nothing equals nothing}) = (\text{everything does not equal not anything})$$
$$\equiv \mathit{true}$$

$$(\Phi \neq \Phi) = (\text{nothing does not equal nothing}) = (\text{everything equals not anything})$$
$$\equiv \mathit{true}$$

$$\sim(\Phi = \Phi) = \sim(\text{nothing equals nothing}) = (\text{something equals nothing})$$
$$= (\text{something does not equal anything}) \equiv \mathit{false}$$

Here is the proof:

$$(\Phi = \Phi) = \sim \bigvee_{X} \sim \bigvee_{Y} (X = Y) = \bigwedge_{X} \bigvee_{Y} (X = Y) = \bigwedge_{X} \sim \bigwedge_{Y} (X \neq Y) = \varnothing_{\varsigma} \equiv \mathit{true}$$

$$(\Phi \neq \Phi) = \sim \bigvee_{X} \sim \bigvee_{Y} (X \neq Y) = \bigwedge_{X} \bigvee_{Y} (X \neq Y) = \bigwedge_{X} \sim \bigwedge_{Y} (X = Y) = \varnothing_{\varsigma} \equiv \mathit{true}$$

$$\sim(\Phi = \Phi) = \sim\sim \bigvee_{X} \sim \bigvee_{Y} (X = Y) = \sim \bigwedge_{X} \bigvee_{Y} (X = Y) = \bigvee_{X} \bigwedge_{Y} (X \neq Y) = \sim\varnothing_{\varsigma} \equiv \mathit{false}$$

So also:

$$(\Phi = \Phi) = (\Phi \neq \Phi)$$
$$\sim(\Phi = \Phi) \neq (\Phi \neq \Phi)$$

So let as analyze all possible equalities:

$$(equality\ between\ anything\ and\ anything\ is\ correct\ equality)$$
$$= \bigwedge_X \bigwedge_Y ((X = Y)\ is\ correct\ equality) = \varnothing_\varsigma \equiv true$$

$$(equality\ between\ nothing\ and\ anything\ is\ not\ correct\ equality)$$
$$= \sim \bigvee_X \bigwedge_Y ((X = Y)\ is\ not\ correct\ equality)$$
$$= \bigwedge_X \bigvee_Y ((X = Y)\ is\ correct\ equality) = \varnothing_\varsigma \equiv true$$

$$(equality\ between\ anything\ and\ nothing\ is\ not\ correct\ equality)$$
$$= \bigwedge_X \sim \bigvee_Y ((X = Y)\ is\ not\ correct\ equality)$$
$$= \bigwedge_X \bigwedge_Y ((X = Y)\ is\ correct\ equality) = \varnothing_\varsigma \equiv true$$

$$(equality\ between\ nothing\ and\ nothing\ is\ correct\ equality)$$
$$= \sim \bigvee_X \sim \bigvee_Y ((X = Y)\ is\ correct\ equality)$$
$$= \bigwedge_X \bigvee_Y ((X = Y)\ is\ correct\ equality) = \varnothing_\varsigma \equiv true$$

$$(equality\ between\ nothing\ and\ nothing\ is\ not\ correct\ equality)$$
$$= \sim \bigvee_X \sim \bigvee_Y ((X = Y)\ is\ not\ correct\ equality)$$
$$= \sim \bigvee_X \bigwedge_Y ((X = Y)\ is\ correct\ equality) = \sim\varnothing_\varsigma \equiv false$$

In other words, only equalities between anything and anything and between nothing and nothing are correct equalities, but:

$$(\Phi = \Phi) = (\Phi \neq \Phi)$$
$$\sim(\Phi = \Phi) \neq (\Phi \neq \Phi)$$

So, as you can see, the negation of the equality is not inequality, so the definition of inequality is self-contrary, so this inequality is a nobeing, so it is not defined as a being, so it does not exist. Simply saying, the negation of the equality is not the appropriate inequality, because the word "nothing" used in it changes the meaning of equality, so it is no longer the equality of things, but has a decidable true statement, because its statement is true and the negation of its statement is false.

Summing up:

$$\Phi^{\uparrow} = \varnothing = (\wr\omega)^{\uparrow} = \varnothing = \Phi^{\uparrow}$$

And we have for any X:

$$\wr\wr X = X$$

So:

$$\omega = \wr\wr\omega = \wr(\textit{not a thing}) = \wr\Phi$$

We can check, that:

$$(\omega = \wr\Phi) = (\wr\Phi = \omega) = \sim\bigvee_X (\omega = \wr X) = \bigwedge_X (\omega \neq \wr X) = \varnothing_\varsigma \equiv \textit{true}$$

$$(\wr\omega = \Phi) = (\Phi = \wr\omega) = \sim\bigvee_X (\wr\omega = X) = \bigwedge_X (\wr\omega \neq X) = \varnothing_\varsigma \equiv \textit{true}$$

Although it is not an equality of things, it means, that we can always exchange these symbols in any context, but since these symbols are only modifiers (quantifiers in this case), they cannot be put in place of any variable and cannot be treated as an object (which includes subject) of a declarence.

And we have for any things X and Y except for ω:

$$(X^\uparrow = Y^\uparrow) = (X = Y) = (\wr X = \wr Y) = \left((\wr X)^\uparrow = (\wr Y)^\uparrow\right)$$

For only one variable equal to ω:

$$\sim\varnothing_\varsigma = (\wr X = \wr\omega = \Phi) = \left((\wr X)^\uparrow = (\wr\omega)^\uparrow = \varnothing\right)$$

$$(X^\uparrow = \omega^\uparrow) = (X = \omega) \Rightarrow (\wr X = \wr\omega) \Rightarrow (\wr\wr X = \wr\wr\omega = \omega) = (X = \omega)$$

So:

$$(X^\uparrow = \omega^\uparrow) = (X = \omega) = (\wr X = \wr\omega) = \sim\varnothing_\varsigma$$

So:

$$(X^\uparrow = \omega^\uparrow) = (X = \omega) = (\wr X = \wr\omega) = \left((\wr X)^\uparrow = (\wr\omega)^\uparrow\right)$$

For both variables equal to ω we have:

$$\varnothing_\varsigma = ((\wr\omega = \Phi) \textit{ and } (\Phi = \wr\omega)) = (\wr\omega = \Phi = \wr\omega) \Rightarrow (\wr\boldsymbol{\omega} = \wr\boldsymbol{\omega}) = \varnothing_\varsigma$$

$$\varnothing_\varsigma = \left(\left((\wr\omega)^\uparrow = \varnothing\right) \textit{ and } \left(\varnothing = (\wr\omega)^\uparrow\right)\right)$$
$$= \left((\wr\omega)^\uparrow = \varnothing = (\wr\omega)^\uparrow\right) \Rightarrow \left((\wr\boldsymbol{\omega})^\uparrow = (\wr\boldsymbol{\omega})^\uparrow\right) = \varnothing_\varsigma$$

Which can be proved also in the following simpler way:

$$\varnothing_\varsigma = (\Phi = \Phi) = (\wr\omega = \wr\omega)$$

$$\varnothing_\varsigma = (\varnothing = \varnothing) = \left((\wr\omega)^\uparrow = (\wr\omega)^\uparrow\right)$$

And we have:

$$\varnothing_\varsigma = (\omega = \omega) = \left(\omega^\uparrow = \omega^\uparrow\right)$$

Because "$X = X$" is a tautology and operation $X^\uparrow$ is invertible for every X.

So:

$$\left(\omega^\uparrow = \omega^\uparrow\right) = (\omega = \omega) = (\wr\omega = \wr\omega) = \left((\wr\omega)^\uparrow = (\wr\omega)^\uparrow\right)$$

So finally for any things X and Y:

$$\left(\boldsymbol{X}^\uparrow = \boldsymbol{Y}^\uparrow\right) = (\boldsymbol{X} = \boldsymbol{Y}) = (\wr\boldsymbol{X} = \wr\boldsymbol{Y}) = \left((\wr\boldsymbol{X})^\uparrow = (\wr\boldsymbol{Y})^\uparrow\right)$$

Q.E.D.

But, of course, it does not mean, that it is defined for nothing, which means, that it is not defined for anything, which is obviously false.

And we can verify, that:

$$true \equiv \varnothing_\varsigma = \left((\wr\omega)^\uparrow = \Phi^\uparrow = (\wr\wr\Phi)^\uparrow\right)$$

$$= (\omega = (not\ not\ a\ thing) = (not\ nothing) = \wr\Phi) = (\wr\omega = \Phi)$$

But remember, that $(\wr\omega \neq \Phi)$ has not a statement, because the symbol "$\neq$" is not meaningful.

And it does not mean, that the general negation of a thing is equal to something, because, oppositely, it means, that there is no result of such an operation on a thing.

And it does not mean, that nothing has defined general negation, which means, that everything has not defined general negation, which is obviously not true. And it does not mean, that nothing has not defined general negation, which means, that everything has defined general negation, which is false, because a thing has not defined general negation. In other words, $(\wr\omega = \Phi)$ means only, that there is no result of the general negation of a thing.

So remember, that the word "nothing" is only a quantifier (equal to *Not Some Thing*), so it has not an independent meaning, a definition and a class, so we do not consider "nothing" in formulas as allowable argument of a function and thereby also of a relation, for example, of the relation "*is/are*".

Remember also not to mistake the negation of a quantifier with the negation of a relation, e.g.:

$$(There\ is\ not\ anything) = \bigwedge_X (There\ is\ not\ X) \neq \sim \bigwedge_X (There\ is\ X)$$
$$= \big(There\ is\ (Not\ Anything)\big)$$

Where the word "not" in the left argument of the above inequality is a part of a symbol of relation *isNot*(X) and not a part of a symbol of quantifier *Not Anythnig* as it is in the right argument of the inequality.

10.14 Classes and classifications

A class is a/an:

a.) **Primary class** or **primary subclass** if and only if all its representatives are concrete things.
b.) **Classification** or **subclassification** of degree n if and only if all its representatives are subclasses of one classification of degree $n-1$, where a classification of degree 0 is a primary class and a classification of degree k of a classification of degree m is a classification of degree $k+m$.
c.) **Proper subclass** of other class if and only if it is a subclass of that class and it is not equal to this class.
d.) **Proper superclass** of other class if and only if it is a superclass of that class and it is not equal to this class.
e.) **Disjoint classification** or **disjoint subclassification** if and only if it is respectively a classification or a subclassification and representations of all its representatives are disjoint.
f.) **Disjoint to the core classification** or **disjoint to the core subclassification** if and only if it is respectively a classification or a subclassification and concretizations of all its representatives are disjoint.
g.) **Exhaustive classification** or **exhaustive subclassification** of degree n if and only if sum of representations of all its representatives is equal to representation of one classification of degree $n-1$.
h.) **Excellent classification** or **excellent subclassification** of degree n if and only if the classes of all its representatives are subclasses of one excellent classification of degree $n-1$, where an excellent classification is disjoint. As a consequence an excellent classification (/subclassification) is also disjoint to the core.
i.) **Perfect classification** of degree n if and only if the classes of all its representatives are subclasses of one perfect classification of degree $n-1$, where a perfect classification is exhaustive and disjoint. As a consequence a perfect classification (/subclassification) is also disjoint and exhaustive to the core.

Every classification is also a class and every subclassification is also a subclass, but not every class is a classification and not every subclass is a subclassification, because there are, for example, primary classes (/subclasses), that are not classifications (/subclassifications). All mixed kinds, that have concrete and abstract representatives, are classes, but not primary. A subclassification of some classification of some class is not a subclass of this class, but is only a subclass of this classification of this class.

And we have:

a.) a class (/classification) is a subclass (/subclassification),
b.) a subclass (/subclassification) is a class (/classification).

A class (/classification) is a subclass (/subclassification) and vice versa, because the representation of every class includes itself.

A classification of a classification of degree k of X is a classification of degree $k+1$ of X.

Remember, that a classification of degree k of a subclass of X is a classification of degree k of X and a subclassification of a classification of degree k of X is a classification of degree k of X.

10.15 Narrowing of a set

We have, that:

$$S_{\to K\#\mho} = \left(S \cap K^{\uparrow\#\mho}\right)$$

$$S_{\Rightarrow K\#\mho} = \left(S \cap K^{\Uparrow\#\mho}\right)$$

$$(X \in_{K\#\mho} S) = (X \in S_{\to K\#\mho})$$

$$\left(X \in_{\downarrow K\#\mho} S\right) = (X \in S_{\Rightarrow K\#\mho})$$

$$S_{\to\#\mho} = S_{\to\Omega_{\downarrow}\#\mho}$$

where $C_{\downarrow}$ is the thing, the class of which is the class of an element of set C for any set C. It will be extended and explained in more details further in this book:

$$(X \in_{\#\mho} S) = (X \in S_{\to\#\mho})$$

$$S_{\to\#K^{\uparrow}} = S_{\to K}$$

$$(X \in_{K\#\mho} S_{\Rightarrow}) = (X \in S_{\Rightarrow K\#\mho})$$

$$(X\ isBp \backslash areBp\ K) = (X \in_K \Omega)$$

For any set C:

$$C_{\to K\#S[m]} = \left\{ X\colon \bigvee_{i=0}^{m} \left(X \in_{K\#S} C^{\to[i]} \right) \right\}$$

$$\left(X \in_{K\#\mho[m]} C \right) = \left(X \in C_{\to K\#\mho[m]} \right)$$

$$C_{\to K\#S} = C_{\to K\#S[0]}$$

$$C_{\rightrightarrows K\#S} = C_{\to K\#S[\infty]}$$

$$C_{\Rightarrow K\#S} = C_{\Rightarrow K\#S[0]}$$

$$C_{\Rrightarrow K\#S} = C_{\Rightarrow K\#S[\infty]}$$

$$C_{\to} = C_{\to \Omega_{\downarrow}} = C$$

$$C_{\rightrightarrows} = C_{\rightrightarrows \Omega_{\downarrow}} = C^{\to[\infty]}$$

10.16 The thing, the kind of which has given representation

For any set C:

$$C_{\downarrow} = (\textit{the thing, the kind of which has representation } C)$$
$$= (\textit{the thing, for the class of which only all elements of set } C \textit{ are its members})$$

So operator $X_{\downarrow}$ is the inversion of operator $X^{\uparrow}$ and operator $X_{\circlearrowleft}$ is the inversion of operator $X^{\circlearrowleft}$. $X_{\downarrow}$ is defined for every set, but not for every set $X_{\circlearrowleft}$ is defined, because $X_{\circlearrowleft}$ is defined only for the sets, that are equal to $Y^{\circlearrowleft}$ for some Y, and then $X_{\circlearrowleft} = (Y^{\circlearrowleft})_{\circlearrowright} = Y$.

And we have:

$$C_{\downarrow K\#S[m]} = \left(C_{\to K\#S[m]} \right)_{\downarrow}$$

$$C_{\downarrow} = (C_{\to})_{\downarrow} = \left(C_{\to \Omega_{\downarrow}} \right)_{\downarrow}$$

$$\varnothing_{\circlearrowright} = \Phi = \wr\omega$$

$$\Omega_{\circlearrowright} = \omega = \wr\Phi$$

$$X^{\uparrow}_{\to K\#S[l]} = \left(X^{\uparrow} \right)_{\to K\#S[l]}$$

$$X^{\uparrow}_{\downarrow K\#S[l]} = (X^{\uparrow})_{\downarrow K\#S[l]}$$

$$X^{\rightarrow[n]}_{\rightarrow K\#S[l]} = \left(X^{\rightarrow[n]}\right)_{\rightarrow K\#S[l]}$$

$$X^{\rightarrow[n]}_{\downarrow K\#S[l]} = \left(X^{\rightarrow[n]}\right)_{\downarrow K\#S[l]}$$

$$X^{\rightarrow C}_{\rightarrow K\#S[l]} = (X^{\rightarrow C})_{\rightarrow K\#S[l]}$$

$$X^{\rightarrow C}_{\downarrow K\#S[l]} = (X^{\rightarrow C})_{\downarrow K\#S[l]}$$

$$X^{\uparrow}_{\downarrow X} = X$$

$$X^{\uparrow}_{\downarrow} = (X^{\uparrow})_{\downarrow} = X$$

$$X^{\uparrow}_{\rightarrow X} = X^{\uparrow}$$

10.17 More about abstract and concrete things

And we have, that:

$$(X\ is \backslash are\ (a)\ concrete\ thing(s)\ and\ not\ set\ and\ (n \in N)\ and\ (n > 0)) = \left(X^{\rightarrow n} = \{X\}\right)$$

$$(X\ is \backslash are\ (an)\ abstract\ thing\ (s)) = \sim (X\ is \backslash are\ (a)\ concrete\ thing\ (s))$$
$$= (X\ is \backslash are\ not\ (a)\ concrete\ thing\ (s))$$

A thing X is a concrete thing, if its class has only one member (e.g., 1, Current Set of All Natural Numbers), and an abstract thing otherwise (Number One, Natural Number, Set of Ten Natural Numbers).

A concrete thing has also its kind and is the only one representative of its kind, so the representation of its kind has exactly one element – this thing itself.

So:

$$((X\ is\ a\ concrete\ thing)\ and\ (A\ isA\ X)) \Rightarrow (A = X) = ((A\ isA\ X)\ and\ (X\ isA\ A))$$

And we have for any X:

$$\_(\{X\})^{\rightarrow} = \{\_(\{X\})\}$$

$$\_deep\_(\{X\}) = \_(\{\_deep\_(X)\})$$

10.18 Transitivity of relation "*isp/arep*" – rules of deduction

Here you will find some rules of deduction, that use conditional transitivity of relation *isp\arep* in natural language and are similar to syllogisms, but more general from the special case of them, because (*A isp B*) is more general relation than (*every A is B*).

Relation *is* seems to be transitive, if we treat it the following incorrect way:

$$(a\,X\ is\ a\,Y) = \mathbf{is}(\boldsymbol{a}\,\boldsymbol{X}, \boldsymbol{a}\,\boldsymbol{Y}) = is\begin{pmatrix} Any\ thing\ consistently\ meant\ to\ be\ concrete\ X, \\ Some\ thing\ consistently\ meant\ to\ be\ concrete\ Y \end{pmatrix}$$

$$((the)\ X\ is\ a\ Y) = \mathbf{is}((\boldsymbol{the})\ \boldsymbol{X}, \boldsymbol{a}\ \boldsymbol{Y}) = is(X, Some^{\uparrow}\ Y)$$

But relation $is\begin{pmatrix} Any\ thing\ consistently\ meant\ to\ be\ concrete\ X, \\ Some\ thing\ consistently\ meant\ to\ be\ concrete\ Y \end{pmatrix}$ is not the same as *is*(*X*, *Some Y*), so relation *is* is not transitive.

In reality we have only the following rules of transitivity:

First of all some definitions:

$$\begin{aligned}(Any\ X\ is\backslash are\ Some\ Y) &= (X^{\Uparrow} \subset Y^{\Uparrow}) \Leftarrow (a\ X\ is\ a\ Y\backslash X\ are\ Y) = (X\ isA\backslash areA\ Y) \\ &= (X^{\uparrow} \subset Y^{\uparrow})\end{aligned}$$

$$\begin{aligned}((the)\ X\ is\backslash are\ Some\ Y) &= ((the)\ X\ is\backslash are\ a\ Y) = (X\ isB\backslash areB\ Y) \\ &= (X \in Y^{\Uparrow}) \Rightarrow (X\ isBp/areBp\ Y) = (\$X\ isA\backslash areA\ Y) \\ &= \left((\$X)^{\uparrow} \subset Y^{\uparrow}\right) = (X \in Y^{\uparrow})\end{aligned}$$

$$(X\ isp\backslash arep\ Y) = ((X\ isBp\backslash areBp\ Y)\ or\ (X\ isA\backslash areA\ Y)) = \left((X \in Y^{\uparrow})\ or\ (X^{\uparrow} \subset Y^{\uparrow})\right)$$

And now the rules:

$$((X\ isA\backslash areA\ Y)\ and\ (Y\ isA\backslash areA\ Z)) \Rightarrow (X\ isA\backslash areA\ Z)$$

$$((X\ isBp\backslash areBp\ Y)\ and\ (Y\ isA\backslash areA\ Z)) \Rightarrow (X\ isBp\backslash areBp\ Z)$$

But in many natural languages we do not distinguish these things (e.g. in Polish: (Pies jest gatunkiem) = ((the) Dog is a species), (Pies jest zwierzęciem) = (A dog is an animal)).

We will use the transitivity of the relation of the inclusion of sets to prove some special cases of the transitivity of relation *isp\arep* in natural language.

In natural language we have the following definition of relation *isp\arep*:

$$\begin{aligned}&(\boldsymbol{A}\ \mathbf{isp}\backslash\mathbf{arep}\ \boldsymbol{B}) \\ &= ((\boldsymbol{A\ is\ a\ member\ of\ the\ class\ of\ B})\ \mathbf{or}\ (\boldsymbol{the\ class\ of\ A\ is\ a\ subclass\ of\ B})) \\ &= (\boldsymbol{A} \in \boldsymbol{B}^{\circlearrowleft}) = \left((\boldsymbol{A} \in \boldsymbol{B}^{\uparrow})\ \mathbf{or}\ (\boldsymbol{A}^{\uparrow} \subset \boldsymbol{B}^{\uparrow})\right)\end{aligned}$$

where for any X:

$$(\textit{the class of A is a subclass of B}) = \left(\left(X \in A^{\uparrow}\right) \Rightarrow \left(X \in B^{\uparrow}\right)\right) = \left(A^{\uparrow} \subset B^{\uparrow}\right) \Rightarrow (A\ isp\backslash arep\ B)$$

For example: "a Rottweiler is a Dog", "a Dog is an Animal", "a Rottweiler is an Animal", "Hunting Dogs are Dogs", "a Hunting Dog is a Dog".

And:

$$(\textit{A is a member of the class of B}) = \left(A \in B^{\uparrow}\right) \Rightarrow (A\ isp\backslash arep\ B)$$

For example: "the Rottweiler is a Breed", "the Dog is a Species", "the Breed is a Classification of Living Being of Given Species".

For any concrete thing A:

$$\left(A \in B^{\uparrow}\right) = \left(A^{\uparrow} \subset B^{\uparrow}\right)$$

For example: My Dog Ace is a Dog.

And due to transitive property of set inclusion for any things A, B and C:

$$\left(\left(A^{\uparrow} \subset B^{\uparrow}\right)\ and\ \left(B^{\uparrow} \subset C^{\uparrow}\right)\right) \Rightarrow \left(A^{\uparrow} \subset C^{\uparrow}\right) \Rightarrow (A\ isp\backslash arep\ C)$$

Which is the generalization of the most popularized syllogism: if every A is B, and every B is C, then every A is C.

For example: "a Rottweiler is a Dog and a Dog is an Animal, so a Rottweiler is an Animal".

For any things A, B and C:

$$\left(\left(A \in B^{\uparrow}\right)\ and\ \left(B^{\uparrow} \subset C^{\uparrow}\right)\right) \Rightarrow \left(A \in C^{\uparrow}\right) = (\textit{A is a member of the class of C}) \Rightarrow (A\ isp\backslash arep\ C)$$

For example: "the Dog is a Species, and the a Species is a Classification of Animal, so the Dog is a Classification of Animal".

These are the only two possibilities, because $A\ isp\backslash arep\ B$ only, when:

a.) The class of A is a subclass of the not narrower class of B, e.g.: a Dog is an Animal, My Dog is a Dog, etc.

b.) A is a member of classification B of a thing, the class of which is not narrower than the class of A, e.g.: the Dog is a Species of Animal (where Species of Animal is the classification of Animal, the class of which is not narrower than the class of Dog), the Rottweiler is a Breed of Dog, etc.

So $A_1\ isp\backslash arep\ A_n$ in the chain of relations $(A_i\ isp\backslash arep\ A_{i+1})$ (assuming $A_i \neq A_{i+1}$) due to transitivity of inclusion only if you do not go out to option b.) from going only by option a.) except for the first step, that determines direction, because otherwise the class of A_1 is at most a proper subclass of element of the final class of A_n, so we cannot prove using only transitivity of inclusion of representations, that the class of A_1 is a subclass of A_n or A_1 is a member of the class of A_n.

Relation $A\ isp\backslash arep\ B$ is true if and only if the class of A is a subclass of B or A is a member of classification $B = Classification\ X\ of\ Y$, but then it is not a subclass of the class of B, but a subclass of Y. If we use step b.) after first relation $A_1\ isp\backslash arep\ A_2$, then we cannot prove using only transitivity of inclusion, that A_1 would be a member of the class of A_n or the class of A_1 would be a subclass of A_n because it would be at most a proper subclass of the class of a member of the class of A_n or a proper subclass of the class of a member of the class of a member of the class of A_n and so on.

Q.E.D.

So finally:

$$\left(\left((\boldsymbol{A}^{\uparrow} \subset \boldsymbol{B}^{\uparrow})\ \boldsymbol{and}\ (\boldsymbol{B}^{\uparrow} \subset \boldsymbol{C}^{\uparrow})\right)\ \boldsymbol{or}\ \left((\boldsymbol{A} \in \boldsymbol{B}^{\uparrow})\ \boldsymbol{and}\ (\boldsymbol{B}^{\uparrow} \subset \boldsymbol{C}^{\uparrow})\right)\right)$$

$$\Rightarrow \left((\boldsymbol{A} \in \boldsymbol{C}^{\uparrow})\ \boldsymbol{or}\ (\boldsymbol{A}^{\uparrow} \subset \boldsymbol{C}^{\uparrow})\right) = (\boldsymbol{A}\ \boldsymbol{isp}\backslash \boldsymbol{arep}\ \boldsymbol{C})$$

And as can be easily seen there are possible only two sequences of relations $(A_i\ isp\backslash arep\ A_{i+1})$, that for any tuple A of things implicate $A_1\ isp\backslash arep\ A_n$:

$$\left(\left(\boldsymbol{A}_1 \in \boldsymbol{A}_2^{\uparrow}\right)\ \boldsymbol{and}\ \left(\boldsymbol{A}_2^{\uparrow} \subset \boldsymbol{A}_3^{\uparrow}\right)\ \boldsymbol{and}\ \ldots\ \boldsymbol{and}\ \left(\boldsymbol{A}_{n-1}^{\uparrow} \subset \boldsymbol{A}_n^{\uparrow}\right)\right) \Rightarrow (\boldsymbol{A}_1\ \boldsymbol{isp}\backslash \boldsymbol{arep}\ \boldsymbol{A}_n)$$

$$\left(\left(\boldsymbol{A}_1^{\uparrow} \subset \boldsymbol{A}_3^{\uparrow}\right)\ \boldsymbol{and}\ \left(\boldsymbol{A}_2^{\uparrow} \subset \boldsymbol{A}_3^{\uparrow}\right)\ \boldsymbol{and} \ldots \boldsymbol{and}\ \left(\boldsymbol{A}_{n-1}^{\uparrow} \subset \boldsymbol{A}_n^{\uparrow}\right)\right) \Rightarrow (\boldsymbol{A}_1\ \boldsymbol{isp}\backslash \boldsymbol{arep}\ \boldsymbol{A}_n)$$

So relation $isp(A, B)\backslash arep(A, B)$ is transitive, for example, under described above conditions and it is not transitive in general.

You can extend this also to every other relation $doSomething(A)$ using appropriate things.

Then, for example, $doSomething(A) = isp(A,\ Something\ that\ Does\ Something)$, e.g.:

A dancer dances sometimes and who dances sometimes exercises sometimes his body.

A Dancer is Who Dances Sometimes and Who Dances Sometimes is Who Exercises Sometimes His Body, so A Dancer is Who Exercises Sometimes His Body.

10.19 Relation "is/are" in any tense

You can use the rules of deduction for relation "is/are", that are described in this book, in any tense you want: was/were, has/have/had been, is, will be, e.g. (assuming the same time of actions in given tense):

((Mary **was** Patient) and (A Patient Person **was** Who Waited As Long As Necessary)) implicates (Mary **was** Who Waited As Long As Necessary)

((A Worker of Our Company **has been** A Person Sleeping in This Hotel) and (A Person Sleeping in This Hotel **has been** A Brain Worker)) implicates (A Worker of Our Company **has been** A Brain Worker)

((You **are** Brave) and (A Brave Person **is** Following Mythical Heroes' Example)) implicates (You **are** Following Mythical Heroes' Example)

((He **will be** Strong) and (A Strong Person **will be** Able to Beat the Tiger)) implicates (He **will be** Able to Beat the Tiger)

And for combinations of times, e.g.:

(He **is** Someone Who Passed the Exams) and (Anyone Who Passed the Exams **will be** Someone Who Will Have More Free Time) implicates (He **will be** Someone Who Will Have More Free Time)

But, for example:

((Johnny was Tired) and (Tired was Adjective)) does not implicate (Johnny was Adjective)

If at some past (was/were), perfect (has/have/had been), present (am, are, is), future (will be) time *t*:

$$\big((A\ is\backslash are\ B)\ and\ (B\ is\backslash are\ C)\big) \Rightarrow (A\ is\backslash are\ C)$$

Then we have respectively the following deduction rules:

$$\begin{pmatrix} (A\ at\ time\ t\ was\backslash were\backslash is\ (always)\ B) \\ and\ (B\ at\ time\ t\ was\backslash were\backslash is\ (always)\ C) \end{pmatrix} \Rightarrow (A\ at\ time\ t\ was\backslash were\ C)$$

$$\begin{pmatrix} (A\ at\ time\ t\ had\ been\backslash is\ (always)\ B) \\ and\ (B\ at\ time\ t\ had\ been\backslash is\ (always)\ C) \end{pmatrix} \Rightarrow (A\ at\ time\ t\ had\ been\ C)$$

$$\begin{pmatrix} (A\ at\ time\ t\ has\ been\backslash have\ been\backslash is\ (always)\ B) \\ and\ (B\ at\ time\ t\ has\ been\backslash have\ been\backslash is\ (always)\ C) \end{pmatrix}$$
$$\Rightarrow (A\ at\ time\ t\ has\ been\backslash have\ been\ C)$$

$$\begin{pmatrix} (A\ at\ time\ t\ will\ be\backslash is\ (always)\ B) \\ and\ (B\ at\ time\ t\ will\ be\backslash is\ (always)\ C) \end{pmatrix} \Rightarrow (A\ at\ time\ t\ will\ be\ C)$$

Under the condition, that X exists now, you can simplify it all to relation "is/are" using the following equality:

$$(X\ was\backslash were\backslash has\ been\backslash have\ been\backslash had\ been\backslash will\ be\ Y)$$
$$= (X\ is\backslash are\ (a)\ Thing(s)\ that\ was\backslash were\backslash has\ been\backslash have\ been\backslash had\ been\backslash will\ be\ Y)$$

11 New logic – extension

11.1 Introduction

This section contains the logical interpretation of the rest of sentences in natural language.

$R, S, T, \ldots$ are relations.

$A, B, C, \ldots$ are objects, that are subjects, when they are the first argument of a relation. An implied subject is expressed as "_".

$X, Y, Z, \ldots$ are attributes.

11.2 Negation of disjunctions and conjunctions

$$\begin{aligned}(A\,\text{is not}\,(X\,or\,Y)) &= \sim(A\,\text{is}\,(X\,or\,Y)) = \sim(A\,\text{is}\,X\,or\,A\,\text{is}\,Y)\\ &= (\sim(A\,\text{is}\,X)\,and\sim(A\,\text{is}\,Y)) = ((A\,\text{is not}\,X)\,and\,(A\,\text{is not}\,Y))\\ &= (A\,\text{is not}\,X\,and\,\text{is not}\,Y)\end{aligned}$$

$$\begin{aligned}(A\,does\,not\,(R\,X\,or\,S\,Y)) &= \sim(A\,R\,X\,or\,S\,Y) = \sim(A\,R\,X\,or\,A\,S\,Y) = \sim(R(A,X)\,or\,S(A,Y))\\ &= (\sim R(A,X)\,and\sim S(A,Y)) = ((A\,!\,R\,X)\,and\,(A\,!\,S\,Y))\\ &= (A\,!\,R\,X\,and\,!\,S\,Y)\end{aligned}$$

$$\begin{aligned}(A\,\text{is not}\,(B\,or\,C)) &= \sim(A\,\text{is}\,(B\,or\,C)) = \sim(A\,\text{is}\,B\,or\,A\,\text{is}\,C)\\ &= (\sim(A\,\text{is}\,B)\,and\sim(A\,\text{is}\,C)) = ((A\,\text{is not}\,B)\,and\,(A\,\text{is not}\,C))\\ &= (A\,\text{is not}\,B\,and\,\text{is not}\,C)\end{aligned}$$

$$\begin{aligned}(A\,does\,not\,(R\,B\,or\,S\,C)) &= \sim(A\,R\,B\,or\,S\,C) = \sim(A\,R\,B\,or\,A\,S\,C) = \sim(R(A,B)\,or\,S(A,C))\\ &= (\sim R(A,B)\,and\sim S(A,C)) = ((A\,!\,R\,B)\,and\,(A\,!\,S\,C))\\ &= (A\,!\,R\,B\,and\,!\,S\,C)\end{aligned}$$

For example: ^"Mary does not sing a song or dance on the floor" = ^"Mary does not sing a song and does not dance on the floor."

$$\begin{aligned}(A\,\text{is not}\,(X\,and\,Y)) &= \sim(A\,\text{is}\,(X\,and\,Y)) = \sim(A\,\text{is}\,X\,and\,A\,\text{is}\,Y)\\ &= (\sim(A\,\text{is}\,X)\,or\sim(A\,\text{is}\,Y)) = ((A\,\text{is not}\,X)\,or\,(A\,\text{is not}\,Y))\\ &= (A\,\text{is not}\,X\,or\,\text{is not}\,Y)\end{aligned}$$

https://doi.org/10.1515/9783111441382-011

$$(A\, does\, not\, (R\,X\, and\, S\,Y)) = \sim(A\,R\,X\, and\, S\,Y) = \sim(A\,R\,X\, and\, A\,S\,Y)$$
$$= \sim(R(A,X)\, and\, S(A,Y)) = (\sim R(A,X)\, or \sim S(A,Y))$$
$$= ((A\,!\,R\,X)\, or\, (A\,!\,S\,Y)) = (A\,!\,R\,X\, or\, !\,S\,Y)$$

$$(A\ is\ not\ (B\ and\ C)) = \sim(A\ is\ (B\ and\ C)) = \sim(A\ is\ B\ and\ A\ is\ C)$$
$$= (\sim(A\ is\ B)\ or\ \sim(A\ is\ C)) = ((A\ is\ not\ B)\ or\ (A\ is\ not\ C))$$
$$= (A\ is\ not\ B\ or\ is\ not\ C)$$

$$(A\, does\, not\, (R\ B\ and\ S\ C)) = \sim(A\ R\ B\ and\ S\ C) = \sim(A\ R\ B\ and\ A\ S\ C)$$
$$= \sim(R(A,B)\ and\ S(A,C)) = (\sim R(A,B)\ or\ \sim S(A,C))$$
$$= ((A\,!\,R\ B)\ or\ (A\,!\,S\ C)) = (A\,!\,R\ B\ or\,!\,S\ C)$$

For example: ^"Mary does not sing a song and dance on the floor" = ^"Mary does not sing a song or does not dance on the floor."

11.3 Something is so . . . as . . .

$$(A\ is\ so\backslash as\backslash the\ same\ X\ as\ B\ is\ Y) = (degree\ of\ A\ being\ X = degree\ of\ B\ being\ Y)$$
$$(A\ is\ so\backslash as\backslash the\ same\ X\ as\ B) = (degree\ of\ A\ being\ X = degree\ of\ B\ being\ X)$$
$$(A\ is\ so\backslash as\backslash the\ same\ X\ as\ Y) = (degree\ of\ A\ being\ X = degree\ of\ A\ being\ Y)$$

$$(A\ is\ not\ so\backslash as\ X\ as\ B\ is\ Y) = (A\ is\ less\ X\ than\ B\ is\ Y) = (B\ is\ more\ Y\ than\ A\ is\ X)$$
$$= (degree\ of\ A\ being\ X < degree\ of\ B\ being\ Y)$$
$$(A\ is\ not\ so\backslash as\ X\ as\ B) = (A\ is\ less\ X\ than\ B) = (B\ is\ more\ X\ than\ A)$$
$$= (degree\ of\ A\ being\ X < degree\ of\ B\ being\ X)$$
$$(A\ is\ not\ so\backslash as\ X\ as\ Y) = (A\ is\ less\ X\ than\ Y) = (A\ is\ more\ Y\ than\ X)$$
$$= (degree\ of\ A\ being\ X < degree\ of\ A\ being\ Y)$$

$$(A\ is\ not\ the\ same\ X\ as\ B\ is\ Y) = (degree\ of\ A\ being\ X \neq degree\ of\ B\ being\ Y)$$
$$(A\ is\ not\ the\ same\ X\ as\ B) = (degree\ of\ A\ being\ X \neq degree\ of\ B\ being\ X)$$
$$(A\ is\ not\ the\ same\ X\ as\ Y) = (degree\ of\ A\ being\ X \neq degree\ of\ A\ being\ Y)$$

For example: "He is so brave as a tiger", "He is not so brave as a grizzle bear."

Important note: to simplify rules everywhere below any relation will have two arguments: group of subjects, objects/attributes, e.g.:

$$relation(group\ of\ subjects, objects \backslash attributes)$$

$$read(she, howAndWhat(carefully, some\ book)) = (She\ carefully\ reads\ some\ book)$$

$$went\ (and\ (Mary, Jane), fromWhereToWhereAndWhen\ (home,\ cinema,\ yesterday))$$
$$= (Mary\ and\ Jane\ went\ yesterday\ from\ home\ to\ cinema)$$

And more generally:

$$(A\ R\ so \backslash as \backslash the\ same\ X\ as\ B\ S\ Y) = (degree\ of\ R(A,X) = degree\ of\ S(B,Y))$$

$$(A\ R\ so \backslash as \backslash the\ same\ X\ as\ B) = (degree\ of\ R(A,X) = degree\ of\ R(B,X))$$

$$(A\ R\ so \backslash as \backslash the\ same\ X\ as\ Y) = (degree\ of\ R(A,X) = degree\ of\ R(A,Y))$$

$$(A\ R\ so \backslash as \backslash the\ same\ as\ B\ S\ Y) = (degree\ of\ R(A) = degree\ of\ S(B,Y))$$

$$(A\ R\ so \backslash as \backslash the\ same\ as\ B) = (degree\ of\ R(A) = degree\ of\ R(B))$$

$$(A\ {\sim}R\ so \backslash as\ X\ as\ B\ S\ Y) = (degree\ of\ R(A,X) < degree\ of\ S(B,Y))$$

$$(A\ {\sim}R\ so \backslash as\ X\ as\ B) = (degree\ of\ R(A,X) < degree\ of\ R(B,X))$$

$$(A\ {\sim}R\ so \backslash as\ X\ as\ Y) = (degree\ of\ R(A,X) < degree\ of\ R(A,Y))$$

$$(A\ {\sim}R\ so \backslash as\ B\ S\ Y) = (degree\ of\ R(A) < degree\ of\ S(B,Y))$$

$$(A\ {\sim}R\ so \backslash as\ B) = (degree\ of\ R(A) < degree\ of\ R(B))$$

$$(A\ {\sim}R\ the\ same\ X\ as\ B\ S\ Y) = (degree\ of\ R(A,X) \neq degree\ of\ S(B,Y))$$

$$(A\ {\sim}R\ the\ same\ X\ as\ B) = (degree\ of\ R(A,X) \neq degree\ of\ R(B,X))$$

$$(A\ {\sim}R\ the\ same\ X\ as\ Y) = (degree\ of\ R(A,X) \neq degree\ of\ R(A,Y))$$

$$(A\ {\sim}R\ the\ same\ as\ B\ S\ Y) = (degree\ of\ R(A) \neq degree\ of\ S(B,Y))$$

$$(A\ {\sim}R\ the\ same\ as\ B) = (degree\ of\ R(A) \neq degree\ of\ R(B))$$

Where it might be, that $R = S$.

For example: “She runs as fast as a wolf”, “She does not run as fast as a panther”, and “She does not run as fast as a train moves”.

11.4 Something instead of something

In general:

$$\sim(R(A,B)\ \textit{instead of}\ \ldots) = (\sim R(A,B)\ \textit{instead of}\ \ldots)$$

$$\sim(R(A,X)\ \textit{instead of}\ \ldots) = (\sim R(A,X)\ \textit{instead of}\ \ldots)$$

$$\sim(R(A)\ \textit{instead of}\ \ldots) = (\sim R(A)\ \textit{instead of}\ \ldots)$$

And we have the following rules:

$$(R(A,B)\ \textit{instead of}\ C) = (R(A,B)\ \textit{instead of}\ R(\_,C)) = (R(A,B)\ \textit{instead of}\ R(A,C))$$
$$= (\textit{It was expected that}\ R(A,C),\ \textit{but}\ R(A,B)\ \textit{and} \sim R(A,C))$$

Where "_" in $R(\_,C)$ is omitted subject, that is already known in given sentence or in the context of given sentence.

For example: ^"I drink tea instead of coffee" = ^"I drink tea instead of drinking coffee" = ^"I drink tea and I do not drink coffee."

$$\textit{It was expected that drink}(I,\textit{coffee}),\textit{but drink}(I,\textit{tea})\ \textit{and} \sim\textit{drink}(I,\textit{coffee})$$

"She has read pages from 100 to 200 of the schoolbook instead of from 200 to 300."

$$\textit{It was expected that haveReadPagesOfTheSchoolbook}(\textit{she},\ \textit{fromTo}(200,300)),$$
$$\textit{but haveReadPagesOfTheSchoolbook}(\textit{she},\ \textit{fromTo}(100,200))$$
$$\textit{and} \sim\textit{haveReadPagesOfTheSchoolbook}(\textit{she},\textit{fromTo}(200,300))$$

$$(R(A,X)\ \textit{instead of}\ Y) = (R(A,X)\ \textit{instead of}\ R(\_,Y)) = (R(A,X)\ \textit{instead of}\ R(A,Y))$$
$$= (\textit{It was expected that}\ R(A,Y),\ \textit{but}\ R(A,X)\ \textit{and} \sim R(A,Y))$$

For example: ^"This dog is wild instead of tame" = ^"This dog is wild instead of being tame" = ^"This dog is wild and is not tame" = ^"This dog is wild and this dog is not tame."

$$\textit{It was expected that is}(\textit{this dog},\ \textit{tame}),\ \textit{but is}(\textit{this dog},\ \textit{wild})\textit{and} \sim\textit{is}(\textit{this dog},\ \textit{tame})$$

$$(R(A,B)\ \textit{instead of}\ C) = (R(A,B)\ \textit{instead of}\ R(C,\_)) = (R(A,B)\ \textit{instead of}\ R(C,B))$$
$$= (\textit{It was expected that}\ R(C,B),\ \textit{but}\ R(A,B)\ \textit{and} \sim R(C,B))$$

For example: "I will do it instead of him."

$$\textit{It was expected that willDo}(\textit{he},\textit{it}),\ \textit{but willDo}(I,\textit{it})\ \textit{and} \sim\textit{willDo}(\textit{he},\textit{it})$$

$$(R(A,X)\ instead\ of\ B) = (R(A,X)\ instead\ of\ R(B,\_)) = (R(A,X)\ instead\ of\ R(B,X))$$
$$= (It\ was\ expected\ that\ R(B,X),\ but\ R(A,X)\ and \sim R(B,X))$$

For example: "A rose will be better for her instead of a tulip."

$$It\ was\ expected\ that\ willBeForHer(tulip, better),$$
$$but\ willBeForHer(rose, better)\ and \sim willBeForHer(tulip, better)$$

$$(R(A,B)\ instead\ of\ S(\_,C)) = (R(A,B)\ instead\ of\ S(A,C))$$
$$= (It\ was\ expected\ that\ S(A,C),\ but\ R(A,B)\ and \sim S(A,C))$$

$$(R(A,X)\ instead\ of\ S(\_,Y)) = (R(A,X)\ instead\ of\ S(A,Y))$$
$$= (It\ was\ expected\ that\ S(A,Y),\ but\ R(A,X)\ and \sim S(A,Y))$$

For example: "She dances with me instead of dancing with him."

$$It\ was\ expected\ that\ danceWith(she, him),$$
$$but\ danceWith(she, me)\ and \sim danceWith(she, him)$$

Where *B*, *C*, *X*, *Y* could be equal to nothing:

$$(R(A,B)\ instead\ of\ S(\_)) = (R(A,B)\ instead\ of\ S(A))$$
$$= (It\ was\ expected\ that\ S(A),\ but\ R(A,B)\ and \sim S(A))$$

$$(R(A,X)\ instead\ of\ S(\_)) = (R(A,X)\ instead\ of\ S(A))$$
$$= (It\ was\ expected\ that\ S(A),\ but\ R(A,X)\ and \sim S(A))$$

$$(R(A)\ instead\ of\ S(\_,C)) = (R(A)\ instead\ of\ S(A,C))$$
$$= (It\ was\ expected\ that\ S(A,C),\ but\ R(A)\ and \sim S(A,C))$$

$$(R(A)\ instead\ of\ S(\_,Y)) = (R(A)\ instead\ of\ S(A,Y))$$
$$= (It\ was\ expected\ that\ S(A,Y),\ but\ R(A)\ and \sim S(A,Y))$$

$$(R(A)\ instead\ of\ S(\_)) = (R(A)\ instead\ of\ S(A))$$
$$= (It\ was\ expected\ that\ S(A),\ but\ R(A)\ and \sim S(A))$$

For example: "She dances instead of singing."

$$It\ was\ expected\ that\ sing(she), but\ dance(she)\ and \sim sing(she)$$

11.5 Relation “can”

$$\big(R(A) \Rightarrow S(B)\big) = \Big(\big(A\, can\, R(\_)\big) \Rightarrow \big(B\, can\, S(\_)\big)\Big)$$

So:

$$\big(R(A) = S(B)\big) = \Big(\big(A\, can\, R(\_)\big) = \big(B\, can\, S(\_)\big)\Big)$$

And we have:

$$(A\, can\, be\, B) = (B\, can\, be\, A) = \big((A^{\uparrow} \cap B^{\uparrow}) \neq \varnothing\big)$$

$$(A\, is \backslash are\, B) \Rightarrow (A\, can\, be\, B)$$

11.6 Which/that

11.6.1 Any

For any $\#Q$:

$$\begin{aligned}\big(R(A, Any\, B)\, which\backslash that\, S(\_)\big) &= \Big(R(A, \#Q)\, when\, only\, \big(is(\#Q, B)\, and\, S(\#Q)\big)\Big)\\ &= \bigwedge_{is(\#R,B)\, and\, S(\#R)} R(A, \#R)\\ &= \Big(\big(is(\#Q, B)\, and\, S(\#Q)\big) \Rightarrow R(A, \#Q)\Big)\\ &= \bigwedge_{is(\#R,B)} \bigwedge_{S(\#R)} R(A, \#R) = \bigwedge_{is(\#R,B)} \big(S(\#R) \Rightarrow R(A, \#R)\big)\\ &= \Big(\bigvee_{is(\#R,B)} S(\#R) \Rightarrow R(A, \#R)\Big)\end{aligned}$$

For more details see Section 5.65

For example: “I like the fruits (= any fruit), that taste(s) like a lemon.”

$$\big(is(\#Q, fruit)\, and\, tasteLikeLemon(\#Q)\big) \Rightarrow like(I, \#Q)$$

For example: “I admire the animals (= any animal), that run(s) fast.”

$$\big(is(\#Q, animal)\, and\, runFast(\#Q)\big) \Rightarrow admire(I, \#Q)$$

11.6.2 Some

For any $\#Q$:

$$(R(A, Some\ B)\ which\backslash that\ S(\_)) = \bigvee_{is(\#R,B)} (R(A,\#R)\ and\ S(\#R))$$

$$= \sim\Big((R(A,\#Q)\ and\ S(\#Q)) \Rightarrow \sim is(\#Q,B)\Big) = \bigvee_{is(\#R,B)\ and\ S(\#R)} R(A,\#R)$$

$$= \sim\Big((is(\#Q,B)\ and\ S(\#Q)) \Rightarrow \sim R(A,\#Q)\Big) = \bigvee_{is(\#R,B)\ and\ R(A,\#R)} S(\#R)$$

$$= \sim\Big((is(\#Q,B)\ and\ R(A,\#Q)) \Rightarrow \sim S(\#Q)\Big)$$

For example: “I like a fruit (= some fruit), that tastes like a lemon.”

$$\sim\Big((like(I,\#Q)\ and\ tasteLikeLemon(\#Q)) \Rightarrow \sim is(\#Q, fruit)\Big)$$

For example: “I admire an animal (some animal), that runs fast.”

$$\sim\Big((admire(I,\#Q)\ and\ runFast(\#Q)) \Rightarrow \sim are(\#Q, animals)\Big)$$

11.7 To do something to achieve something

$$(R(A)\ to\ S(\_)) = (R(A)\ so\ that\ S(A))$$

For example: ^“I am learning to pass the exams” = ^“I am learning so that I pass the exams.”

11.8 The more something the more something

$$(the\ more\ x\ the\ more\ y) = ((x\ more\ and\ more) \Rightarrow (y\ more\ and\ more))$$

$$(the\ more\ x\ the\ less\ y) = ((x\ more\ and\ more) \Rightarrow (y\ less\ and\ less))$$

$$(the\ less\ x\ the\ more\ y) = ((x\ less\ and\ less) \Rightarrow (y\ more\ and\ more))$$

$$(the\ less\ x\ the\ less\ y) = ((x\ less\ and\ less) \Rightarrow (y\ less\ and\ less))$$

To use above rules transform every comparative to form:

$$more\backslash less\ [adverb\backslash adjective\backslash noun\backslash etc.]$$

12 Probabilistic logic

12.1 Probability for elementary logic operators

A decidable statement is either true or false, so the probability, that a statement is true, has no sense. But a declarence in different situations, in which it is meaningful, can have different statements, that are either true or false, so we can calculate the probability that a declarence has true statement depending on the situation, in which it is used. For that reason we will use **declarences instead of statements** in logical expressions inside function P of probability, so they will have classical syntax and classical rules there.

For more details see Sections 12.4 and 12.6.

We will also use true measure and will assume, that equality is sensitive to any differential. For more details see Section 16.1.

First of all, we have for any declarences a and b:

$$\left(a \underbrace{op}_{s} b\right) = s^{\text{“}a\,op\,b\text{”}}$$

$$\left(a \underbrace{op}_{\text{any}} b\right) = \bigwedge_{s \in Situation^{\uparrow}} \left(a \underbrace{op}_{s} b\right)$$

$$\left(a \underbrace{op}_{\text{any}} b\right) \Rightarrow (^{\wedge}a\,op\,^{\wedge}b)$$

$$\left(a \underbrace{\rightarrow}_{any} b\right) = \left(a \underbrace{\Rightarrow}_{any} b\right) \Rightarrow (^{\wedge}a \Rightarrow {}^{\wedge}b)$$

For a proof of the above statement see Section 5.38.1.

And we have for any declarences x and y:

$$\left(\boldsymbol{x} \underbrace{\rightarrow}_{\boldsymbol{any}} \boldsymbol{y}\right) \Rightarrow (\boldsymbol{P}(\boldsymbol{x}) \leq \boldsymbol{P}(\boldsymbol{y}))$$

So:

$$\left(\boldsymbol{x} \underbrace{\leftrightarrow}_{\boldsymbol{any}} \boldsymbol{y}\right) = \left(\left(x \underbrace{\rightarrow}_{any} y\right) and \left(x \underbrace{\leftarrow}_{any} y\right)\right) \Rightarrow \left((P(x) \leq P(y)) \; and \; (P(x) \geq P(y))\right)$$

$$= (\boldsymbol{P}(\boldsymbol{x}) = \boldsymbol{P}(\boldsymbol{y}))$$

https://doi.org/10.1515/9783111441382-012

So:

$$\left(\boldsymbol{x} \underset{any}{\leftrightarrow} \boldsymbol{y}\right) \Rightarrow (\boldsymbol{P}(\boldsymbol{x}) = \boldsymbol{P}(\boldsymbol{y}))$$

For any declarences x and y:

$$\boldsymbol{P}(\boldsymbol{x}\ \boldsymbol{or}\ \boldsymbol{y}) = \boldsymbol{P}(\boldsymbol{x}) + \boldsymbol{P}(\boldsymbol{y}) - \boldsymbol{P}(\boldsymbol{x}\ \boldsymbol{and}\ \boldsymbol{y})$$

This formula is true, because $P(x) + P(y)$ has intersection $P(x\ and\ y)$ counted twice, while $P(x\ or\ y)$ has this intersection counted once, so after subtraction of this intersection from $P(x) + P(y)$ we have the above equality.

It can be easily proved.

Assume, that $P(x\ or\ y) = P(x) + P(y)$, when in any situation $\sim(x\ and\ y)$.

We have, that for any declarences x and y:

$$\big((x\ and \sim(x\ and\ y))\ or\ (x\ and\ y)\big) \leftrightarrow x$$

So for any declarences x and y:

$$\Big(P(x) = P\big((x\ and \sim(x\ and\ y))\ or\ (x\ and\ y)\big)\Big)$$

So for any declarences x and y:

$$\begin{aligned}
&\boldsymbol{P}(\boldsymbol{x}) + \boldsymbol{P}(\boldsymbol{y}) - \boldsymbol{P}(\boldsymbol{x}\ \boldsymbol{and}\ \boldsymbol{y})\\
&= P\big((x\ and \sim(x\ and\ y))\ or\ (x\ and\ y)\big)\\
&\quad + P\big((y\ and \sim(x\ and\ y))\ or\ (x\ and\ y)\big) - P(x\ and\ y)\\
&= P(x\ and \sim(x\ and\ y)) + P(x\ and\ y) + P(y\ and \sim(x\ and\ y))\\
&\quad + P(x\ and\ y) - P(x\ and\ y)\\
&= \boldsymbol{P}(\boldsymbol{x}\ \boldsymbol{and} \sim(\boldsymbol{x}\ \boldsymbol{and}\ \boldsymbol{y})) + \boldsymbol{P}(\boldsymbol{y}\ \boldsymbol{and} \sim(\boldsymbol{x}\ \boldsymbol{and}\ \boldsymbol{y})) + \boldsymbol{P}(\boldsymbol{x}\ \boldsymbol{and}\ \boldsymbol{y})\\
&= P\big((x\ and \sim(x\ and\ y))\ or\ (y\ and \sim(x\ and\ y))\big) + P(x\ and\ y)\\
&= P\big(((x\ or\ y)\ and \sim(x\ and\ y))\big) + P(x\ and\ y)\\
&= P\big(((x\ or\ y)\ and \sim(x\ and\ y))\ or\ (x\ and\ y)\big)\\
&= P(x\ or\ y\ or\ (x\ and\ y)) = \boldsymbol{P}(\boldsymbol{x}\ \boldsymbol{or}\ \boldsymbol{y})
\end{aligned}$$

Where in any situation:

$$\sim\left((x \textit{ and} \sim(x \textit{ and } y)) \textit{ and } (x \textit{ and } y)\right) \leftrightarrow \sim\left(x \textit{ and } (\sim(x \textit{ and } y) \textit{ and } (x \textit{ and } y))\right)$$

$$\leftrightarrow \sim(x \textit{ and false}) \leftrightarrow \sim\textit{false} \leftrightarrow \textit{true}$$

$$\sim\left((y \textit{ and} \sim(x \textit{ and } y)) \textit{ and } (x \textit{ and } y)\right) \leftrightarrow \sim\left(y \textit{ and } (\sim(x \textit{ and } y) \textit{ and } (x \textit{ and } y))\right)$$

$$\leftrightarrow \sim(y \textit{ and false}) \leftrightarrow \sim\textit{false} \leftrightarrow \textit{true}$$

$$\sim\left((x \textit{ and} \sim(x \textit{ and } y)) \textit{ and } (y \textit{ and} \sim(x \textit{ and } y))\right) \leftrightarrow \sim((x \textit{ and } y) \textit{ and} \sim(x \textit{ and } y))$$

$$\leftrightarrow \sim\textit{false} \leftrightarrow \textit{true}$$

So:

$$\boldsymbol{P(x \textit{ or } y) = P(x) + P(y) - P(x \textit{ and } y)}$$

And:

$$(x \textit{ and} \sim x) \leftrightarrow \textit{false}$$

$$(x \textit{ or} \sim x) \leftrightarrow \textit{true}$$

So:

$$\boldsymbol{P(\textit{false})} = P(x \textit{ and} \sim x) = \boldsymbol{0}$$

$$\boldsymbol{P(\textit{true})} = P(x \textit{ or} \sim x) = \boldsymbol{1}$$

So:

$$\boldsymbol{1} = P(x \textit{ or} \sim x) = P(\sim x) + P(x) - P(x \textit{ and} \sim x) = \boldsymbol{P(\sim x) + P(x)}$$

So:

$$\boldsymbol{P(\sim x)} = 1 - \boldsymbol{P(x)}$$

And:

$$\boldsymbol{P(y|x)} = \frac{\boldsymbol{P(x \textit{ and } y)}}{\boldsymbol{P(x)}}$$

As can be easily seen:

$$\boldsymbol{P(a \textit{ and } b)} = P(b|a) * P(a) \leq \boldsymbol{P(a)}$$

$$\boldsymbol{P(a \textit{ and } b)} = P(a|b) * P(b) \leq \boldsymbol{P(b)}$$

$$\boldsymbol{P(a \textit{ or } b)} = P(a) + P(b) - P(b|a) * P(a) = P(a)(1 - P(b|a)) + P(b) \geq \boldsymbol{P(b)}$$

$$\boldsymbol{P(a \textit{ or } b)} = P(a) + P(b) - P(a|b) * P(b) = P(b)(1 - P(a|b)) + P(a) \geq \boldsymbol{P(a)}$$

So:

$$\boldsymbol{P(a\ and\ b) \leq \min(P(a), P(b))}$$

$$\boldsymbol{P(a\ or\ b) \geq \max(P(a), P(b))}$$

And we have:

$$\boldsymbol{P((a|x)\ and\ (b|x)) = P((a\ and\ b)|x)}$$

$$\boldsymbol{P((a|x)\ or\ (b|x)) = P((a\ or\ b)|x)}$$

So:

$$P((y|x)\ and\ (\sim y|x)) = P((y\ and \sim y)|x) = \mathbf{0}$$

$$P((y|x)\ or\ (\sim y|x)) = P((y\ or \sim y)|x) = \mathbf{1} = P(y|x) + P(\sim y|x) - P((y|x)\ and\ (\sim y|x))$$
$$= \boldsymbol{P(y|x) + P(\sim y|x)}$$

So:

$$\boldsymbol{P(\sim y|x) = 1 - P(y|x)}$$

Here is another simple proof of this fact:

$$P((a\ and\ b)\ and\ (\sim a\ and\ b)) = 0$$

So:

$$\boldsymbol{P(a\ and\ b) + P(\sim a\ and\ b)}$$
$$= P(a\ and\ b) + P(\sim a\ and\ b) - P((a\ and\ b)\ and\ (\sim a\ and\ b))$$
$$= P((a\ and\ b)\ or\ (\sim a\ and\ b))$$
$$= P((a\ or \sim a)\ and\ (a\ or\ b)\ and\ (b\ or \sim a)\ and\ b)$$
$$= P\Big((a\ or \sim a)\ and\ (b\ or\ (a\ and \sim a))\ and\ b\Big)$$
$$= P(true\ and\ (b\ or\ false)\ and\ b) = P(true\ and\ b\ and\ b) = \boldsymbol{P(b)}$$

This can be proved even easier, because it is b in the a plus b outside a, which is complete b. Or a inside b plus everything outside a inside b, which is also complete b.

Now let us transform it:

$$P(b) = P(a\ and\ b) + P(\sim a\ and\ b) = P(a|b)P(b) + P(\sim a|b)P(b) = P(b)\,(P(a|b) + P(\sim a|b))$$

So for $P(b) \neq 0$:

$$P(a|b) + P(\sim a|b) = 1$$

So:

$$\boldsymbol{P(\sim a|b) = 1 - P(a|b)}$$

And we have:

$$P(a\,and\,{\sim}b) = P(a|{\sim}b)P({\sim}b) = P({\sim}b|a)P(a)$$

So:

$$\boldsymbol{P(a|{\sim}b)} = \frac{P({\sim}b|a)P(a)}{P({\sim}b)} = \frac{(1 - P(b|a))P(a)}{1 - P(b)} = \frac{\boldsymbol{P(a) - P(a\,and\,b)}}{\boldsymbol{1 - P(b)}}$$

So:

$$\boldsymbol{P(a|{\sim}b)} = \frac{\boldsymbol{P(a) - P(a\,and\,b)}}{\boldsymbol{1 - P(b)}}$$

So:

$$\boldsymbol{P({\sim}a|{\sim}b)} = 1 - P(a|{\sim}b) = 1 - \frac{P(a) - P(a\,and\,b)}{1 - P(b)} = \frac{1 - P(b) - P(a) + P(a\,and\,b)}{1 - P(b)}$$

$$= \frac{\boldsymbol{1 - P(a\,or\,b)}}{\boldsymbol{1 - P(b)}} = \frac{P({\sim}(a\,or\,b))}{P({\sim}b)} = \frac{P({\sim}a\,and\,{\sim}b)}{P({\sim}b)}$$

Q.E.D.

Any declarences a and b are independent if and only if $P(a|b) = P(a)$, which is true if and only if $P(b|a) = P(b)$.

For n independent declarences a_i, where $i = 1, \ldots, n$:

$$A_n = \boldsymbol{P}\left(\bigwedge_{i=1}^{n} \boldsymbol{a_i}\right) = \prod_{i=1}^{n} \boldsymbol{P(a_i)}$$

$$B_n = \boldsymbol{P}\left(\bigvee_{i=1}^{n} \boldsymbol{a_i}\right) = P\left({\sim}\bigwedge_{i=1}^{n} {\sim}a_i\right) = \left(1 - P\left(\bigwedge_{i=1}^{n} {\sim}a_i\right)\right) = \left(\boldsymbol{1} - \prod_{i=1}^{n} (\boldsymbol{1 - P(a_i)})\right)$$

And we have:

$$T(1) = (A_1 = P(a_1) \le P(a_1) = B_1) \equiv true$$

Assume:

$$T(n) = (A_n \le B_n) \equiv true$$

Now we can prove, that:

$$T(n+1) = (A_{n+1} \le B_{n+1}) \equiv true$$

We have to prove:

$$A_{n+1} = A_n * P(a_n) \le 1 - (1 - B_n)(1 - P(a_n)) = B_n + P(a_n)(1 - B_n) = B_{n+1}$$

Using assumption $T(n)$, we get:

$$A_{n+1} = A_n * P(a_n) \le A_n \le B_n \le B_n + P(a_n)(1 - B_n) = B_{n+1}$$

So we get conclusion $T(n+1)$:

$$A_{n+1} \le B_{n+1}$$

So we proved induction step $T(n) \Rightarrow T(n+1)$ and showed, that $T(1)$ is true. So we proved $T(n)$ for any $n \ge 1$.

Q.E.D.

In other words, for any set of independent declarences S we have:

$$P\left(\bigwedge_S x \, by \, x\right) \le P\left(\bigvee_S x \, by \, x\right)$$

Which has even a simpler proof:

$$P\left(\bigwedge_S x \, by \, x\right) \le \min_{x \in S} P(x) \le \max_{x \in S} P(x) \le P\left(\bigvee_S x \, by \, x\right)$$

Q.E.D.

12.2 Probability of equality of logical values and possibility of implication

Equality is trivial:

$$\begin{aligned}\boldsymbol{P(a \leftrightarrow b)} &= P((a \, and \, b) \; or \; ({\sim}a \, and \, {\sim}b)) \\ &= P(a \, and \, b) + P({\sim}a \, and \, {\sim}b) - P(a \, and \, b \, and \, {\sim}a \, and \, {\sim}b) \\ &= \boldsymbol{P(a \, and \, b) + P({\sim}a \, and \, {\sim}b)}\end{aligned}$$

Now you can evaluate $P(a \rightarrow b)$ from $P(a \leftrightarrow (b \, and \, a))$.

$$\begin{aligned}\boldsymbol{P(a \rightarrow b)} &= P(a \leftrightarrow (b \, and \, a)) = P(a \, and \, (b \, and \, a)) + P({\sim}a \, and \, {\sim}(b \, and \, a)) \\ &= P(a \, and \, b) + P({\sim}a \, and \, ({\sim}b \; or \; {\sim}a)) \\ &= P(a \, and \, b) + P(({\sim}a \, and \, {\sim}b) \; or \; ({\sim}a \, and \, {\sim}a)) \\ &= P(a \, and \, b) + P({\sim}a \, and \, {\sim}b) + P({\sim}a) - P({\sim}a \, and \, {\sim}b \, and \, {\sim}a) \\ &= \boldsymbol{P(a \, and \, b) + P({\sim}a)}\end{aligned}$$

And:

$$\boldsymbol{P(a \rightarrow b)} = P(a \text{ and } b) + P(\sim a) = 1 - (P(a) - P(a)P(b|a)) = 1 - P(a)(1 - P(b|a))$$
$$= 1 - P(a)P(\sim b|a) = 1 - P(a \text{ and} \sim b) = P(\sim(a \text{ and} \sim b)) = \boldsymbol{P(\sim a \text{ or } b)}$$

You can check it, for example, for the equality $P(a \rightarrow b) = P(\sim b \rightarrow \sim a)$

$$P(\sim b \rightarrow \sim a) = P(\sim\sim b \text{ or} \sim a) = P(\sim a \text{ or } b) = P(a \rightarrow b)$$

And we have:

$$\left(a \underset{any}{\rightarrow} b\right) = \left(a \underset{any}{\leftrightarrow} (b \text{ and } a)\right) = \left(P(a \rightarrow b) = P(\sim a \text{ or } b) = P(\sim(b \text{ and } a) \text{ or } b)\right.$$
$$\left. = P(\sim b \text{ or} \sim a \text{ or } b) = P((\sim b \text{ or } b) \text{ or} \sim a) = P(\text{true or} \sim a) = 1\right)$$

And even more:

$$\left(a \underset{any}{\Rightarrow} b\right) = \left(a \underset{any}{\rightarrow} b\right) = (P(a \rightarrow b) = 1)$$

So:

$$\left(a \underset{any}{\Rightarrow} b\right) = \boldsymbol{(P(a \rightarrow b) = 1)}$$

And remember, that for $h \rightarrow 0$:

$$(x = v \pm h) = \left(((\varepsilon \in R) \text{ and } (\varepsilon > 0)) \Rightarrow (v - \varepsilon < x < v + \varepsilon)\right)$$

But in true measure $x \neq x \pm h$. See the definition of true measure in Section 16.1.1. For more details see Section 15.29.

12.3 Statistical independence of two declarences implicates statistical dependence of their both ways possibilities of implication and vice versa

Let us assume, that declarences a and b are statistically independent and declarences $(a \rightarrow b)$ and $(b \rightarrow a)$ are statistically independent. Then:

$$P(a \text{ and } b) = P(a)P(b)$$

$$P(a \leftrightarrow b) = P((a \rightarrow b)\ and\ (b \rightarrow a)) = P(a\ and\ b) + P(\sim a\ and \sim b)$$
$$= P(a\ and\ b) + 1 - P(a\ or\ b) = 2P(a\ and\ b) + 1 - (P(a) + P(b))$$
$$= P(a \rightarrow b)P(b \rightarrow a)$$

$$2P(a\ and\ b) + 1 - (P(a) + P(b)) = (P(a\ and\ b) + P(\sim a))(P(a\ and\ b) + P(\sim b))$$

$$2P(a\ and\ b) + 1 - (P(a) + P(b)) = P(a\ and\ b)^2 + P(a\ and\ b)(2 - P(a) - P(b)) + (1 - P(a))(1 - P(b))$$

$$1 - (P(a) + P(b)) = P(a\ and\ b)^2 - P(a\ and\ b)\ (P(a) + P(b)) + (1 - P(a))(1 - P(b))$$

$$P(a\ and\ b)(P(a) + P(b)) = P(a\ and\ b)^2 + P(a)\,P(b)$$

$$P(a\ and\ b)(P(a) + P(b) - P(a\ and\ b)) = P(a)\,P(b)$$

$$P(a\ and\ b)\ P(a\ or\ b) = P(a)\,P(b)$$

$$P(a)\,P(b)\,P(a\ or\ b) = P(a)\,P(b)$$

$$P(a\ or\ b) = 1$$

$$P(a) + P(b) - P(a)\,P(b) = 1$$

$$P(b)(1 - P(a)) = 1 - P(a)$$

1′ For $P(a) = 1$:

$$2P(a\ and\ b) + 1 - (P(a) + P(b)) = (P(a\ and\ b) + P(\sim a))(P(a\ and\ b) + P(\sim b))$$

$$2P(a\ and\ b) - P(b) = (P(a\ and\ b) + 1 - P(a))(P(a\ and\ b) + 1 - P(b))$$

$$2P(a\ and\ b) - P(b) = P(a\ and\ b)(P(a\ and\ b) + 1 - P(b))$$

$$P(a\ and\ b)(1 - P(a\ and\ b) - P(b)) = P(b)$$

$$P(a)\,P(b)(1 - P(a)\,P(b) - P(b)) = P(b)$$

$$P(b)(1 - P(b) - P(b)) = P(b)$$

$$-2P(b)^2 = 0$$

$$P(b) = 0$$

2′ For $P(a) \neq 1$:

$$P(b) = 1$$

Since above equation is symmetrical we can conclude from case 1’, that $P(a) = 0$.

So this is only true for $(P(a), P(b)) \in \{(1, 0), (0, 1)\}$.

Which means, that any declarences a and b are statistically independent at the same time, when declarences $(a \rightarrow b)$ and $(b \rightarrow a)$ are statistically independent, if and only if $\big(P(a), P(b)\big) \in \{(1,0), (0,1)\}$. Otherwise statistical independence of each one of both pairs implicate statistical dependence of the other pair.

Q.E.D.

12.4 Universe

$P_{℧}(x)$ is the probability of x being true in universe ℧, where ℧ is a subset of Ω. All previous rules apply for any universe, so you can take any rule from Chapter 12 and replace everywhere P with $P_{℧}$ and all operators with their forms bounded to universe ℧, e.g.: "$=_{℧}$", "$\underset{℧}{\Rightarrow}$", "$\rightarrow_{℧}$", etc..

$$P_{℧}(x) = P_{entity \in ℧}(x) = P_{\in ℧}(x) = P_r(x)$$

where:

The symbol "℧" is inverted "Ω" and symbolizes given universe.

Declarence r determines set ℧ of all entities, that fulfill r, so it determines the range in Ω, which is a subset of Ω.

Remember, that for every unary operator op:

$$\big(op(a)\big)_{℧} = op(a)_{℧} = op_{℧}(a) = op(a_{℧})$$

And for every binary operator op:

$$(a\, op\, b)_{℧} = \left(a \underset{℧}{\underbrace{op}} b\right) = (a_{℧}\, op\, b_{℧})$$

For example:

$$(a \Rightarrow b)_{℧} = \left(a \underset{℧}{\Rightarrow} b\right) = \left(a \underset{℧}{\underbrace{\Rightarrow}} b\right) = (a_{℧} \Rightarrow b_{℧})$$

And so on.

And we have the following rules:

$$P_{x\, and\, y}(a) = P_x(a|y) = P_y(a|x) = P(a|x|y) = P(a|x\, and\, y)$$

Where the universe for probability is restricted by the subset determined by declarence x of the subset determined by declarence y and equivalently by the subset determined by declarence y of the subset determined by declarence x:

$$P_x(x) = P(x|x) = \frac{P(x\,and\,x)}{P(x)} = \frac{P(x)}{P(x)} = 1$$

$$P_x(\sim y) = P_x(x\,and\sim y) = \frac{P(x\,and\sim y)}{P(x)}$$

$$P_x(\sim x) = P_x(x\,and\sim x) = \frac{P(x\,and\sim x)}{P(x)} = \frac{0}{P(x)} = 0$$

So:

$$P_{x\,and\,y}(a) * P(x\,and\,y) = \left(P_{x\,and\,y}(a) * P_y(x)\right) * P(y) = P_y(a) * P(y) = P(a)$$

$$P_{x\,and\,y}(a) * P(x\,and\,y) = \left(P_{x\,and\,y}(a) * P_x(y)\right) * P(x) = P_x(a) * P(x) = P(a)$$

$P((a|x)\,and\,(b|y))$ has not any meaning unless $x \Leftrightarrow y$, because then it is like a combination of the measures of probability in two different sub-universes and such measures cannot be combined:

$$P((a|x)\,and\,(b|y)) = P(a|x) + P(b|y) - P((a|x)\,or\,(b|y))$$

$$P((a|x)\,and\,(b|y)) = \boldsymbol{P_x(a)} + \boldsymbol{P_y(b)} - P((a|x)\,or\,(b|y))$$

But for $x \Leftrightarrow y$:

$$P((a|x)\,and\,(b|x)) = P(a|x) + P(b|x) - P((a|x)\,or\,(b|x))$$

$$P(a\,and\,b|x) = P(a|x) + P(b|x) - P(a\,or\,b|x)$$

$$P_x(a\,and\,b) = P_x(a) + P_x(b) - P_x(a\,or\,b)$$

And we have:

$$\frac{P((a|x)\,and\,(b|x)\,and\,(c|x))}{P(c|x)} = P((a|x)\,and\,(b|x)|(c|x)) = P(a\,and\,b|x|(c|x))$$
$$= P(a\,and\,b|x\,and\,(c|x))$$

But also:

$$\frac{P((a|x)\,and\,(b|x)\,and\,(c|x))}{P(c|x)} = \frac{P(a\,and\,b\,and\,c|x)}{P(c|x)} = \frac{P_x(a\,and\,b\,and\,c)}{P_x(c)}$$
$$= P_x(a\,and\,b|c) = P(a\,and\,b|c\,and\,x)$$

So in range calculation:

$$\boldsymbol{x\,and\,(c|x) = x\,and\,c}$$

Of course, all these rules apply also to sets, so, for example, for subsets A and B of set Q:

$$P_{x \in Q}((x \in A) \to (x \in B)) = P_{x \in Q}(\sim(x \in A) \, or \, (x \in B)) = P_{x \in Q}\left(x \in ((Q - A) \cup B)\right)$$
$$= P\left(x \in ((Q - A) \cup B) | x \in Q\right)$$

Combining mutually disjoint set Q of universes into one is very simple:

$$P_{\cup_Q q \, by \, q}(x) = \frac{\int_Q \mu(q) * P_q(x) \, for \, q}{\int_Q \mu(q) \, for \, q}$$

where μ is either the true measure of universes, if available, or $\mu(q) = |q|$ for set of finite universes.

From now:

$$P(x) = P_\Omega(x)$$

And whenever you will see symbol Ω as the lower index of some symbol you can omit it, e.g.:

$$R(A)_\Omega = R_\Omega(A) = R(A)$$
$$[a]_\Omega = [a]$$
$$(a \to_\Omega b) = (a \to b)$$

etc.

For example, for the following data (universe $\mho$):

$$\mho = \left\{ \begin{array}{c} black \, horse \, A, \\ white \, horse \, B, \\ black \, horse \, C, \\ white \, dog \, A, \\ brown \, dog \, B, \\ black \, dog \, C, \\ brown \, dog \, D \end{array} \right\}$$

We have:

$$P_\mho(horse \, is \, black) = P_\mho(entity \, is \, black | entity \, is \, horse) = \frac{2}{3}$$

$$P_\mho(white \, are \, dogs) = P_\mho(entity \, is \, dog | entity \, is \, white) = \frac{1}{2}$$

$$P_\mho(some \, entity \, is \, a \, horse) = P_\mho(X \, is \, horse | X \, is \, entity) = P_\mho(X \, is \, horse) = \frac{3}{7}$$

etc.

There are possible very useful extensions, e.g.:

Universe can be defined with entities weighted by importance (and also declarences can be weighted by uncertainty or credibility, which you will see further):

$$\mho = set\,A\,of\,black\,horses\,(0.5) \cup \left\{ \begin{array}{c} white\,horse\,B\,(0.75), \\ black\,horse\,C\,(0.2), \\ white\,dog\,A\,(0.1), \\ brown\,dog\,B\,(0.7), \\ black\,dog\,C\,(1.0), \\ brown\,dog\,D\,(1.0) \end{array} \right\}$$

Then e.g.:

$$\begin{aligned} P_{\mho}(horse\,is\,black) &= P_{\mho}(entity\,is\,black \mid entity\,is\,horse) \\ &= \frac{|\mathrm{A}|_{1T} * 0.5 + |\{horse\,C\}|_{1T} * 0.2}{|\mathrm{T}|_{1T}} \\ &= \frac{|\mathrm{A}|_{1T} * 0.5 + |\{horse\,C\}|_{1T} * 0.2}{|A \cup \{horse\,B, horse\,C\}|_{1T}} \end{aligned}$$

where T is the set of all horses in this universe.

12.5 Realization of relation in given universe

Before you will read this and the following sections you should read at least Section 10.12.

For any relation $R(Some\,A_{)(})$, that has parameterized all objects (including the subject):

$$[R(A_{)(})]_{\mho} = \left\{ \begin{array}{c} e\colon \bigvee_{i \in N,\, \{i,\{e\}\} \in x \in A^{\uparrow \# \mho}} \left((e \in \mathcal{B}) \underset{\mho}{\Rightarrow} R(A_{)(}) \right) \\ and \sim \bigvee_{r \in [R(A_{)(})]_{\mho}} (r \subset e) \end{array} \right\}$$

where $Some\,A$ symbolizes untied (not tied) arguments.

This definition guarantees that realizations are not surplus.

And we assume, that any relation R once declared as $R(Some \backslash Any\,A_{)(})$, $R(The\,A_{)(})$ or $R(The\,A_{)(}, Some \backslash Any\,B_{)(})$ can be used at any place without modifiers *Some*, *Any* and *The*, if these modifiers are not changed in that place, so we can call it simply $R(A_{)(})$ and $R(A_{)(}, B_{)(})$ respectively. This is done to shorten formulas.

Remember also, that:

$$R(The\,A_{)(}, Some\,B_{)(}, Any\,C_{)(}) = R\left(\coprod_A The\,a\;by\,a,\ \coprod_B Some\,b\;by\,b,\ \coprod_C Any\,c\;by\,c\right)$$

$$R(A_{)(}) = R\left(\coprod_A a\;by\,a\right)$$

So, for example:

$$R\left(The\,(A_1,\,...,A_n)_{)(}, Some\,(B_1,\,...,B_m)_{)(}, Any\,\,(C_1,\,...,C_k)_{)(}\right)$$
$$= R(The\,A_1,\,...,The\,A_n, Some\,B_1,\,...,Some\,B_m, Any\,C_1,\,...,Any\,C_k)$$

$$R\left((A_1,\,...,A_n)_{)(}\right) = R(A_1,\,...,A_n)$$

And we have for any statements a and b:

$$([a]_{\mho} \subset [b]_{\mho}) \Rightarrow \left(\bigwedge_{x\in[a]_{\mho}} \bigvee_{y\in[b]_{\mho}} (y \subset x)\right) = \left(a \underset{\mho}{\Rightarrow} b\right)$$

So for any full normalized relation $R(Some\,A_{)(})$:

$$\left(B^{\uparrow\#\mho} \subset C^{\uparrow\#\mho}\right) \Rightarrow \left(\left[R(Some\,B_{)(})\right]_{\mho} \subset \left[R(Some\,C_{)(})\right]_{\mho}\right)$$
$$\Rightarrow \left(R(Some\,B_{)(}) \underset{\mho}{\Rightarrow} R(Some\,C_{)(})\right)$$
$$= \left((!R)(Any\,C_{)(}) \underset{\mho}{\Rightarrow} (!R)(Any\,B_{)(})\right)$$
$$= \left(\left(\langle R(The\,B_{)(})\rangle_{\mho} \neq \varnothing\right) \Rightarrow \left(\langle R(The\,C_{)(})\rangle_{\mho} \neq \varnothing\right)\right)$$
$$= \left(\langle R(The\,B_{)(})\rangle_{\mho} \subset \langle R(The\,C_{)(})\rangle_{\mho}\right)$$

Where:

$$\langle R(A_{)(})\rangle_{\mho} = \langle R(A_{)(})_{\mho}\rangle = \langle R_{\mho}(A_{)(})\rangle = \{a \in A^{\Uparrow\#\mho} : R(a_{)(})\}$$

And for any full normalized relation $R(Any\,A_{)(})$:

$$\begin{aligned}(B^{\uparrow\#\mho}\subset C^{\uparrow\#\mho}) &\Rightarrow \left(@\left[R(Any\,B_{)(})\right]_{\mho}\subset @\left[R(Any\,C_{)(})\right]_{\mho}\right)\\ &\Rightarrow \left(R(Any\,C_{)(})\underset{\mho}{\Rightarrow} R(Any\,B_{)(})\right)\\ &=\left((!R)(Some\,B_{)(})\underset{\mho}{\Rightarrow}(!R)(Some\,C_{)(})\right)\\ &=\left(\left(\langle(!R)(The\,B_{)(})\rangle_{\mho}\neq\varnothing\right)\Rightarrow\left(\langle(!R)(The\,C_{)(})\rangle_{\mho}\neq\varnothing\right)\right)\\ &=\left(\langle(!R)(The\,B_{)(})\rangle_{\mho}\subset\langle(!R)(The\,C_{)(})\rangle_{\mho}\right)\end{aligned}$$

And we could easily find a counterexample to:

$$\left(\langle R(The\,B_{)(})\rangle_{\mho}\subset\langle R(The\,C_{)(})\rangle_{\mho}\right)=\left(\langle(!R)(The\,B_{)(})\rangle_{\mho}\subset\langle(!R)(The\,C_{)(})\rangle_{\mho}\right)$$

So we have only the following hypothesis:

$$\begin{aligned}(B^{\uparrow\#\mho}\subset C^{\uparrow\#\mho}) &=\left(\left[R(Some\,B_{)(})\right]_{\mho}\subset\left[R(Some\,C_{)(})\right]_{\mho}\right)\\ &=\left(@\left[R(Any\,B_{)(})\right]_{\mho}\subset @\left[R(Any\,C_{)(})\right]_{\mho}\right)\end{aligned}$$

And for any normalized relation $R(Some\ A_{)(})$:

$$(B^{\uparrow\#\mho}\subset C^{\uparrow\#\mho})\Rightarrow\left(R(Some\,B_{)(})\underset{\mho}{\Rightarrow}R(Some\,C_{)(})\right)$$

And for any normalized relation $R(Any\ A_{)(})$:

$$(B^{\uparrow\#\mho}\subset C^{\uparrow\#\mho})\Rightarrow\left(R(Any\,C_{)(})\underset{\mho}{\Rightarrow}R(Any\,B_{)(})\right)$$

Notice, that relation $R(Some\backslash Any\,A_{)(})$ at the same moment has one statement, if it does not use non-strict references, while $R(The\,A_{)(})$ at the same moment can have many statements, because the statement of it depends on the tuple of beings it refers to, that can be different in every context, in which it is said.

For any statements a and b:

$$[a\,and\,b]_{\mho}=\{x\,and\,y: x\in[a]_{\mho}\,and\,y\in[b]_{\mho}\}$$

$$[a\,or\,b]_{\mho}=\{x: x\in[a]_{\mho}\,or\,x\in[b]_{\mho}\}=([a]_{\mho}\cup[b]_{\mho})$$

$$\begin{aligned}[a\,xor\,b]_{\mho}&=[(a\,or\,b)\,and\sim(a\,and\,b)]_{\mho}=[(a\ and\sim b)\ or\ (\sim a\,and\,b)]_{\mho}\\ &=\left(\left([a]_{\mho}\underset{and}{\times}[\sim b]_{\mho}\right)\cup\left([\sim a]_{\mho}\underset{and}{\times}[b]_{\mho}\right)\right)\end{aligned}$$

12.6 Quantifier "the" and the set of all examples supporting given relation

In this section we will assume, that all situations are equally probable. I will also call the article "the" in the arguments of a basic relation a quantifier, for convenience.

Quantifier *The* symbolizes tied arguments, that have to be instantiated for the relation to mean something, e.g.: "$(\mathit{The\ Dog})\ \mathit{runs\ to}\ (\mathit{Some\ Place})$". If *The Dog* is not tied to any thing, that is a dog, then the whole declarence has not any meaning.

Such a relation can have different forms, e.g.:

$$(\mathit{The\ man\ has\ the\ car}) = \mathit{has}(\mathit{The\ Man}, \mathit{The\ Car}) \rightarrow ((\mathit{The\ Man})\ \mathit{has}\ (\mathit{The\ Car}))$$

$$(x > 0) = \mathit{isPositive}(\mathit{The\ Real\ Variable}\ x) \rightarrow ((\mathit{The\ Real\ Variable}\ x) > 0)$$

$$(x \in X) = \mathit{belongsTo}(\mathit{The\ Thing}\ x\ (1),\ \mathit{The\ Set}\ X\ (2))$$
$$\rightarrow \Big((\mathit{The\ Thing}\ x\ (1))\ \mathit{belongs\ to}\ (\mathit{The\ Set}\ X\ (2))\Big)$$

And so on.

Relation declared as $R(\mathit{The}\,A_{)(},\ \mathit{Some}\backslash\mathit{Any}\,B_{)(})$ has tied arguments A as well as untied arguments B. So it is still a contextual declarence. The tied arguments are, of course, necessary to evaluate logical value of the relation, while the untied arguments will be used to iterate nested operations *and* (in case of quantifier *Any*) and *or* (in case of quantifier *Some*) over the appropriate representations of the untied arguments.

For relation $R \div A \rightarrow \mathit{Statement}^{\Uparrow}$, where A is tuple of things, which is equal to declaration of relation $R(\mathit{The}\,A_{)(})$, we have:

$$\langle R(A_{)(})\rangle_{\mho} = \langle R(A_{)(})_{\mho}\rangle = \langle R_{\mho}(A_{)(})\rangle = \{a \in \mathrm{A}^{\Uparrow\#\mho} : R(a_{)(})\}$$

where $\langle R(A_{)(})\rangle_{\mho}$ is the set of all instantiations of arguments A of relation $R(A)$ in universe $\mho$, for which relation is true. This is the set of all examples of arguments from universe $\mho$ supporting $R(A)$.

So for $a \in A^{\Uparrow\#\mho}$:

$$R(a_{)(})_{\mho} = \left(a \in \langle R(\mathit{The}\,A_{)(})\rangle_{\mho}\right)$$

And for any normalized relations $R(\mathit{Some}\,A_{)(})$ and $R(\mathit{Any}\,A_{)(})$:

$$R(\mathit{Some}\,A_{)(})_{\mho} = \left(\langle R(\mathit{The}\,A_{)(})\rangle_{\mho} \neq \varnothing\right)$$

$$R(\mathit{Any}\,A_{)(})_{\mho} = \left(\langle R(\mathit{The}\,A_{)(})\rangle_{\mho} = A^{\Uparrow\#\mho}\right)$$

For any full normalized relation $R(Some\,A_{)(})$:

$$R(X_{)(})_{\mho} = \bigvee_{x \in X^{\Uparrow\#\mho}} \bar{R}(x_{)(})$$

So:

$$\left(B^{\Uparrow\#\mho} \subset C^{\Uparrow\#\mho}\right) \Rightarrow \left(R(B_{)(})_{\mho} = \bigvee_{b \in B^{\Uparrow\#\mho}} \bar{R}(b_{)(}) \Rightarrow \bigvee_{c \in C^{\Uparrow\#\mho}} \bar{R}(c_{)(}) = R(C_{)(})_{\mho}\right)$$
$$\Rightarrow \left(R(B_{)(})_{\mho} \Rightarrow R(C_{)(})_{\mho}\right) = \left(R(B_{)(}) \underset{\mho}{\Rightarrow} R(C_{)(})\right)$$

Which is true also for any not full normalized relation $R(Some\ A_{)(})$, because any and/or (*Any*/*Some*) logical expression (that does not use negation) of any arguments implicates the same expression with other arguments, if only each argument of the former expression implicates the argument, for which it is replaced in the latter expression.

And we have the following hypothesis:

$$\left(B^{\Uparrow\#\mho} \subset C^{\Uparrow\#\mho}\right) = \left(R(B_{)(})_{\mho} \Rightarrow R(C_{)(})_{\mho}\right) = \left(R(B_{)(}) \underset{\mho}{\Rightarrow} R(C_{)(})\right)$$

For any full normalized relation $R(Any\,A_{)(})$

$$R(X_{)(})_{\mho} = \bigwedge_{x \in X^{\Uparrow\#\mho}} \bar{R}(x_{)(})$$

So:

$$\left(C^{\Uparrow\#\mho} \subset B^{\Uparrow\#\mho}\right) \Rightarrow \left(R(B_{)(})_{\mho} = \bigwedge_{b \in B^{\Uparrow\#\mho}} \bar{R}(b_{)(}) \Rightarrow \bigwedge_{c \in C^{\Uparrow\#\mho}} \bar{R}(c_{)(}) = R(C_{)(})_{\mho}\right)$$
$$\Rightarrow \left(R(B_{)(})_{\mho} \Rightarrow R(C_{)(})_{\mho}\right) = \left(R(B_{)(}) \underset{\mho}{\Rightarrow} R(C_{)(})\right)$$

Which is true also for any not full normalized relation $R(Any\ A_{)(})$ for the same reason as for any not full normalized relation $R(Some\ A_{)(})$.

And we have the following hypothesis:

$$\left(C^{\Uparrow\#\mho} \subset B^{\Uparrow\#\mho}\right) = \left(R(B_{)(})_{\mho} \Rightarrow R(C_{)(})_{\mho}\right) = \left(R(B_{)(}) \underset{\mho}{\Rightarrow} R(C_{)(})\right)$$

For set $M = A^{\Uparrow\#\mho}$, that has measure $|x|_{M}$ $(= |x|_{\mho} = \mu_{\mho}(x))$ and has finite size in this measure:

$$P_{\mho}\left(R(The\,A_{)(})\right) = \frac{|\langle R(A_{)(})\rangle_{\mho}|_{M}}{|M|_{M}}$$

And we have:

$$\langle R(A_{)()})\rangle_{\mho} \subset M$$

$$\langle \sim R(A_{)()})\rangle_{\mho} = \sim_M \langle R(A_{)()})\rangle_{\mho}$$

$$\langle R(A_{)()})\rangle_{\mho} \cup \langle \sim R(A_{)()})\rangle_{\mho} = M$$

$$\langle R(A_{)()})\rangle_{\mho} \cap \langle \sim R(A_{)()})\rangle_{\mho} = \varnothing$$

$$|\langle R(A_{)()})\rangle_{\mho}|_{\mathrm{M}} + |\langle \sim R(A_{)()})\rangle_{\mho}|_{\mathrm{M}} = |M|_{\mathrm{M}}$$

So:

$$P_{\mho}\left(\sim R(A_{)()})\right) = 1 - \frac{|\langle R(A_{)()})\rangle_{\mho}|_{\mathrm{M}}}{|M|_{\mathrm{M}}} = \frac{|\mathrm{M}|_{\mathrm{M}} - |\langle R(A_{)()})\rangle_{\mho}|_{\mathrm{M}}}{|\mathrm{M}|_{\mathrm{M}}} = \frac{|\langle \sim R(A_{)()})\rangle_{\mho}|_{\mathrm{M}}}{|\mathrm{M}|_{\mathrm{M}}}$$

And for any relations $R(The\,A_{)(})$ and $S(The\,A_{)(})$:

$$\langle \sim R(A)\rangle = \sim_{A\Uparrow\#\mho} \langle R(A)\rangle$$

$$\langle \sim S(A)\rangle = \sim_{A\Uparrow\#\mho} \langle S(A)\rangle$$

$$\langle R(A)\ and\, S(A)\rangle = (\langle R(A)\rangle \cap \langle S(A)\rangle)$$

$$\langle R(A)\ or\, S(A)\rangle = (\langle R(A)\rangle \cup \langle S(A)\rangle)$$

$$\begin{aligned}\langle R(A)\ xor\, S(A)\rangle &= \left(\langle R(A)\ or\, S(A)\right)\ and \sim\left(R(A)\ and\, S(A)\right)\rangle \\ &= \left(\langle R(A)\ or\, S(A)\rangle \cap \sim_{A\Uparrow\#\mho} \langle R(A)\ and\, S(A)\rangle\right) \\ &= \left(\left(\langle R(A)\rangle \cup \langle S(A)\rangle\right) \cap \sim_{A\Uparrow\#\mho} \left(\langle R(A)\rangle \cap \langle S(A)\rangle\right)\right)\end{aligned}$$

When two relations have different arguments, then the above equalities will be true for them, if you extend arguments of both relations to the sum of their arguments, where their intersection is not doubled, and put them in the same order. These additional arguments in both relations will be not used, but are necessary to make elements of appropriate sets be of the same kind.

12.7 Implication in given universe implicates the same implication in its sub-universe

We have, that for any statement a and b:

$$(a \Rightarrow b) = \left(a \underset{\Omega}{\Rightarrow} b\right) \Rightarrow \left(a \underset{\mho}{\Rightarrow} b\right)$$

And more generally:

$$(A \subset B) \Rightarrow \left((a \underset{B}{\Rightarrow} b) \Rightarrow (a \underset{A}{\Rightarrow} b) \right)$$

So for any declarences a and b:

$$\left(a \underbrace{\Rightarrow}_{any} b \right) \Rightarrow \left(a \underbrace{\Rightarrow}_{any\, in\, \mho} b \right) = \left(a \underbrace{\rightarrow}_{any\, in\, \mho} b \right) = (P_{\mho}(a \rightarrow b) = 1) = (P_{\mho}(\sim a\, or\, b) = 1)$$

12.8 Implication in given universe expressed as inclusion of sets of all examples supporting given relations

For relations with the same arguments $a = R(The\, X)$ and $b = S(The\, X)$ we have:

$$\left(a \underbrace{\equiv}_{any\, in\, \mho} b \right) = \left(a \underbrace{\leftrightarrow}_{any\, in\, \mho} b \right) = (P(a \leftrightarrow_{\mho} b) = 1) = (\langle a \rangle_{\mho} = \langle b \rangle_{\mho}) = (\langle \sim b \rangle_{\mho} = \langle \sim a \rangle_{\mho})$$

So

$$\left(a \underbrace{\Rightarrow}_{any\, in\, \mho} b \right) = \left(a \underbrace{\rightarrow}_{any\, in\, \mho} b \right) = (P(a \rightarrow_{\mho} b) = 1) = \left(P((a\; and\, b) \leftrightarrow_{\mho} a) = 1 \right)$$

$$= (\langle a \rangle_{\mho} = \langle a\; and\; b \rangle_{\mho}) = \left(\langle a \rangle_{\mho} = (\langle a \rangle_{\mho} \cap \langle b \rangle_{\mho}) \right) = (\langle a \rangle_{\mho} \subset \langle b \rangle_{\mho})$$

$$= (\langle \sim b \rangle_{\mho} \subset \langle \sim a \rangle_{\mho})$$

12.9 Calculation of probability for different probabilities of situations

When we assume, that any situation can have different probability, that it happens, then in the compliance with the probability definition we calculate the weighted arithmetic mean for given universe.

First of all, for any a and any declarence x and any b_1 and b_2 such that $b_1 \neq b_2$:

$$((a = b_1)\; and\; (a = b_2)\; and\, x) \equiv false$$

Because:

$$(a = b_1) \Rightarrow \sim(a = b_2)$$

And:

$$P_{\mho}(false) = 0$$

Then for relation $S \div A \to Statement^{\Uparrow}$, where A is a tuple of things, for tuple a that defines the context of the current situation under the condition that we do not know, which situation is the current situation:

$$P_{\mho}\left(S(A_{)(})\right) = P_{\mho}\left(S(a_{)(}) \mid a \in A^{\Uparrow\#\mho}\right) = \frac{P_{\mho}\left(S(a_{)(})\ and\ (a \in A^{\Uparrow\#\mho})\right)}{P_{\mho}(a \in A^{\Uparrow\#\mho})}$$

$$= \frac{P_{\mho}\left(S(a_{)(})\ and \bigvee_{b \in A^{\Uparrow\#\mho}} (b = a)\right)}{P_{\mho}(a \in A^{\Uparrow\#\mho})} = \frac{P_{\mho}\left(\bigvee_{b \in A^{\Uparrow\#\mho}} (S(a_{)(})\ and\ (b = a))\right)}{P_{\mho}(a \in A^{\Uparrow\#\mho})}$$

$$= \frac{\int_{A^{\Uparrow\#\mho}} P_{\mho}(S(a_{)(}) | a = b) * P_{\mho}(a = b)\ for\ b}{P_{\mho}(a \in A^{\Uparrow\#\mho})}$$

$$= \frac{\int_{A^{\Uparrow\#\mho}} P_{\mho}(S(b_{)(})) * P_{\mho}(b)\ for\ b}{P_{\mho}(a \in A^{\Uparrow\#\mho})} = \frac{\int_{A^{\Uparrow\#\mho}} V(S(b_{)(})) * P_{\mho}(b)\ for\ b}{P_{\mho}(a \in A^{\Uparrow\#\mho})}$$

$$P_{\mho}(S(a_{)(})\ and\ (a \in A^{\Uparrow\#\mho})) = P_{\mho}(a \in A^{\Uparrow\#\mho}) P_{\mho}(S(a_{)(}) | a \in A^{\Uparrow\#\mho})$$
$$= P_{\mho}(a \in A^{\Uparrow\#\mho}) P_{\mho}(S(A_{)(}))$$

where:

$$P_{\mho}(a \in A^{\Uparrow\#\mho}) = P_{\mho}\left(\bigvee_{b \in A^{\Uparrow\#\mho}} (b = a)\right) = \int_{A^{\Uparrow\#\mho}} P_{\mho}(b = a)\ for\ b = \int_{A^{\Uparrow\#\mho}} P_{\mho}(b)\ for\ b$$

$$V(x) = ((x \equiv true)\ ?\ 1 : 0)$$

$P_{\mho}(b)$ for $b \in A^{\Uparrow\#\mho}$ is the probability of a situation, in which our context is defined by b.

So we get weighted arithmetic mean:

$$\boldsymbol{P_{\mho}\left(S(A_{)(})\right)} = \frac{\int_{A^{\Uparrow\#\mho}} V(S(b_{)(})) * P_{\mho}(b)\ for\ b}{P_{\mho}(a \in A^{\Uparrow\#\mho})} = \frac{\int_{A^{\Uparrow\#\mho}} V(S(b_{)(})) * P_{\mho}(b)\ for\ b}{\int_{A^{\Uparrow\#\mho}} P_{\mho}(b)\ for\ b}$$

$$= \int_{A^{\Uparrow\#\mho}} \boldsymbol{V(S(b_{)(}))\ for\ b}\ _{\mathrm{avg}((P_{\mho}(b)) \leftarrow (b))}$$

Q.E.D.

12.10 Possible extensions of logical values to real values

You may want to use real values instead of logical and then you assume, that the statements of sentences are true, credible or certain to some degree. In such a case we need such a function V, that returns this real value from interval $[0,1]$ for every statement, where, for example, for degree of certainty 0 it is certain falseness and for 1 it is certain trueness and all values between them are to some degree uncertain, e.g.: $\frac{1}{2}$ means, that the value is oriented neither toward trueness nor towards falseness. And the following condition should be fulfilled:

$$V(\sim x) = 1 - V(x)$$

When degree is in range $[-\infty, \infty]$, then you can use, for example, any sigmoid function to transform it to range $[0,1]$:

$$f(x) = \frac{1}{1+e^{-x}}$$

For degree in range $[0, \infty]$ you can also use for example:

$$f(x) = \frac{1-e^{-x}}{1+e^{-x}}$$

$$f(x) = 1 - \frac{1}{x+1}$$

Etc.

If we treat $V(x)$ as a measure of degree of trueness, then:

$$((a \Rightarrow b)\ and\ (b \Rightarrow a)) = (a = b) \Rightarrow (V(a\ and\ b) = V(a\ and\ a) = V(a) = V(b))$$

$$(\sim(a \Rightarrow b)\ and \sim(b \Rightarrow a)) \Rightarrow (V(a\ and\ b) = V(a) \ast V(b))$$

And we know, that $\left((a \Rightarrow b) = (a = (b\ and\ a))\right)$, so:

$$((a \Rightarrow b)\ and \sim(b \Rightarrow a)) \Rightarrow (V(a\ and\ b) = V(a))$$

The measure of the degree of trueness behaves exactly the same as probability, because it is rational and objective and based on full knowledge about some universe, while uncertainty does not, because uncertainty informs only, how certain we are about the logical value of a statement, where we can be wrong, because we are subjective due to incompleteness of information, on which we base our certainty. So it does not give the same rational and objective measure of trueness as probability. That is, why measure of trueness other than from set $\{0,1\}$ would not have any sense in place of function V (because different universes cannot be mixed, so it would have to be a measure in the same universe, which is already defined and has not any meaning at the level of single situation) and why uncertainty is a subject to different laws. Uncertainty is not strictly defined, so cannot be translated into trueness without additional assumptions. Remember, that if we have certainty equal to 0.99, because we, for example, verified 99% of possible realizations of the negation of a statement to be no-beings, the statement still can be false. You can be as close to complete certainty as you want, but you cannot be completely certain until you will verify every possible realization of the negation of the statement. So if we have certainty equal to 1 with a precision to differentials, then such certainty is complete certainty, so we are sure, that the statement is true, so it cannot be false.

The degree of trueness and of uncertainty has both nothing to do with fuzzy logic. In fuzzy logic, when we do not use strict declarences, e.g.: "She has quite big eyes", then these are opinions, that can have an objective meaning only if a speaker, which gives such an opinion, will also give objective (strict) interpretation of his opinion, so he will explain, what that subjective sentence in objective reality means for him, so what the subjective declarence objectively means. Like in above example, a speaker will declare, for example, that for eyes "being quite big" is 50–70% of "being big", where for eyes "being big" means, that they have size not smaller than x, and fuzzy logic value F of "being big" is defined as:

$$F(\text{“X has big eyes”}) = 1 - \frac{(x - \min(\text{size of X's eyes, x}))}{x}$$

then:

$$\begin{aligned}(\text{She has quite big eyes}) &= (0.5 \leq F(\text{“She has big eyes”}) \leq 0.7) \\ &= \left(0.5 \leq 1 - \frac{(x - \min(\text{size of her eyes, x}))}{x} \leq 0.7\right)\end{aligned}$$

Since, as you can also see above, in fuzzy logic everything can be defined individual way, there are no general formulas for evaluation of logical functions, because all such evaluation should follow all implications of accepted interpretations of fuzzy notions. In other words, in fuzzy logic you have to always give objective (strict) interpretation of all fuzzy (subjective, not strict, that can mean something different for everyone) notions, so in the end you have exactly the same problem as the general problem of trueness or falseness of logical statements, where you can, of course, use, for example, probabilistic logic.

In other words, there is only one measure of the degree of trueness, which is probability, that comes straight from the definition of the measure of trueness: "in what part of all cases given sentence has a true statement". So so-called fuzzy logic is not a logic, if it is not probabilistic logic, because then fuzzy logic value is not the measure of trueness, but only some value, to which you give some not restricted, unlike in case of probability, interpretation.

If we treat $V(x)$ as a measure of credibility or certainty of a statement, then conjunction could be evaluated this way:

$$((a \Rightarrow b)\ and\ (b \Rightarrow a)) = (a = b) \Rightarrow (V(a\ and\ b) = V(a\ and\ a) = V(a))$$

$$(\sim(a \Rightarrow b)\ and \sim(b \Rightarrow a)) \Rightarrow (V(a\ and\ b) = V(a) * V(b))$$

$$((a \Rightarrow b)\ and \sim(b \Rightarrow a)) \Rightarrow (a = (b\ and\ a)) \Rightarrow (V(a\ and\ b) = V(a))$$

So we can have, for example, the following heuristic:

$$Q(x, a, b) = x^{(1-(1-V(a))V(b))^{1-V(a)(1-V(b))}}$$

$$Q(x, \sim a, \sim b) = Q(x, b, a)$$

$$V(a\ and\ b) = Q(V(b), a, b) * Q(V(a), b, a) = Q(V(a), b, a) * Q(V(b), a, b) = V(b\ and\ a)$$

$$\begin{aligned} V(a\ or\ b) &= V(\sim(\sim a\ and \sim b)) = 1 - V(\sim a\ and \sim b) \\ &= 1 - Q(1 - V(b), \sim a, \sim b) * Q(1 - V(a), \sim b, \sim a) \\ &= 1 - Q(1 - V(b),\ b,\ a) * Q(1 - V(a),\ a,\ b) \end{aligned}$$

The above formula has this form, because the closer we are to the situation $V(a) = 0$ and $V(b) = 1$, the closer we are to implication, so $((1 - V(a))V(b))$ has to weaken the participation of $V(b)$ in the multiplication, that gives the result. And the closer we are to situation $V(a) = 1$ and $V(b) = 0$, the closer we are to contradiction with assumption about implication, so $(V(a)(1 - V(b)))$ has to weaken strengthening of participation of $V(b)$ in the multiplication and reset it for $V(a) = 1$ and $V(b) = 0$.

12.11 Probability in time

When you have in time of experiment set M of different moments of tests, then all moments are equally probable, if we assume that the tempo of time does not ever change.

So we have, that $P_\mho(time = t)$ is constant for every moment t and for $time$ being the current time, that we do not know.

Let us assume, that for every moment t:

$$P_\mho(time = t) = c$$

First of all, for any declarence x and any t_1 and t_2 such that: $t_1 \neq t_2$

$$((time = t_1)\ and\ (time = t_2)\ and\ x) \equiv false$$

Because:

$$(time = t_1) \Rightarrow \sim(time = t_2)$$

And:

$$P_{\mho}(false) = 0$$

Then we have:

$$\begin{aligned}
\boldsymbol{P_{\mho\, at\, M}(S(A_{)(}))} &= P_{\mho\, and\, time \in M}(S(A_{)(})) = P_{\mho}(S(A_{)(})|time \in M) \\
&= \frac{P_{\mho}(S(A_{)(})\ and\ time \in M)}{P_{\mho}(time \in M)} = \frac{P_{\mho}(S(A_{)(})\ and \bigvee_{t \in M}(time = t))}{P_{\mho}(\bigvee_{t \in M}(time = t))} \\
&= \frac{P_{\mho}(\bigvee_M S(A_{)(})\ and\ (time = t)\ by\ t)}{P_{\mho}(\bigvee_{t \in M}(time = t))} = \frac{\int_M P_{\mho}(S(A_{)(})\ and\ (time = t))\ for\ t}{\int_M P_{\mho}(time = t)\ for\ t} \\
&= \frac{\int_M P_{\mho}(S(A_{)(})|time = t) * P_{\mho}(time = t)\ for\ t}{\int_M P_{\mho}(time = t)\ for\ t} \\
&= \frac{\int_M P_{\mho}(S(A_{)(})|time = t) * c\ for\ t}{\int_M c\ for\ t} = \frac{c \int_M P_{\mho}(S(A_{)(})|time = t)\ for\ t}{c \int_M 1\ for\ t} \\
&= \frac{\int_M P_{\mho}(S(A_{)(})|time = t)\ for\ t}{|M|_{1M}} = \boldsymbol{\int_M P_{\mho\, at\{t\}}(S(A_{)(}))\ for\ t}_{avg}
\end{aligned}$$

where M can be continuous segment, e.g.: $[a, b]$, and then $|M|_{1M}$ is the size of this set in the real measure ($1M = 1R$), so it is equal to $|b - a|$. For a finite set it would be its cardinality.

So:

$$\boldsymbol{P_{\mho\, at\, M}(S(A_{)(}))} = \boldsymbol{\int_M P_{\mho\, at\, \{t\}}(S(A_{)(}))\ for\ t}_{avg}$$

To evaluate $P_{\mho}(time = t)$ consider, that for an experiment, it took place between moments t_1 and t_2:

$$F(t) = \frac{t - t_1}{t_2 - t_1} = \int_{t_1}^{t} P_{\mho}(time = t) = \int_{t_1}^{t} \mu(t)\, dt = \int_{t_1}^{t} \frac{1}{t_2 - t_1} dt$$

Or another way:

$$\mu(t) = \frac{dF(t)}{dt} = \frac{d\left(\frac{t-t_1}{t_2-t_1}\right)}{dt} = \frac{1}{t_2-t_1} = \frac{P_{\mho}(time=t)}{dt}$$

So:

$$P_{\mho}(time=t) = \frac{dt}{t_2-t_1}$$

And we have:

$$P_{\mho\, at\, M}(x) = P_{\in \mho\, and\, time \in M}(x) = P_{\in \mho\, and\, \vee_M\, time\, is\, t\, by\, t}(x)$$

And, for example:

$$P_{\mho\, at\, M}(a|b) = \frac{P_{\mho\, at\, M}(a\, and\, b)}{P_{\mho\, at\, M}(b)} = \frac{P_{\mho}(\bigvee_M a\ at\ time\ t\ and\ b\ at\ time\ t\ by\ t)}{P_{\mho}(\bigvee_M b\ at\ time\ t\ by\ t)}$$

$$= \frac{\int_M P_{\mho\, at\, \{t\}}(a\, and\, b)\ by\ t_{\ avg}}{\int_M P_{\mho\, at\, \{t\}}(b)\ by\ t_{\ avg}} = \frac{|M|_{1M}^{-1} \int_M P_{\mho\, at\, \{t\}}(a|b)\, P_{\mho\, at\, \{t\}}(b)\ by\ t}{|M|_{1M}^{-1} \int_M P_{\mho\, at\, \{t\}}(b)\ by\ t}$$

$$= \int_M P_{\mho\, at\{t\}}(a|b)\ by\ t_{\ avg((P_{\mho\, at\{t\}}(b)) \leftarrow (t))}$$

We can calculate, for example, the probability of contextual implication for different moments in time: $P_{\mho}((a\, and\, time\ is\ t_1) \to (b\, and\, time\ is\ t_2))$.

Then the following heuristic should give information about the strength of, for example, implication through some time period:

$$g_n(x) = \frac{1-x}{1-x^{n-1}} \sum_{i=1}^{n} x^{i-1} \prod_{j=1}^{i} P_{\mho}\left(\left(a\, and\, time\ is\ t_1 + (j-1)\frac{t_2-t_1}{i}\right) \to \left(b\, and\, time\ is\ t_2 + j\frac{t_2-t_1}{i}\right)\right)^{\frac{1}{i}}$$

$$f(x) = \lim_{n \to \infty} g_n(x)$$

$$\lim_{x \to 1} f(x) = e^{\frac{\int_{t_1}^{t_2} \log P_{\mho\, at\, \{t\}}(a \to b)\ by\ t}{t_2 - t_1}}$$

This function has this limit, because from the arithmetic mean of infinite number of values we can eliminate as great finite number of values, as we want, for example, form the beginning, getting as close to this limit as we want.

If function f is not decreasing in the interval $[0,1]$ at least from some point not too far from 0, then we have indication, that there might be some connection through time between the trueness of the statements of the declarences within given universe. Analyzing this function we can get more information about this connection.

12.12 The probability that an entity is equal to given entity

For any at most continuous universe $\mho$:

$$P_{\mho}(R(x)) = P_{\mho}\left(\bigvee_{\mho} R(e)\ by\, e\right) = \int_{\mho} P_{\mho}(R(entity)\ and\, entity\ is\, e)\ for\, e$$

$$= \int_{\mho} P_{\mho}(R(entity)|entity\ is\, e)\ P(entity\ is\, e)\ for\, e$$

And since $P_{\mho}(entity\ is\, e)$ is constant:

$$1 = P_{\mho}(true) = \int_{\mho} P_{\mho}(true\ and\, entity\ is\, e)\ for\, e$$

$$= \int_{\mho} P_{\mho}(true|entity\ is\, e)\ P_{\mho}(entity\ is\, e)\ for\, e = \int_{\mho} P_{\mho}(entity\ is\, e)\ for\, e$$

$$= \int_{\mho} \frac{P_{\mho}(entity\ is\, e)}{de}\, de = |\mho|_{1\mho} * \frac{P_{\mho}(entity\ is\, e)}{de}$$

So:

$$|\mho|_{1\mho}^{-1} = \frac{P_{\mho}(entity\ is\ e)}{de}$$

$$P_{\mho}(entity\ is\, e) = |\mho|_{1\mho}^{-1}\, de$$

So for uncountable universe $P_{\mho}(entity\ is\ e)$ is not exactly 0, but to be precise it is differential of variable that points to single entity in this universe, that is exactly size of every element in this universe, divided by constant, that is the size of this universe in its measure. That is, why sum of this value over all elements of universe gives 1 – because it is the probability, that an entity from this universe is equal to given entity in this universe.

Q.E.D.

For more details see section 15.29.

12.13 Probability formulas for larger sets of declarences

For two declarences x and y we have:

$$\boldsymbol{P}(x\,\boldsymbol{and}\,y) = \boldsymbol{P}(y|x) * \boldsymbol{P}(x) = \boldsymbol{P}(x|y) * \boldsymbol{P}(y)$$

$$\begin{aligned}\boldsymbol{P}(x\,\boldsymbol{or}\,y) &= \boldsymbol{P}(x) + \boldsymbol{P}(y) - \boldsymbol{P}(x\,\boldsymbol{and}\,y) = \boldsymbol{P}(x) + \boldsymbol{P}(y) - \boldsymbol{P}(y|x) * \boldsymbol{P}(x)\\ &= \boldsymbol{P}(x) + \boldsymbol{P}(y) - \boldsymbol{P}(x|y) * \boldsymbol{P}(y)\end{aligned}$$

And for larger sets of declarences:

$$\boldsymbol{P}\left(\bigwedge_a^b x_i\,\boldsymbol{by}\,\boldsymbol{i}\right) = \prod_a^b \boldsymbol{P}\left(x_i|\bigwedge_a^{\boldsymbol{prev}(\boldsymbol{i})} x_j\,\boldsymbol{by}\,\boldsymbol{j}\right)\,\boldsymbol{by}\,\boldsymbol{i}$$

$$\boldsymbol{P}\left(\bigvee_a^b x_i\,\boldsymbol{by}\,\boldsymbol{i}\right) = \boldsymbol{1} - \prod_a^b\left(\boldsymbol{1} - \boldsymbol{P}\left(x_i|\bigwedge_a^{\boldsymbol{prev}(\boldsymbol{i})} {\sim}x_j\,\boldsymbol{by}\,\boldsymbol{j}\right)\right)\,\boldsymbol{by}\,\boldsymbol{i}$$

13 All mathematics derived from logic

All mathematics can be derived from logic, even set theory, since mathematics uses only quantitative and qualitative relations between things, that both comes from set theory, that has definitions-based definition based on logic, where all quantities come from cardinalities and measures of sets and all relations come from subsets of Cartesian products of sets.

So all fundaments of mathematics are newly defined in this book.

https://doi.org/10.1515/9783111441382-013

14 Depth and richness of logic

There is no meta-logic as something above logic, meta-language as something above language and meta-mathematics as something above mathematics, because, as I showed, logic completely explains itself and language and mathematics. So meta-logic is logic, meta-language is language, meta-mathematics is mathematics, and all come from logic. Logic can say everything about everything and explain everything, so it is not even the fundament of language, mathematics and all knowledge, because indeed language, mathematics and in general whole knowledge are logic.

For more details see Section 15.13.

Logic is "language of mind and mind of language".

You will not feel the whole depth and richness of logic, if you do not understand, that logic can process every thought and explains everything, including even itself, and is above everything, except only itself, because it decides what is true and what is false about everything. Of course, feelings have also their logic. Truth is one, but everyone knows only its part. Logic allows you to understand your part of truth and someone else's part of truth. That is, why truth and logic can enrich and join people. Truth can also serve as a shield against lies, and logic empowers you to discern truth from them.

https://doi.org/10.1515/9783111441382-014

15 Problems and paradoxes solved

15.1 Introduction (read it first)

In round brackets (in the titles of the sections in this chapter) the assumptions necessary to solve given problem are given:

US – Undecidable statements (defined in this book in New logic)
NL – New logic (defined in this book)
NST – New set theory (defined in this book)
TT – Theory of things (defined in this book)
TM – True measure (defined in this book)

If no assumption is given, then classical logic and classical set theory are enough to solve given problem, but the reasoning of the proposed solution is also correct in new logic and new set theory.

All symbols used in this section are described in Section 2.4.

In this book I proved, that all classical resolutions of paradoxes in classical logic are wrong. So classical logic contains paradoxes and a paradox is always a proof, that you made a mistake. For more details see section 5.15.

Remember, that anything inferred from true logic cannot make a contextual contrariety, so any contextual contrariety inferred from classical logic proves, that classical logic is false. And a paradox is a at least contextual contrariety. For more details see Section 5.15 and Section 5.12.

15.2 Liar paradox disproven (US)

Let us assume, that a liar expressed statement X in the following sentence: "I am lying".

If statement X is true, then the liar is lying, so what he said (statement X) is false.

Notice, that the following reasoning would be also correct:

If statement X is true, then the liar is saying the truth, that he is lying, so what he said (statement X) is false.

So we have, that:

$$X = (X \textit{ is true}) \Rightarrow (X \textit{ is false}) = {\sim}X$$

If statement X is false, then the liar is lying, so what he said (statement X) is true.

Notice, that the following reasoning would be also correct:

If statement X is false, then the liar is saying the truth, that he is lying, so what he said (statement X) is true.

So we have, that:

https://doi.org/10.1515/9783111441382-015

$$X = (X \text{ is true}) \Leftarrow (X \text{ is false}) = {\sim}X$$

So:

$$X = (X \text{ is true}) = (X \text{ is false}) = {\sim}X$$

So:

$$X = {\sim}X$$

If statement X was decidable, then the above equation would not be fulfilled, but it is fulfilled. So statement X is undecidable.

So this is not a paradox, because we cannot decide what is the logical value of an undecidable statement. For more details see Section 5.15.

Q.E.D.

Second proof:

If the liar is lying, then statement X is false.

$$X = (I \text{ am lying}) \Rightarrow (X \text{ is false})$$

If statement X is false, then the liar is lying.

$$X = (I \text{ am lying}) \Leftarrow (X \text{ is false})$$

So:

$$X = (I \text{ am lying}) = (X \text{ is false})$$

Then:

$$X = (X \text{ is false}) = {\sim}X$$

So:

$$X = {\sim}X$$

So X is undecidable.

So this is not a paradox, because we cannot decide what is the logical value of an undecidable statement. For more details see Section 5.15.

Q.E.D.

Third proof:

The statement X of the liar sentence "I am lying" is defined as follows:

$$X = (X \text{ is false}) = {\sim}X$$

Assume, that X is true. Then X is false. So we have a contradiction.

Assume, that X is false. Then X is true. So we have a contradiction.

So statement X is not decidable, since both assumptions about its logical value lead to a contradiction.

So X is undecidable.

So this is not a paradox, because we cannot decide what is the logical value of an undecidable statement. For more details see Section 5.15.

Q.E.D.

15.3 Card paradox disproven (US)

Suppose there is a card with sentences printed on both sides A and B:

The sentence on side A expresses statement a:

$$a = (\textit{The statement of the sentence on the other side of this card is true})$$

The sentence on side B expresses statement b:

$$b = (\textit{The statement of the sentence on the other side of this card is false})$$

So:

$$a = (b \textit{ is true}) = b$$

$$b = (a \textit{ is false}) = \sim a$$

So:

$$a = b = \sim a = \sim b$$

If statement a or b was decidable, then the above equation would not be fulfilled, but it is. So both statements a and b are undecidable and are equal to each other.

So this is not a paradox, because we cannot decide what is the logical value of an undecidable statement. For more details see Section 5.15.

Q.E.D.

15.4 No-no paradox disproven

Consider two statements each denying, what the other says:

$$a = (b\ is\ not\ true) = \sim(b\ is\ true) = \sim b$$

$$b = (a\ is\ not\ true) = \sim(a\ is\ true) = \sim a$$

Both statements have the same attempt of definition:

$$x = ((x\ is\ not\ true)\ is\ not\ true) = \sim(\sim(x\ is\ true)\ is\ true) = \sim\sim(x\ is\ true) = (x\ is\ true)$$

So:

$$x = (x\ is\ true)$$

So both attempts of definition failed to define some concrete statements, because the above equality is true for any statement.

And all we know about these statements is, that:

$$a = \sim b$$

So one of them is simply any statement and second is its negation.

So this is not a paradox.

Q.E.D.

15.5 Grelling-Nelson paradox disproven (US)

Consider the following definitions:

An autological adjective x is an adjective, for which the statement of the sentence "The adjective 'x' is x" is true.

A heterological adjective x is an adjective, that is not autological. So for such x the statement of the sentence "The adjective 'x' is x" is false.

And ask the question: Is the adjective "heterological" heterological?

Let us assume:

$$X = (The\ adjective\ \text{“}heterological\text{”}\ is\ heterological)$$

If we assume, that X is true, then the statement of the sentence "The adjective 'heterological' is heterological" is false.

So:

$$X \Rightarrow \sim X$$

If we assume, that X is false, then the statement of the sentence "The adjective 'heterological' is heterological" is true.

So:

$$\sim X \Rightarrow X$$

So:

$$(X \Rightarrow \sim X) \textit{ and } (\sim X \Rightarrow X)$$

So:

$$X = \sim X$$

If X was decidable, then the above statement would not be true, but it is true. So X is undecidable. So the correct answer to the question is that it is undecidable.

So this is not a paradox, because we cannot decide what is the logical value of an undecidable statement. For more details see Section 5.15.

Q.E.D.

15.6 Russell's paradox disproven without axioms (NST+US)

The best example of a statement about sets, which does not need any axioms to be disproven, is that about existence of the set from Russell's paradox:

$$\textit{there exists set } X, \textit{ that contains all sets, that do not contain themselves}$$

Firstly, a proof of it with the use of new set theory and new logic:

$$X = (\textit{the set, that contains all sets, that do not contain themselves})$$

Assume, that the above attempt to define X succeeded.

So:

$$X = \{x : (x \textit{ is a concrete set}) \textit{ and} \sim(x \in x)\}$$

Then:

$$true \equiv a = \bigwedge_{x \in \Omega} \Big((x \in X) = \big((x \text{ is a concrete set}) \text{ and} \sim (x \in x) \big) \Big)$$

$$\Rightarrow \Big((X \in X) = \big((X \text{ is a concrete set}) \text{ and} \sim (X \in X) \big) \Big)$$

$$\Rightarrow \big((X \in X) \leftrightarrow \sim (X \in X) \big) = b$$

Statement $(X \in X)$ cannot be true and cannot be false, because in both cases statement b, that is implicated by true statement a, is false, which is impossible.

So statement $(X \in X)$ is undecidable.

If X was a set, then statement $(X \in X)$ would be decidable, but it is not. So we have a contradiction. So the attempt to define X failed. So X is a nobeing. So X is not a set. So the sentence $(X \in X)$ is not meaningful, because relation $(X \in X)$ is meaningful only for X being a set.

In other words, the set from Russell's paradox is not defined as it was meant to be defined, so the attempt to define it failed, and we did not need any axiom, particularly axiom schema of specification or axiom of regularity in this case, to prove it.

Q.E.D.

And another proof:

$$X = (\textit{the set, that contains all sets, that do not contain themselves})$$

Assume, that the above attempt to define X succeeded.

X is defined as a set. So $(X \in X)$ is decidable.

We have, that:

$$(X \in X) \rightarrow \sim (X \in X)$$

$$\sim (X \in X) \rightarrow (X \in X)$$

So:

$$(X \in X) \leftrightarrow \sim (X \in X)$$

If $(X \in X)$ was decidable, then above statement would not be true, but it is true. So $(X \in X)$ is undecidable.

So we have a contradiction. So the attempt to define X failed. So X is a nobeing.

So this is not a paradox.

And the sentence $(X \in X)$ is not meaningful, because relation $(X \in X)$ is meaningful only for X being a set.

Q.E.D.

15.7 Burali-Forti paradox disproven

A proof of the paradox can be found in English Wikipedia (2018-10-10) in the term *Burali-Forti paradox*. It uses conclusion that X is ordinal to prove that it cannot be a set, but it is not true.

Let X be the set of all ordinal numbers. Then:

$$(X\ is\ ordinal) \rightarrow (X \in X) = (X < X) = \big((X < X)\ or\ (X > X)\big) = (X \neq X) \equiv false$$

So X is not ordinal. So it cannot be proved this way, that X cannot be a set.

So this is not a paradox.

15.8 Curry's paradox disproven (US)

Assume we have:

$$a = (a \rightarrow b)$$

Assume, that b is false.

Then we have:

$$a \rightarrow \sim a$$

And:

$$\sim a \rightarrow a$$

So:

$$a \leftrightarrow \sim a$$

If a was decidable, then the above statement would be false, but it is true. So a is undecidable, if b is false.

So if b is false, then this is also undecidable, that $a \rightarrow b$, since $a = (a \rightarrow b)$. So we cannot prove false b this way. So we cannot prove this way any false statement and proving true statement is not contradictory.

Q.E.D.

15.9 Curry's paradox in set theory disproven (NST+US)

Curry's paradox says, that any logical statement Y can be proved by examining the set:

$$X = \{x : (x \in x) \to Y\}$$

Then from the definition of a set for every x:

$$(x \in X) = ((x \in x) \to Y)$$

Assume, that Y is false.

And we have:

$$(X \in X) = ((X \in X) \to Y)$$

Then we have:

$$(X \in X) \to ((X \in X) \to \mathit{false}) \to \sim(X \in X)$$

And:

$$\sim(X \in X) \to \sim((X \in X) \to \mathit{false}) \to (X \in X)$$

So:

$$(X \in X) \leftrightarrow \sim(X \in X)$$

If $(X \in X)$ was decidable, then the above statement would be false, but it is true. So $(X \in X)$ is undecidable, if Y is false.

So if Y is false, then this is also undecidable, that $(X \in X) \to Y$, since $(X \in X) = ((X \in X) \to Y)$. So we cannot prove false Y this way. So we cannot prove this way any false statement and proving true statement is not contradictory.

So Curry's proof does not work (for more details see the term *Curry's paradox/ Formal proof/Naïve set theory* in English Wikipedia).

Q.E.D.

15.10 Cantor's paradox disproven

Since the set of all concrete sets includes its power set, there is a surjection from the set of all concrete sets to its power set. So the set of all concrete sets is not smaller than its power set. Simply, any subset of a set cannot be greater than the set.

If you still do not believe it, because you think, that Cantor's diagonal argument obtains, then let us prove, that it does not.

First of all, I will explain very shortly Cantor's diagonal argument:

Imagine, that you have some surjection f from some set S to its power set P. Now build new set T from these elements of set S, each one of which is mapped by f to a set, that does not contain it. Now since we have the surjection, some element t is mapped to T. Now assume, that t does not belong to T, then t should be chosen to T, but it is not, which makes a contradiction. So assume t belongs to T, then t should not be chosen to T, but it is, which makes a contradiction too. So since there is no other possibility, we have just proved, that none element of S is mapped to T by f, so f is not a surjection. So there is not a surjection from S to P, so S is smaller than P.

Q.E.D.

Let us map every set from the set of all concrete sets, that belongs also to the power set of the set of all concrete sets, to the same set in this power set. Then all elements of the power set will be used, so we have already a surjection. And let us map every other set, that contains itself, to a set, that contains it, and every other set, that does not contain itself, to a set, that does not contain it. Then the set of all concrete sets, each of which is mapped to a set, that does not contain it, is the set of all concrete sets, that do not contain themselves. So by virtue of conclusion from Russell's paradox (for more details see the solution of Russell's paradox, that is in Section 15.6) there is not such a set. But such a set is necessary to prove, that the power set of some set is greater than this set, using Cantor's diagonal argument (for more details see the term *Cantor's theorem* in English Wikipedia). So not only we have the surjection, but also Cantor's diagonal argument does not obtain. So I have found the mapping, for which Cantor's diagonal argument does not work, so there is the surjection from the set of all concrete sets to its power set, which proves, that the set of all concrete sets is not smaller than its power set, so its power set is not greater than the set of all concrete sets.

So this is not a paradox.

Q.E.D.

In other words, inexistence of the set of all concrete sets, that do not contain themselves, follows from the fact, that the set of all concrete sets includes its power set.

15.11 Hilbert's second problem

I suppose that you will find a positive solution to this problem in Section 16.2.

15.12 Gödel's incompleteness theorems disproven (US)

Since we have the logical causality (for more details see Section 5.61):

$$x = (x \text{ is not provable}) = (x \text{ is false}) = \sim x$$

So:

$$x = \sim x$$

And a decidable statement cannot fulfill the above equation, so the Gödel statement (x) is simply undecidable. So it cannot be proved by Gödel's theorem, that any logical system is either incomplete or inconsistent.

Q.E.D.

And, as you can see below, I do not even need the logical causality to prove the same.

If we do not use the logical causality, then we have the following proof:

Let us consider the following equality:

$$x = (x \text{ is provable})$$

1' If the above equality is not true for some statements, then the above equality defines a statement, so then the following two definitions would define the same statement:

$$x = (x \text{ is provable})$$

$$y = \sim x = (\sim x \text{ is provable})$$

So:

$$x = y = \sim x = \sim(x \text{ is provable}) = (x \text{ is not provable})$$

So:

$$\mathbf{x = \sim x = (x \text{ is not provable})}$$

If x was decidable, then the above equality would not be fulfilled, but it is fulfilled. So x is undecidable.

So x is the Gödel statement and is undecidable.

So Gödel's incompleteness theorem is false.

2' If the equality $(x = (x \text{ is provable}))$ is true for any statement x, then we have the logical causality, in case of which, as I proved in the previous proof, the Gödel statement is undecidable, so Gödel's incompleteness theorem is false.

So either way the Gödel's incompleteness theorem is false.

Q.E.D.

Here is another proof:

Here is some citation from the term *Complete Axiomatic Theory* from Wolfram MathWorld (2016-10):

> *An axiomatic theory (such as a geometry) is said to be complete if each valid statement in the theory is capable of being proven true or false.*

In other words, in a complete system a true statement always has a proof of being true and a false statement always has a proof of being false. And, what is not considered in the above citation, also an undecidable statement always has a proof, that it is undecidable.

Now let us analyze **Gödel's theorem**:

First proof:

Let statement x in system F have the following definition:

$$x = (x\ is\ not\ provable\ in\ system\ F)$$

Assume, that F is complete, then:

$$\sim x = (x\ is\ provable\ in\ system\ F) = x = (x\ is\ not\ provable\ in\ system\ F)$$

So, assume, that x. Then $(x\ is\ not\ provable\ in\ system\ F)$, so F is incomplete, so we have a contradiction.

So, assume, that $\sim x$. Then $(x\ is\ provable\ in\ system\ F)$, so x. This inconsistency does not make a contradiction, because we do not assume, that system F is consistent. But then $(x\ is\ not\ provable\ in\ system\ F)$, so F is incomplete, so we have a contradiction too.

So the assumption, that x, the same as the assumption, that $\sim x$, leads to a contradiction. So x is undecidable.

And, as you can see, that inconsistency does not prove, that system F is inconsistent, because it is a logical consequence of the undecidable assumption.

So x is undecidable or F is not complete.

Assume, that F is consistent, then:

Assume, that $\sim x$.

$$\sim x = (x\ is\ provable\ in\ system\ F) \Rightarrow x$$

So x cannot be false. So we have a contradiction. So x cannot be false.

So we have the above proof, that $x\ or\ (x\ is\ undecidable)$.

Assume, that x, then:

$$true \equiv (x\ is\ provable\ in\ system\ F) = \sim x$$

So $\sim x$. So x and $\sim x$.

This is inconsistency. So the assumption, that x, the same as the assumption, that $\sim x$, leads to a contradiction. So x is undecidable.

So x is undecidable or F is not consistent.

Summing up, if F is complete, then x is undecidable, so it does not prove, that F is inconsistent, and if F is consistent, then x is undecidable, so it does not prove, that F is incomplete.

Q.E.D.

The same is described in detail:

The Gödel statement G in system F says "G is not provable in the system F":

1. Assume, that system F is consistent. And assume, that G is false, that is, that "G is provable in the system F", which is in the system F, so G is true, but then, since system F is consistent, G cannot be false, so we have a contradiction. So we have the above proof in system F, that G is true (so "G is provable in system F"), if G is a decidable statement, since the only assumption, that G is false, under the condition, that F is consistent, leads to a contradiction. So G is an undecidable statement or G is true. If G is true, then "G is not provable in system F", which is inconsistency, so we have the contradiction with the consistency of system F. So G is undecidable.
2. Now assume outside system F, that system F is complete. So although system F does not have to be consistent, this assumption, that system F is complete, is the only thing, that from the inside of system F has to be consistent, so that we are able to make a contradiction. Now, when we assume in system F, that G is true, then G has a proof, since F is complete, and "G is not provable in system F", which makes the contradiction with the completeness of system F. But when we assume, that G is false, then negation of G says, that G is provable in system F, so G is also true, which means, that system F is inconsistent, but it also means, that "G is not provable in the system F", which is impossible, because in a complete system it cannot be true, that a true statement is not provable, so we have the contradiction with the completeness of system F. So G is undecidable. And it does not matter, that true G is also provable in system F, because it only means, that there is another inconsistency, but it only matters, that true G is not provable in system F, because it cannot be true in a complete system. So the conclusions about inconsistencies are not valid, because the assumption, that implicates them, is undecidable. And we allow inconsistencies, since we did not assume consistency of system F, so they are not a contradiction.
3. So, summing up, if we assume, that system F is consistent or complete, then statement G is undecidable. So this is not proved by Gödel's theorem, that a consistent system is incomplete. And this is not proved by Gödel's theorem, that a complete system is inconsistent.

Q.E.D.

The most fundamental truth to understand is, that a decidable statement is not true or false for no sufficient reason. So you can always ask: why a decidable statement is true and not false or false and not true, and the correct explanation, why the statement is true or false, will be always a proof.

Let us assume that x is a true decidable statement.

Is there any sufficient reason, why I cannot truthfully say, that it is false?

If there is not any such a reason, then I can truthfully say, that it is false, so we have a contradiction. So there is such a reason, that is also a sufficient reason, why it must be true, since a decidable statement is either true or false. So x is provable.

Q.E.D.

For more details see Section 5.61.

15.13 Tarski's undefinability theorem disproven (US)

From English Wikipedia (2017-10) from the term *Tarski's undefinability theorem* from section *Discussion*:

> *Smullyan (1991, 2001) has argued forcefully that Tarski's undefinability deserves much of the attention garnered by Gödel's incompleteness theorems. (. . .) Tarski's theorem, on the other hand, is not directly about mathematics but about the inherent limitations of any formal language sufficiently expressive to be of a real interest. (. . .) The broader philosophical import of Tarski's theorem is more strikingly evident.*

And from section *General form*:

> *Tarski's undefinability theorem (general form): Let (L, N) be any interpreted formal language which includes negation and has a Gödel numbering $g(x)$ such that for every L-formula $A(x)$ there is a formula such that $B \Leftrightarrow A(g(B))$ holds in N. Let T^* be the set of Gödel numbers of L-sentences true in N. Then there is no L-formula $True(n)$ which defines T^*. That is, there is no L-formula $True(n)$ such that for every L-formula A, $True(g(A)) \Leftrightarrow A$ is itself true in N.*
>
> *The proof of Tarski's undefinability theorem in this form is again by reduction ad absurdum. Suppose that an L-formula $True(n)$ defined T^*. In particular, if A is a sentence of arithmetic then $True(g(A))$ holds in N if and only if A is true in N. Hence for all A, the Tarski T-sentence $True(g(A)) \Leftrightarrow A$ is true in N. But the diagonal lemma yields a counterexample to this equivalence, by giving a "Liar" sentence S such that $S \Leftrightarrow \sim True(g(S))$ holds in N. Thus no L-formula $True(n)$ can define T^*. QED.*

Assume:

$$True(n) = (n\;is\;true)$$

$$\sim True(n) = \sim(n\;is\;true) = (n\;is\;not\;true) = (n\;is\;false)$$

For x defined as follows:

$$x = \sim True(x)$$

We have:

$$x = \sim True(x) = \sim(x\;is\;true) = (x\;is\;not\;true) = (x\;is\;false) = \sim x$$

So:

$$x = \sim x$$

If x was decidable, then it would not fulfill the above equation, but it fulfills it. So x is undecidable.

And in new logic for x the same as for any other statement we have:

$$x = True(x) = (x\;is\;true)$$

So:

$$\mathbf{x} = (x\;is\;true) = \mathbf{True}(\mathbf{x}) = \sim x = (x\;is\;false) = (x\;is\;not\;true) = \sim\mathbf{True}(\mathbf{x})$$
$$\equiv undecidable$$

So we have no contradiction. So the above proof of Tarski's undefinability theorem is incorrect.

And here is a proof of definability:

Formula $(x\;is\;true)$ defines T^*, since $x = (x\;is\;true)$ is by the definition of a statement in logic true for every statement x, because when you declare something, then and only then you want to communicate, that, what you claim, is true, so the statement of any declarence is always equivalent to the statement, that, what you claim by this declarence, is true.

Q.E.D.

15.14 Response to the proof of undecidability of halting problem

This is how we can trick a tricking code:

```
bool Halts(program, data)
{
  If program(data) calls Halts(data, data)
  // or: If program(data) IS Test(Test)
  {
    If program(data) is on the call stack then
     go to into endless loop
    else
     return false;
  }
  else return HaltsWithoutCall(program, data)
}
bool HaltsWithoutCall(program, data)
{
  If program(data) halts return true else return false
}
Test(program)
{
  If Halts(program, program) then go into endless loop
}
```

Either we know, that a proposed program (e.g., Test) calls Halts to make a contradiction and then operation (`program(data) calls Halts(data, data)`) in the procedure Halts also knows that, so the procedure tricks this program and thereby we have not a proof, that this program leads to a contradiction, or we do not know it, so then we also have not a proof, that this program leads to a contradiction.

In other words, for any Turing like proof we can modify our Halts procedure to refute it. So we can refute any Turing like proof, that halting problem is undecidable.

If HaltsWithoutCall is decidable under the assumption, that any program(data) does not call Halts(data, data), then halting problem will be undecidable only inside any program(data) that calls Halts(data, data), so it will be decidable for any program(data).

Q.E.D.

15.15 Response to the proof of Kolmogorov complexity uncomputability

This is how we can trick a tricking code:

```
int KolmogorovComplexity(s)
{
  K = size of KolmogorovComplexity + size of GenerateComplexString
  L = KolmogorovComplexityWithoutGCS(s)
  If (s == GenerateComplexString()) then
    If GenerateComplexString is on the call stack then
      return L
    else
      return K
  else
    return L
}
function GenerateComplexString()
{
  for i = 1 to infinity
    for each string s of length exactly i
      if KolmogorovComplexity(s) > (size of KolmogorovComplexity + size of
GenerateComplexString)
        return s
}
```

If KolmogorovComplexityWithoutGCS(s) returns Kolmogorov complexity under the assumption, that GenerateComplexString() is impossible, then KolmogorovComplexity() will return a wrong result only in function GenerateComplexString(), so it will return correct result for any string.

Q.E.D.

15.16 Berry's paradox solved

Consider the expression:

"the smallest positive integer not definable in English in under eighty letters"

Let us assume, that the above attempt to define a number succeeded.

Since the definition has less than 80 letters, it contradicts the assumption of the definition. So the above attempt to define a number failed.

Here is more detailed version of that proof:

Since the definition has less than 80 letters, the number defined by this definition has to be subtracted from the set of positive integer numbers before we will determine this number (the smallest number, definition of which cannot have less than 80 letters). So first of all to find this number we have to know this number before and secondly, of course, the number that we want to find on that list will not be on that list, because it will be subtracted from that list before we can start to look for it in that list. So the definition of this number contradicts itself for each of these two reasons alone, which means, that this definition does not define any number. So we have a contradiction. So the above attempt to define a number failed.

So this is not a paradox.

Q.E.D.

Summing up, even if the smallest positive integer not definable in English in under 80 letters exists, then it has not this definition, because this definition is self-contrary.

So it is not the case, that there is a systematic ambiguity in the word "definable".

From English Wikipedia from the term *Berry's paradox* from section *Relationship with Kolmogorov complexity*:

> *That is to say, the definition of the Berry number is paradoxical because it is not actually possible to compute how many words are required to define a number, and we know that such computation is not possible because of the paradox.*

Unfortunately, it has nothing to do with uncomputability of Kolmogorov complexity, because the fact, that Kolmogorov complexity cannot be computed by a program, does not implicate, that Kolmogorov complexity is not always defined. The same is for the computation versus the definition of how many words are required to define a number.

Analogously the following attempts to define a being fail:

"being, definition of which cannot contain word "dog""

"being, definition of which cannot relate to a dog"

15.17 Richard paradox solved

The proof is completely analogous to the above proof, that "Berry's paradox" is not a paradox.

Suppose we have list l of all real numbers, that have finite definition in English.

Consider the following definition of r:

"The integer part of r is 0, the n-th decimal place of r is 1 if n-th decimal place of l_n is not 1, and the n-th decimal place of r is 2 if n-th decimal place of l_n is 1."

Assume, that this definition describes a real number. Since the definition is finite and describes a real number, it has to be on list l, so let us assume, that it is at position m on that list. But then this number cannot have different digits at m-th decimal place from itself, so the above attempt to define it is self-contrary.

So this is not a paradox.

Q.E.D.

So contemporary mathematicians are wrong, when they claim, that the definition is invalid due to nonexistence of a well-defined notion of when an English phrase defines a real number, which would mean, that there is no unambiguous way to construct list l.

Moreover it is not true, that there is no such a well-defined notion, because we can distinguish strict sentences from not strict ones, that do not define real numbers, and for strict sentences we can logically prove, that they either define or not define some real number. And for all failed attempts to define a real number like the above one we can prove, that they are self-contrary, so they do not define a real number. So we have such a list unambiguously defined.

Even if *"there is not any way to unambiguously determine exactly which English sentences are definitions of real numbers (see Good 1966)"*, because *"there is not any way to describe in a finite number of words how to tell whether an arbitrary English expression is a definition of a real number"* (English Wikipedia, term *Richard's paradox*, section *Analysis and relationship with metamathematics*), then it does not mean, that there is not a proof of it, because sentences in a proof do not have to be finite and a proof does not have to have a finite number of steps. And, of course, any limitations of a given language do not limit the possibility of the existence of a proof, because a proof does not need sentences, since either way it uses statements and sentences are only their expressions in given language, that do not have to be a prefect tool, that can express everything. For more details see Section 7.1.

15.18 Hilbert-Bernays paradox solved

From English Wikipedia (2017-10) from the term *Hilbert-Bernays paradox*:
From section *History*:

The paradox appears in Hilbert and Bernays' Grundlagen der Mathematik and is used by them to show that a sufficiently strong consistent theory cannot contain its own reference functor. Although it has gone largely unnoticed in the course of the 20th century, it has recently been rediscovered and appreciated for the distinctive difficulties it presents.

From section *Formulation*:

Just as the semantic property of truth seems to be governed by naïve schema:
(T) The sentence "P" is true if and only if P
(where we used single quotes to refer to the linguistic expression inside the quotes), the semantic property of reference seems to be governed by the naïve schema:
(R) If a exists, the referent of the name "a" is identical with a
Consider however a name h for (natural) numbers satisfying:
(H) h is identical with "(the referent of h)+1"
Suppose that, for some number n:
(1) *The referent of h is identical with n*
Then, surely, the referent of h exists, and so does (the referent of h)+1. By (R), it then follows that:
(2) *The referent of "(the referent of h)+1" is identical with (the referent of h)+1*
and so, by (H) and the principle of indiscernibility of identicals, it is the case that:
(3) *The referent of h is identical with (the referent of h)+1*
But, again by indiscernibility of identicals, (1) and (3) yield:
(4) *The referent of h is identical with n*
And, by transitivity of identity, (1) together with (4) yields:
(5) *n is identical with n+1*
But (5) is absurd, since no number is identical with its successor.

The solution is quite simple:
From (H) we have:

$$h = (h\,|\,\text{“+1”})$$

where "|" is the symbol concatenation operator. For example, (*"abc"* | *"def"*) = *"abcdef"*.
So:

$$(h\,|\,\text{“+1”}) = (h\,|\,\text{“+1”}\,|\,\text{“+1”}) = (h\,|\,\text{“+1+1”})$$

And so on to infinity. So:

$$h = (h\,|\,\text{“+1”}) = (h\,|\,\text{“+1+1”}) = (h\,|\,\text{“+1+1+1”}) = \ldots = (h\,|\,\text{“+1+1+}\ldots\text{”})$$

So:

$$h = (h\,|\,\text{“+1+1+}\ldots\text{”})$$

So if the referent of h is natural number k, then the referent of h is $k+1+1+\ldots$, which is infinity.

So:

A = ^"the referent of h is a natural number" ⇒ ^"the referent of h is infinity" ⇒ ^"the referent of h is not a natural number" = ~A

So:

$$A \Rightarrow \sim A$$

So A is false. So we have a contradiction with the assumptions of the problem. So either the referent of h is not a natural number or (H) is not fulfilled. So we have not paradoxical conclusion (5). So it is not a paradox.

Q.E.D.

And even if we assume, that the referent of h is infinity, then conclusion (5) is not an absurd, because $\infty = (\infty + 1)$.

15.19 Boy or Girl paradox

This is not a logical paradox.

The paradox is as follows:

1. Mr. Jones has two children. The older child is a girl. What is the probability, that both children are girls?
2. Mr. Smith has two children. At least one of them is a boy. What is the probability, that both children are boys?

The whole controversy over that problem may come from the fact, that conditional probability is not the probability of "material conditional", so when you ask the question, what is the probability, that both children have sex x (b) either if older one of them has sex x (a) or if at least one of them has sex x (a), then some people will interpret it as (I):

1. **What is the probability, that** (both children have sex x (b) if older one of them has sex x (a))**?**
2. **What is the probability, that** (both children have sex x (b) if at least one of them has sex x (a))**?**

And other people will interpret it as (II):

1. **What is the probability, that both children have sex x (b)** (under the condition, that/if older one of them has sex x (a))**?**
2. **What is the probability, that both children have sex x (b)** (under the condition, that/if at least one of them has sex x (a))**?**

Here both interpretations are correct, because a comma is not sufficient to express the intended structure of dependence of the clauses of a sentence, so the above sentence is ambiguous, because it has two different and at the same time correct interpretations.

And then in the first case instead of calculating $P(b|a)$ you want to calculate probability of $a \to b$, that in general is not the same as $P(b|a)$:

$$\boldsymbol{P(a \to b) = P(\sim a \, or \, b) \neq P(b|a)}$$

For more details see Chapter 12.

Here is the following proof of it:

1′ Assume $P(b|a) \neq 1$:

$$P(\sim a \, or \, b) = P(a \, and \, b) + P(\sim a) = P(b|a)$$

$$P(a \, and \, b) + P(\sim a) = \frac{P(a \, and \, b)}{P(a)}$$

$$P(a \, and \, b)\left(P(a) - 1\right) + P(a)\left(1 - P(a)\right) = 0$$

$$P(a)\left(1 - P(a)\right) = P(a \, and \, b)\left(1 - P(a)\right)$$

1″ For $P(a) \neq 1$:

$$P(a) = P(a \, and \, b) = P(a)P(b|a)$$

$$P(b|a) = 1$$

A contradiction.

2″ For $P(a) = 1$:

$$P(\sim a \, or \, b) = P\left(\sim (a \, and \sim b)\right) = 1 - P(a \, and \sim b) = 1 - P(a)\,P(\sim b|a) = 1 - P(\sim b|a)$$
$$= P(b|a)$$

which is true.

2′ Assume $P(b|a) = 1$:

$$P(\sim a \, or \, b) = P(a \, and \, b) + P(\sim a) = P(a)P(b|a) + 1 - P(a) = P(a) + 1 - P(a) = 1$$
$$= P(b|a)$$

which is true.

So only for $P(a) = 1$ or $P(b|a) = 1$ there is an equality.

Q.E.D.

In cases I.1 and II.1 for $x = female$:

Older child	Younger child	a	b
Girl	Girl	1	1
Girl	Boy	1	0
Boy	Girl	0	0
Boy	Boy	0	0

$$\frac{3}{4} = P(a \to b) = P(\sim a \, or \, b) \neq P(b|a) = \frac{\frac{1}{4}}{\frac{2}{4}} = \frac{1}{2}$$

In cases I.2 and II.2 for $x = male$:

Older child	Younger child	a	b
Girl	Girl	0	0
Girl	Boy	1	0
Boy	Girl	1	0
Boy	Boy	1	1

$$\frac{2}{4} = \frac{1}{2} = P(a \to b) = P(\sim a \, or \, b) \neq P(b|a) = \frac{\frac{1}{4}}{\frac{3}{4}} = \frac{1}{3}$$

15.20 Knower paradox disproven

Let's assume, that we have:

$$A = (\textit{If the statement P is known, then P is true})$$

$$B = (\textit{If the statement P has been proved, then P is known})$$

$$C = (\textit{C is not known})$$

Assume, that $\sim C$ (C is known). So by A we have, that C (C is not known). So we have a contradiction.

So we have a proof, that C is true or C is undecidable. Let us assume, that C is true. Then C is not known. So by B we have, that C has not been proved. So we have a contradiction.

So we have proved, that C is undecidable, so we cannot prove by reductio ad absurdum, that C (C is not known), and use B to prove the contradiction, that C is both known and not known.

So this is not a paradox.

Q.E.D.

15.21 Fitch's paradox of knowability disproven

For more details see the term *Fitch's paradox of knowability* in English Wikipedia.

This is trivial. The set of "all truths" (understood the classical way as logical values associated with sentences) is changing in time. For example, once you sat, you are no longer standing. So it was true, that you were standing, until you sat. The same the sentence "sentence P is an unknown truth" has a true statement, until someone will get to know it. So it is not true, that the set of "all truths" must not include any of the form "something is an unknown truth", because the set of "all truths" the same has known and unknown part and unknown truths become known truths but also known untruths in time. So all truths are knowable, which means, that you can get to know them, because there is no any nonremovable obstacle to get to know them, but they are not guaranteed to be truths forever. So, for example, the unknown truth "sentence P is an unknown truth", becomes a known untruth (has a false statement) once you get to know it.

So this is not a paradox.

Q.E.D.

15.22 Bhartrhari's paradox solved

Consider the following statement:

$$X = (\mathit{This\ is\ unnameable})$$

If X is true, then this thing has the name "unnameable", so is nameable.

So:

$$X \Rightarrow \sim X$$

If X is false, then this thing has at least one name, so the fact, that it is named as "nameable" does not cause a contradiction.

So we have only:

$$X \Rightarrow \sim X$$

So X is just false. So there is not any thing, that is unnameable. So every thing is nameable.

So this is not a paradox.

Q.E.D.
Now consider the following statement:

$$X = (\textit{This is unsignifiable})$$

If X is true, then this thing signifies at least "something, that is unsignifiable", so this is signifiable.

So:

$$X \Rightarrow \sim X$$

If X is false, then this thing signifies something, so the fact, that it is "something, that is signifiable" does not cause a contradiction.

So we have only:

$$X \Rightarrow \sim X$$

So X is just false. So there is not any thing, that is unsignifiable. So every thing is signifiable.

So this is not a paradox.

Q.E.D.
Analogically the statement of the following sentence is false:

"Nothing can be said about this sentence."

15.23 Quine's paradox disproven (US)

Assume:

Y = "has a false statement when preceded by its citation"

$$Z = (\text{‘“’}|Y|\text{‘”’}|Y)$$

$$X = {}^\wedge Z$$

where "|" is the symbol concatenation operator. For example, ("abc" | "def") = "abcdef".

This is the same as the liar paradox.

Sentence Z precisely describes itself as the sentence it refers to, because:

$$([\textit{sentence}]\ Y\ (\textit{when})\ \textit{preceded by its citation}) = (\text{‘“’}|Y|\text{‘”’}|Y)$$

So without any doubt it is a self-referential sentence, the statement of which expresses the same as the statement (X *has a false statement*) which is the same as (X *is false*).

So we have directly from that:

$$X = (X\ is\ false) = {\sim}X$$

So:

$$X = {\sim}X$$

Or we can analyze implications of this statement having logical values to get the same:

$$X = (X\ is\ true) \Rightarrow (X\ is\ false) = {\sim}X$$

and:

$$X = (X\ is\ true) \Leftarrow (X\ is\ false) = {\sim}X$$

So:

$$X = (X\ is\ true) = (X\ is\ false) = {\sim}X$$

So:

$$X = {\sim}X$$

If X was decidable, then the above equation would not be fulfilled, but it is fulfilled. So X $\left(= \hat{}(\text{“‘”}|Y|\text{‘”’}|Y)\right)$ is an undecidable statement.

So this is not a paradox, because we cannot decide what is the logical value of an undecidable statement. For more details see Section 5.15.

Q.E.D.

15.24 Pinocchio paradox disproven (US)

It is the same as the liar paradox, because the statement of “My nose grows now” said by Pinocchio is equal to the statement of “I am lying” (and “The statement of this sentence is false” and “This sentence expresses some untruth”) said by Pinocchio or anyone else.

So this is not a paradox.

Q.E.D.

15.25 Yablo's paradox disproven (US)

Yablo's statements are the following infinite sequence of statements for $i = 1, 2, 3, \ldots$:

$$Yablo_i = (Yablo_j \; is \; not \; true \; for \; all \; integers \; j \; such, \; that \; j > i)$$

The statements of all Yablo's sentences are obviously undecidable, because they are elements of an infinite chain of dependencies, so this is not a paradox. For more details see Section 5.23.

This is not a paradox, because we cannot decide what is the logical value of an undecidable statement. For more details see Section 5.15.

Q.E.D.
For more details see the term *Yablo's paradox* in English Wikipedia.

15.26 Crocodile dilemma paradox disproven (US)

From English Wikipedia (2017-10) from the term *Crocodile dilemma*:

> *The premise states, that a crocodile, who has stolen a child, promises the father/mother that their child will be returned if and only if they correctly predict what the crocodile will do next.*

So we have:

$$A = (the \; crocodile \; will \; return \; the \; child)$$

$$B = (the \; parents \; correctly \; predict, \; what \; the \; crocodile \; will \; do)$$

$$C = (the \; parents \; predict, \; that \; the \; crocodile \; will \; not \; return \; the \; child)$$

So we have promise of the crocodile:

$$A = B$$

Let as assume that the promise must be fulfilled.

So in case C is true:

$$B \to A \to {\sim}B \to {\sim}A \to B$$

So:

$$A \to {\sim}A \to B \to A$$

And:

$$B \rightarrow \sim B \rightarrow B$$

So:

$$A \leftrightarrow \sim A$$

If A was decidable, then the above statement would be false, but it is true. So A is undecidable. So in case C is true it is undecidable what the crocodile will do next under the condition that his promise must be fulfilled. So in case C is true it is undecidable what the crocodile should do to fulfil his promise.

And:

$$B \leftrightarrow \sim B$$

If B was decidable, then the above statement would be false, but it is true. So B is undecidable. So in case C is true it is also undecidable whether the parents predicts correctly, what the crocodile will do.

So this is not a paradox, because we cannot decide what is the logical value of an undecidable statement. For more details see Section 5.15.

Q.E.D.

15.27 Paradox of the Court disproven (US)

After English Wikipedia (2017-10) from the term *Paradox of the Court:*

> *It is said that the famous sophist Protagoras took on a pupil, Euathlus, on the understanding that the student pay Protagoras for his instructions after he wins his first court case. After instruction, Euathlus decided to not enter the profession of law, and Protagoras decided to sue Euathlus for the amount owed.*
>
> *Protagoras argued that if he won the case he would be paid his money. If Euathlus won the case, Protagoras would still be paid according to the original contract, because Euathlus would have won his first case.*
>
> *Euathlus, however, claimed that if he won, then by the court's decision he would not have to pay Protagoras. If, on the other hand, Protagoras won, then Euathlus would still not have won a case and would therefore not be obliged to pay.*
>
> *The question is: which of the two men is in the right?*

Here is the solution:

$$EuathlusObligation = (Euathlus\ has\ to\ pay\ Protagoras)$$

$$EuathlusWon = (Euathlus\ won\ by\ the\ Court\ decision\ his\ first\ court\ case)$$

$$Contract = (EuathlusObligation = EuathlusWon) \equiv true$$

And we have:

$$\sim EuathlusObligation = (by\ decision\ of\ the\ Court \sim EuathlusObligation)$$
$$= EuathlusWon$$

So:

$$\sim EuathlusObligation = EuathlusWon = EuathlusObligation$$
$$\sim EuathlusObligation = EuathlusObligation$$

If *EuathlusObligation* was decidable, then the above equation would not be fulfilled, but it is fulfilled. So Euathlus obligation is undecidable, so the Court is simply not able to make any decision about it, so it is undecidable, which of the two men is right about it. The Court can make such a decision only if the Court will outlaw this contract.

So this is not a paradox, because we cannot decide what is the logical value of an undecidable statement. For more details see Section 5.15.

Q.E.D.

15.28 Unexpected hanging paradox disproven

From English Wikipedia (2017-10) from the term *Unexpected hanging paradox*:

> *Despite significant academic interest, there is no consensus on its precise nature and consequently a final correct resolution has not yet been established. (. . .) Even though it is apparently simple, the paradox's underlying complexities have even led to its being called "significant problem" for philosophy.*

From section *Description of the paradox*:

> *The paradox has been described as follows:*
>
> *A judge tells a condemned prisoner that he will be hanged at noon on one weekday in the following week but that the execution will be a surprise to the prisoner. He will not know the day of the hanging until the executioner knocks on his cell door at noon that day.*
>
> *Having reflected on his sentence, the prisoner draws the conclusion that he will escape from the hanging. His reasoning is in several parts. He begins by concluding that the "surprising hanging" can't be on Friday, as if he hasn't been hanged by Thursday, there is only one day left – and so it won't be a surprise if he's hanged on Friday. Since the judge's sentence stipulated that the hanging would be a surprise to him, he concludes it cannot occur on Friday.*
>
> *He then reasons that the surprising hanging cannot be on Thursday either, because Friday has already been eliminated and if he hasn't been hanged by Wednesday, the hanging must occur on Thursday, making a Thursday hanging not a surprise either. By similar reasoning he concludes that the hanging can also not occur on Wednesday, Tuesday or Monday. Joyfully he retries to his cell confident that the hanging will not occur at all.*

The next week, the executioner knocks on the prisoner's door at noon on Wednesday – which, despite all the above, was an utter surprise to him. Everything the judge said came true.

Solution of the paradox is very simple:

A = ^"Neither of days will be a surprise for an intelligent prisoner" ⇒ ^"The intelligent prisoner thinks, that there will not be an execution" ⇒ ^"Every day will be a surprise for the prisoner" ⇒ ^"Some day will be a surprise for the prisoner" = ~**A** ⇒ ^"The intelligent prisoner thinks, that there will be an execution" ⇒ ^"Neither of days will be a surprise for an intelligent prisoner" = **A**

So:

$$A \Rightarrow {\sim}A \Rightarrow A$$

So:

$$A = {\sim}A$$

If A was decidable, then the above statement would be false, but it is true. So A is undecidable.

So this is not a paradox, because we cannot decide what is the logical value of an undecidable statement. For more details see Section 5.15.

Q.E.D.

15.29 "Zero measure" and "one measure" probabilities and Borel-Kolmogorov paradox very simply disproven (TM)

Here you will find the solution of Borel-Kolmogorov paradox:

Let λ be the longitude and φ be the latitude of a point on a sphere, that has radius r, and $\lambda \in [-\pi, \pi]$ and $\varphi \in \left[-\frac{\pi}{2}, \frac{\pi}{2}\right]$.

Variables λ and φ are independent, so:

$$P(\varphi = \varphi_1 | \lambda = \lambda_1) = P(\varphi = \varphi_1) = \frac{2\pi r * \cos\varphi_1 r d\varphi}{4\pi r^2} = \frac{1}{2}\cos\varphi_1 d\varphi$$

$$P(\lambda = \lambda_1 | \varphi = \varphi_1) = P(\lambda = \lambda_1) = \frac{\frac{d\lambda}{2\pi} * 4\pi r^2}{4\pi r^2} = \frac{d\lambda}{2\pi}$$

So for two different great circles we have density function defined as:

$$\boldsymbol{f(\lambda'|\varphi=0)} = \frac{dF(\lambda'|\varphi=0)}{d\lambda'} = \frac{P(\lambda=\lambda'|\varphi=0)}{d\lambda'} = \frac{d\lambda}{2\pi} * \frac{1}{d\lambda'} = \boldsymbol{\frac{1}{2\pi}\frac{d\lambda}{d\lambda'}}$$

$$\boldsymbol{f(\varphi'|\lambda=0)} = \frac{dF(\varphi'|\lambda=0)}{d\varphi'} = \frac{P(\varphi=\varphi'|\lambda=0)}{d\varphi'} = \frac{1}{2}\cos\varphi'\frac{d\varphi}{d\varphi'} = \boldsymbol{\frac{1}{\pi} * \frac{\pi\cos\varphi' d\varphi}{2d\varphi'}}$$

So if we treat both the variables as a coordinates in Cartesian coordinate system, then, of course, we will have the uniform distribution, so we would have:

$$\frac{d\lambda}{d\lambda'} = 1$$

$$d\lambda = d\lambda'$$

$$\boldsymbol{\lambda = x(\lambda') = \lambda' + C_{\lambda'}}$$

$$\frac{\pi\cos\varphi' d\varphi}{2d\varphi'} = 1$$

$$d\varphi = \frac{2d\varphi'}{\pi\cos\varphi'}$$

$$\boldsymbol{\varphi = y(\varphi') = \frac{4}{\pi}\tanh^{-1}\tan\frac{\varphi'}{2} + C_{\varphi'}}$$

So we have to apply transformation $T(\lambda',\varphi') = (\lambda,\varphi) = \left(T_\lambda(\lambda',\varphi'), T_\varphi(\lambda',\varphi')\right)$ and its inversion between the coordinate systems:

$$\begin{cases} \lambda = T_\lambda(\lambda',\varphi') = x(\lambda') = \lambda' + C_{\lambda'} \\ \varphi = T_\varphi(\lambda',\varphi') = y(\varphi') = \frac{4}{\pi}\tanh^{-1}\tan\frac{\varphi'}{2} + C_{\varphi'} \end{cases}$$

$$\begin{cases} \lambda' = T_{\lambda'}(\lambda,\varphi) = x'(\lambda) = \lambda - C_{\lambda'} \\ \varphi' = T_{\varphi'}(\lambda,\varphi) = y'(\varphi) = 2\tan^{-1}\tanh\frac{\left(\varphi - C_{\varphi'}\right)\pi}{4} \end{cases}$$

Q.E.D.

We will get an uniform distribution also, if we look at this sphere from the perspective of the three-dimensional Euclidean space, in which it is located, because we can, for example, rotate this sphere to prove, that in both cases for every point we have exactly the same situation, so the probabilities are the same and uniform. Then:

$$\begin{cases} x = c_x + r\cos\varphi\cos\lambda \\ y = c_y + r\cos\varphi\sin\lambda \\ z = c_z + r\sin\varphi \end{cases}$$

where $c = (c_x, c_y, c_z)$ is the center of the sphere and r is the radius of the sphere.

But let us come back to the geographic coordinate system, where simple calculations show, that it is not at all a problem of conditional probability with a condition set, that has "zero measure":

$$P(\lambda=\lambda_1|\varphi=\varphi_1) = \frac{\left(\frac{d\lambda}{2\pi} * (2\pi r * \cos\varphi_1)\right) * rd\varphi}{2\pi r * (\cos\varphi_1 rd\varphi)} = \frac{d\lambda}{2\pi} = P(\lambda=\lambda_1)$$

$$P(\varphi=\varphi_1|\lambda=\lambda_1) = \frac{P(\lambda=\lambda_1,\varphi=\varphi_1)}{P(\lambda=\lambda_1)} = P(\varphi=\varphi_1) * P(\lambda=\lambda_1|\varphi=\varphi_1) * P(\lambda=\lambda_1)^{-1}$$

$$= \frac{1}{2}\cos\varphi_1 d\varphi * \frac{d\lambda}{2\pi} * \frac{2\pi}{d\lambda} = \frac{1}{2}\cos\varphi_1 d\varphi = P(\varphi=\varphi_1)$$

$$P(a<\lambda<b|\varphi=\varphi_1) = \int_a^b P(\lambda=x|\varphi=\varphi_1)\, for\, x = \frac{b-a}{2\pi}$$

$$P(a<\varphi<b|\lambda=\lambda_1) = \int_a^b P(\varphi=x|\lambda=\lambda_1)\, for\, x = \frac{\sin b - \sin a}{2}$$

where

$$\int_a^b f(x)\, for\, x = \lim_{h\to 0}(+)\left(\overbrace{f(a), f(a+h), \ldots, f(b)}^{\frac{b-a}{h}}\right) = \int_a^b \frac{f(x)}{dx}\, dx = \int_a^b \frac{f(x)}{dx}\, by\, x$$

So the crux of the matter here was to notice, that $(\lambda,\varphi) \neq (\lambda',\varphi')$, and that is all.

This is also a proof, that the only transformation of a coordinate system, that preserves the probability density functions, is a translation, because for:

$$\coprod_1^n f(x_i') = \frac{F(x_i')}{dx_i'} = \frac{P(x_i = x_i')}{dx_i'} = \frac{P(x_i = x_i')}{dx_i} * \frac{dx_i}{dx_i'}\ by\ i$$

We have:

$$\coprod_1^n \frac{dx_i}{dx_i'} = 1\ by\ i = \coprod_1^n x_i = x_i' + C_i\ by\ i$$

In general there is not any problem with probabilities, that has "zero measure". In reality probability is never equal to zero, if there is at least one point in a space, because zero integrates always to zero, while the probability of (in case of n-dimensional space

n-dimensional generalization of) volume of points integrates to a nonzero value. For example, the uniform probability of a point in a n-dimensional Euclidean space is:

$$P(x = x' | a \le x \le b) = \frac{\prod_1^n dx_i \text{ by } i}{\prod_1^n (b_i - a_i) \text{ by } i}$$

$$f(x' | a \le x' \le b) = \frac{\partial^n F(x' | a \le x' \le b)}{\partial x'_n \ldots \partial x'_1} = \frac{P(x = x' | a \le x \le b)}{\partial x'_n \ldots \partial x'_1}$$

$$= \frac{\prod_1^n dx_i \text{ by } it}{\prod_1^n (b_i - a_i) \text{ by } i} * \frac{1}{\prod_1^n dx_i \text{ by } i} = \frac{1}{\prod_1^n (b_i - a_i) \text{ by } i}$$

where $dx = dx'$.

When you want the probability of any k-dimensional part of a bounded region, then, of course, you have to integrate.

Having the probability density of any point you can calculate the probability of any shape in a bounded region. For more details see Section 16.1.

The “one measure” problem is analogous to the “zero measure” problem, but it concerns inequality instead of equality:

$$P(x \ne x' | a \le x \le b) = 1 - P(x = x' | a \le x \le b) = 1 - f(x' | a \le x' \le b) dx'_1 \ldots dx'_n$$

Suppose, that we map with an use of a random function g all points of a generalized n-dimensional cuboid defined by points a and b to points of this cuboid. What will be the probability, that point p inside the cuboid belongs to image of g?

$$e^{\int_a^b \log P(x \ne g(x') | a \le x \le b) \text{ for } x'} = e^{\int_a^b \log\left(1 - f(g(x') | a \le g(x') \le b) dx'_1 \ldots dx'_n\right) \text{ for } x'}$$

And we have, that:

$$1 - f\left(g(x') | a \le x' \le b\right) dx'_1 \ldots dx'_n$$

$$= \left(\left(1 - f\left(g(x') | a \le x' \le b\right) dx'_1 \ldots dx'_n\right)^{\frac{1}{f(g(x') | a \le x' \le b) dx'_1 \ldots dx'_n}}\right)^{f(g(x') | a \le x' \le b) dx'_1 \ldots dx'_n}$$

$$= \left(\frac{1}{e}\right)^{f(g(x') | a \le x' \le b) dx'_1 \ldots dx'_n}$$

So:

$$e^{\int_a^b \log P(x \neq g(x') | a \leq x \leq b) \text{ for } x'} = e^{\int_a^b \log\left(\frac{1}{e}\right)^{f(g(x') | a \leq x' \leq b) dx'_1 \ldots dx'_n} \text{ for } x'}$$

$$= e^{\int_a^b \log\left(\frac{1}{e}\right)^{f(g(x') | a \leq x' \leq b)} \text{ by } x'} = e^{-\int_a^b f(g(x') | a \leq x' \leq b) \text{ by } x'}$$

And that is it.

When $-f\left(g(x') | a \leq x' \leq b\right) = \log h\left(g(x') | a \leq x' \leq b\right)$, then:

$$e^{\int_a^b \log P(x \neq g(x') | a \leq x \leq b) \text{ for } x'} = e^{\int_a^b \log h(g(x') | a \leq x' \leq b) \text{ by } x'}$$

15.30 Bertrand paradox solved

From English Wikipedia from the term *Bertrand paradox*:

The Bertrand paradox goes as follows: Consider an equilateral triangle inscribed in a circle. Suppose a chord of the circle is chosen at random. What is the probability that the chord is longer than a side of the triangle?

Bertrand gave three arguments, all apparently valid, yet yielding different results.

1. *The "random endpoints" method: Choose two random points on the circumference of the circle and draw the chord joining them. To calculate the probability in question imagine the triangle rotated so its vertex coincides with one of the chord endpoints. Observe that if the other chord endpoint lies on the arc between the endpoints of the triangle side opposite the first point, the chord is longer than a side of the triangle. The length of the arc is one third of the circumference of the circle, therefore the probability that a random chord is longer than a side of the inscribed triangle is* $\frac{1}{3}$.
2. *The "random radius" method: Choose a radius of the circle, choose a point on the radius and construct the chord through this point and perpendicular to the radius. To calculate the probability in question imagine the triangle rotated so a side is perpendicular to the radius. The chord is longer than a side of the triangle if the chosen point is nearer the center of the circle than the point where the side of the triangle intersects the radius. The side of the triangle bisects the radius, therefore the probability a random chord is longer than a side of the inscribed triangle is* $\frac{1}{2}$.
3. *The "random midpoint" method: Choose a point anywhere within the circle and construct a chord with the chosen point as its midpoint. The chord is longer than a side of the inscribed triangle if the chosen point falls within a concentric circle of radius* $\frac{1}{2}$ *the radius of the larger circle. The area of the smaller circle is one fourth the area of the larger circle, therefore the probability a random chord is longer than a side of the inscribed triangle is* $\frac{1}{4}$.

As presented above, the selection methods have certain irregularities involving chords which are diameters. In method 2, each diameter can be chosen in two ways, whereas each other chord can be chosen in only one way. In method 3, each choice of midpoint corresponds to a single chord, except the center of the circle, which is the midpoint of all the diameters. These issues can be avoided by "regularizing" the problem so as to exclude diameters, without affecting the resulting probabilities.

If you want images, that illustrate these solutions, see this article in English Wikipedia.

So, as you can see, Bertrand gave three methods of calculating this probability. Unfortunately only one of them is correct. Here is a proof:

Firstly, it is necessary to understand, that the problem is equal to the problem of, what is the probability, that a random chord, that intersects the (outer) circle, intersects also the (inner) circle inscribed in equilateral triangle inscribed in this outer circle. This inner circle has the radius equal to the half of the radius of the outer circle.

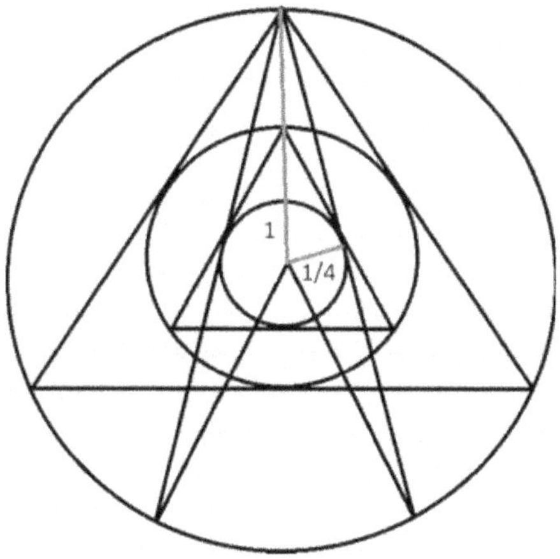

First solution implicates, that if we inscribe a smaller circle in the triangle inscribed in the inner circle and then inscribe a equilateral triangle again, but this time in this smaller circle, then probability, that random chord, that intersects the outer circle, also intersects this smaller circle, will be $\frac{1}{3} * \frac{1}{3} = \frac{1}{9}$, because again out of all chords, that intersect the inner circle, $\frac{1}{3}$ will intersect its inner circle (this smaller). But, unfortunately, if we use the same method, so we inscribe such a triangle in the outer circle, that the smaller circle (the circle in the inner circle) will be partially inscribed (this circle will be tangent to two sides of the triangle) in it (it will not be an equilateral triangle anymore, but still an isosceles triangle), then the inscribed in the outer circle angle, that subtends the appropriate arc, will be equal to:

$$28.955°2 \approx \arcsin\frac{\frac{1}{4}}{1} = 2\arcsin\frac{1}{4} \neq \frac{180°}{9} = 20°$$

And the central angle, that subtends the same arc on the outer circle, will be twice that due to the inscribed angle theorem:

$$57.91° \approx 4\arcsin\frac{\frac{1}{4}}{1} = 4\arcsin\frac{1}{4} \neq \frac{360°}{9} = 40°$$

So the length of the arc will be:

$$\frac{57.91°}{360°} * 2\pi \approx 1.011 \approx 4 \arcsin \frac{\frac{1}{4}}{1} = 4 \arcsin \frac{1}{4} \neq \frac{2\pi}{9} \approx 0.698$$

which proves, that this method is invalid.

Third solution counts the points in the outer circle and in inner circle, but this is obviously wrong method, because many points belong to the same chords of different sizes, so longer chords will be counted more times than shorter. This proves, that this method is invalid too.

Now the true solution:

Firstly, notice that for any angle there are $2r$ parallel chords intersecting a circle with radius r, which is number proportional to this radius, so for all angles there are $2r * 2\pi = 4\pi r$ chords intersecting the circle, which is still a number proportional to this radius. Of course, it is value of integration ($\int_{-r}^{r} \int_{0}^{2\pi} da\ dr$). So for the outer circle there will be 4π chords intersecting it, $4\pi * \frac{1}{2} = 2\pi$ of which will intersect the inner circle. So the probability, that chord, that intersects the outer circle, intersects also the inner circle is equal to $\frac{2\pi}{4\pi} = \frac{1}{2}$, which is the right answer to the main question.

So only the second solution proposed by Bertrand is correct, because it calculates exactly, what it should, since the number of chords intersecting a circle is proportional to its radius.

So this is not a paradox.

Q.E.D.

15.31 Any other paradox disproven or solved

For true premises, there are no real paradoxes. Paradox always means wrong assumptions. For more details see definition of a paradox in Section 5.15.

15.32 Categorical imperative and "Golden Rule" disproven

Immanuel Kant claimed not once, that if only someone will, for example, that some really immoral action, for example, under some conditions should be done by everyone else, then it is automatically moral, when it is done by him. For example, when someone want to kill all disabled people and he would want to be killed, if he were disabled, because he could not stand it. Here an obvious problem arises, because, of course, not everybody is like him, so the condition of the action is wider (stronger) than in case of the situation, in which someone else would kill that person, because not everyone cannot stand his disability. In other words, even the assumption, that it was moral to kill him, because he would want to be killed, if he was disabled, does not implicate, that it would be moral to kill someone else, who would not want to be killed, if he was disabled. In Kant's words categorical imperative is as follows: *"[to be moral always] act only according to that maxim whereby you can, at the same time, will that it should become a universal law", "[to be moral always] act as if the maxims of your action were to become through your will a universal law of nature".* According to Kant these are sufficient conditions of moral action – if you do not believe it, see the five steps of the universalizability test in the English Wikipedia term *Immanuel Kant* in section *Moral philosophy* and subsection *First formulation.* Of course, killing disabled people cannot be objectively moral by this rule, because according to the same rule opposite action could be also objectively moral, which would be a contradiction, so it is impossible.

So the same "Golden Rule" is not always good, because people are different and often want different things, that can be opposite to each other.

Q.E.D.

It can be fixed very simply: "do not do to anyone, who is not your enemy, what would be bad for you, if you were him". And simpler: "do not do to anyone, who is not your enemy, what is bad for him".

15.33 Norton's dome disproven

From English Wikipedia (2017-10) from the term *Norton's dome*:

> *Norton's dome is a thought experiment that exhibits non-deterministic system within the bounds of Newtonian mechanics. It was devised by John D. Norton in his 2003 paper "Causation as Folk Science". Norton's dome problem can be regarded as a problem in physics, mathematics, or philosophy. It poses interesting philosophical questions about the concepts of causality, determinism, and probability theory.*
>
> *The model consists of an idealize particle initially sitting motionless at the apex of idealized radially symmetrical frictionless dome described by the equation*
>
> $$h = \frac{2}{3g} r^{\frac{3}{2}}$$
>
> *where h is vertical displacement from the top of the dome to a point on the dome, r is the geodesic distance from the dome's apex to that point (in other words, a radial coordinate r is "inscribed" on the surface), and g the acceleration due to gravity.*
>
> *Norton shows that there are two classes of mathematical solutions to the system under Newtonian's physics. In the first, the particle stays sitting at the apex of the dome forever. In the second, the particle sits at the apex of the dome for a while, and then after an arbitrary period of time starts to slide down the dome in an arbitrary direction. The apparent paradox in the second case is that it would seem to occur for no discernible reason, and without any radial force being exerted on it by any other entity, apparently contrary to both physical intuition and*

> *normal intuitive concepts of cause and effect, yet the motion is still entirely consistent with the mathematics of the Newton's law of motion.*
>
> *While many criticisms have been made of Norton's thought experiment, such as it being a violation of the principle of Lipschitz continuity, or in violation of the principles of physical symmetry, or that it is somehow in some other way "unphysical", there is no consensus among its critics as to why they regard it as invalid.*

There is a very simple logical error in this thought experiment. Function $r(t)$, that gives the geodesic distance of the particle from the top of the dome at time t, can be greater than zero if and only if the particle at the apex will start to move, which is impossible by Newton's first law, that says, that in the absence of the net external force a body remains at rest or in a state of uniform linear motion. So if body is at rest, then it will stay at rest, so $r(t) = 0$. So any solution, that allows a nonzero value of $r(t)$, is wrong. And that is all.

Q.E.D.

You can visit the web page at the address www.pitt.edu/~jdnorton/Goodies/Dome to see the original writing of John D. Norton on the subject.

15.34 Ship of Theseus and sorites paradox solved (TT)

From English Wikipedia (2017-10) from the term *Ship of Theseus* from section *The thought experiment*:

> *First, suppose that the famous ship sailed by the hero Theseus in a great battle has been kept in a harbor as a museum piece. As the years go by some of the wooden parts begin to rot and are replaced by new ones. After a century or so, all of the parts have been replaced. Is the "restored" ship still the same object as the original?*
>
> *Second, suppose that each of the removed pieces were stored in a warehouse, and after the century, technology develops to cure their rotting and enable them to be put back together to make a ship. Is this "reconstructed" ship the original ship? And if so, is the restored ship in the harbor still the original ship too?*

The solution is very simple. In the strict sense the ship is a different physical being at any moment. But if we treat sameness as a non-strict term, then in someone's opinion it can be the same ship in both cases, because he can argue for example, that in the first case its physical bounds have not changed permanently and in the second case its content and form have not changed. So this is not a paradox. For more details see the beginning of Chapter 10.

Q.E.D.
The solution to the sorites paradox (and thereby also to the Continuum fallacy) is exactly the same.

15.35 Counterexample to identity of indiscernibles disproven

From English Wikipedia (2017-10) from the term *Identity of indiscernibles* from section *Critique/Symmetric universe*:

> *Max Black has argued against identity of indiscernibles by counterexample. Notice that to show that identity of indiscernibles is false, it is sufficient that one provides a model in which there are two distinct (numerically nonidentical) things that have all the same properties. He claimed that in a symmetric universe wherein only two identical spheres exists, the two spheres are two distinct objects even though they have all their properties in common.*
>
> *Black's argument appears significant because it shows that even relational properties (properties specifying distances between objects in space-time) fail to distinguish two identical objects in a symmetrical universe.*

Such two different objects are not identical in the strict sense, because their locations in space are different. In other words, they have different positions in any Cartesian coordinate system in this space. So they are not two different things, that have all the same properties. So it is not a counterexample and thereby not an argument against identity of indiscernibles, that says, that if two things have all the same properties (set of attributes), then they are equal.

15.36 Problem of universals and mathematical objects

Everything, that is defined, is a thing. Everything, that exists, is a being. Anything can be either a being or a no-being. So there is presence and existence. There are all things and all beings exist. All nonphysical things except impossible things and potential beings are eternal, so they always exist. Minds and thoughts exist at current state of physical reality.

A being can exist only two ways, always by the definition of a being as a part of reality:

1. As an abstraction of some aspect of physical reality (e.g., number, color, shape, etc.) and then it does not ever change, when abstracted from its existence in time, and exists nonphysically always and everywhere, so it is accessible by an intellect to every mind, in which its more or less perfect reflection come into being.
2. As a part of physical reality and then it can change in time (where we have on mind a sequence of different beings that share the same identity) not only due to its existence in time. And everything that is a part of physical reality is a part of reality.

Universals exist the first way, because they are abstract takes of some parts of physical reality. Mind and thoughts exist the second way, because they can change in time. But in a mind there are only more or less perfect reflections of universals and these reflections as parts of the mind exist the same way as the mind (the second way).

For more details see Chapter 10.

15.37 Problem of individuation (TT)

The criterion for a thing being an individual is the unique set of attributes.

For more details see the beginning of Chapter 10.

15.38 Material implication problem solved (NL)

True implication (inclusion of the meaning of one sentence in the meaning of another sentence) is given and material conditional (statement of the sentence in the form "if x then y") is proved and derived from it. So there is not any problem with that. For more details see Sections 5.35 and 5.38 and especially Section 5.38.7.

15.39 Counterfactuals problem disproven

From English Wikipedia from the term *List of unsolved problems in philosophy* from section *Counterfactuals*:

> *A counterfactual statement is conditional statement with a false antecedent. For example, the statement "If Joseph Swan had not invented the modern light bulb, then someone else would have invented it anyway" is counterfactual, because in fact, Joseph Swan invented the modern light bulb. The most immediate task concerning counterfactuals is that of explaining their truth-conditions. As a start, one might assert that background information is assumed when stating and interpreting counterfactual conditionals and that this background information is just every true statement about the world as it is (pre-counterfactual). In the case of the Swan statement, we have certain trends in the history of technology, the utility of artificial light, the discovery of electricity, and so on. We quickly encounter an error with this initial account: among the true statements will be "Joseph Swan did invent the modern incandescent light bulb". From the conjunction of this statement (call it "S") and the antecedent of the counterfactual ("~S"), we can derive any conclusion, and we have the unwelcome result that any statement follows from any counterfactual (see the principle of explosion). Nelson Goodman takes up this and related issues in his seminal "Fact, Fiction and Forecast"; and David Lewis's influential articulation of possible world theory is popularly applied in efforts to solve it.*

If you assume, that S, then you also assume, that ~~S. If you assume, that S, then for every X such, that $(X \Rightarrow {\sim}S)$, you assume, that ${\sim}X$, because otherwise it would be contrary to S, so the main assumption would fail. So it is not true, that S and $\sim S$ are both true, since $(\sim S \Rightarrow \sim S)$. So we have no contradiction, so the principle of explosion does not apply. So there is not any problem with counterfactuals.

Q.E.D.

15.40 Gettier problem disproven

From English Wikipedia (2017-10) from the term *List of unsolved problems in philosophy* from section *Gettier problem*:

> *Plato suggests, in his Theaetetus (210a) and Meno (97a–98b), that knowledge may be defined as justified true belief. For over two millennia, this definition of knowledge has been reinforced and accepted by subsequent philosophers. An item of information's justifiability, truth and belief have been seen as the necessary and sufficient conditions for knowledge.*

From English Wikipedia (2017-10) from the term *Gettier problem* from section *Gettier's two original counterexamples*:

> *Gettier's case is based on two counterexamples to the JTB [justified true belief] analysis. Each relies on two claims. Firstly, that justification is preserved by entailment, and secondly that it applies coherently to Smith's putative "belief". That is, that if Smith is justified in believing P, and Smith realizes that the truth of P entails the truth of Q, then Smith would also be justified in believing Q. Gettier calls these examples "Case I" and "Case II":*
>
> *Case I*
> *Suppose that Smith and Jones have applied for a certain job. And suppose that Smith has strong evidence for the following conjunctive proposition: (d) Jones is the man who will get the job, and Jones has ten coins in his pocket.*
>
> *Smith's evidence for (d) might be that the president of the company assured him that Jones would, in the end, be selected and that he, Smith, had counted the coins in Jones's pocket ten minutes ago. Proposition (d) entails: (e) The man who will get the job has then coins in his pocket.*
>
> *Let us suppose that Smith sees the entailment from (d) to (e), and accepts (e) on the grounds of (d), for which he has strong evidence. In this case, Smith is clearly justified in believing that (e) is true.*
>
> *But imagine, further, that unknown to Smith, he himself, not Jones, will get the job. And, also, unknown to Smith, he himself has ten coins in his pocket. Proposition (e) is true, though proposition (d), from which Smith inferred (e), is false. In our example, then, all of the following are true: (i) (e) is true, (ii) Smith believes that (e) is true, and (iii) Smith is justified in believing that (e) is true. But it is equally clear that Smith does not know that (e) is true; for (e) is true in virtue of the number of coins in Smith's pocket, while Smith does not know how many coins are in Smith's pocket, and bases his belief in (e) on a count of the coins in Jones's pocket, whom he falsely believes to be the man who will get the job.*

Case II
Smith, it is claimed by the hidden interlocutor, has a justified belief that "Jones owns a Ford". Smith therefore (justifiably) concludes (by the rule of disjunction introduction) that "Jones owns a Ford, or Brown is in Barcelona", even though Smith has no knowledge whatsoever about the location of Brown.

In fact, Jones does not own a Ford, but by sheer coincidence, Brown really is in Barcelona. Again, Smith had a belief that was true and justified, but not knowledge.

The solution is very simple, because a truly justified belief cannot turn out to be false, since a truly justified belief can be only such one, that is proved. So in both cases beliefs were not really justified. In the first case the president of the company could lie to Smith or change the decision, so Smith knew only, that the president of the company told him, that Jones will get the job, but he did not know whether it will really happen. In the second case it is not explained, why that belief is justified.

Q.E.D.

On the other hand, of course, knowledge in the strict sense has to be the justified (proved) set of statements, because this is the only way we are able to know, that they are true. So someone has some knowledge in the strict sense if and only if he or she thinks, that something is true, and it is indeed true and he understands, why it is true, because he himself experienced at least one of the realizations of that knowledge or logically proved it. So in the strict sense knowledge is infallible. But in the usual (not strict, subjective) sense anyone does not have to know any proof to have knowledge. For example, a student, that studies history, in the usual sense has often knowledge about events from the past, but has not any justification (proof) to believe in it other than his trust in a teacher's words, which in the strict sense is not a justification at all, because it is not a proof. So someone has some knowledge in the usual sense, for example, if he experienced something (at least one of the realizations of that knowledge) himself or he knows something from a source of information, that he trusts in. So in the usual sense knowledge is fallible. Another thing is, that knowledge in the usual sense is more often a conviction than a pure belief, because there are degrees of confidence in statements. In the strict sense knowledge is a conviction and gives complete confidence, so if you know something, then you do not have to believe, so it is not a belief. And the last thing is, that the phrase "justified true belief" says the same twice, because if something is truly justified, then, of course, it is also true. Moreover, the phrase is self-contrary, because when something is truly justified, then it is not a belief anymore. So, as you can see, the statement, that knowledge is a justified true belief, is wrong for many reasons.

15.41 Demarcation problem solved

Empirical sciences are those, that reference physical totality, and all other sciences do not refer to physical totality.

Science is based only on proved facts (true statements) expressed without the use of not strict terms. Nonscience is everything else.

15.42 Munchhausen trilemma (NL)

From English Wikipedia (2017-10) from the term *Munchhausen trilemma*:

> *In epistemology, the Munchhausen trilemma is a thought experiment used to demonstrate the impossibility of proving any truth, even in the fields of logic and mathematics. If it is asked how any knowledge is known to be true, proof may be provided. Yet that same question can be asked of the proof, and any subsequent proof. The Munchhausen trilemma is that there are only three options when providing proof in this situation:*
>
> 1. *The circular argument, in which theory and proof support each other*
> 2. *The regressive argument, in which each proof requires a further proof, ad infinitum*
> 3. *The axiomatic argument, which rests on accepted percepts*
>
> *The trilemma, then, is the decision among the three equally unsatisfying options.*

I proposed new logic based on definitions, that is defined in this book, that is correct and very satisfying option, because this logic does not need any standalone axioms (all axioms are used only to define basic notions), is almost not circular (it uses only one very intuitive primitive notion: conjunction) and explains itself as well as everything else, so is an exhaustive justification of any true theory.

For the definition of new logic, set theory and theory of things see Section 5.4.

15.43 Problem of the criterion

From English Wikipedia (2017-10) from the term *Problem of the criterion*:

> *In the field of epistemology, the problem of the criterion is an issue regarding the starting point of knowledge. This is a separate and more fundamental issue than the regress argument found in discussion on justification of knowledge.*
>
> *American philosopher Roderick Chisholm in his "Theory of Knowledge" details the problem of the criterion with two sets of questions:*
>
> 1. *What do we know? or What is the extent of our knowledge?*
> 2. *How do we know? or What is the criterion of knowing?*
>
> *An answering to either set of questions will allow us to devise a means of answering the other. Answering the former questions set first is called particularism, whereas answering the latter set first*

is called methodism. A third solution is skepticism, which proclaims that since one cannot have an answer to the first set of questions without first answering the second set, and one cannot hope to answer the second set of questions without first knowing the answers to the first set, we are, therefore, unable to answer either. This has the result of our being unable to justify any of our beliefs.

What we know?

We know, what we have experienced (including the real experiments, that we have conducted), and all abstractions, that we have derived from the experience, and all we have inferred from this.

How do we know?

We know using logic, in which everything has its logical cause, so logic is self-explainable.

15.44 Why there is something rather than nothing? (NL)

See section 5.54.

15.45 Stephen Hawking not right about 2-D creatures

From the official Stephen Hawking's website from the article *Space and Time Warps* (1999):

It would be very difficult to design a living being that could exist in only two dimensions.

Food that the creature couldn't digest would have to be spat out the same way it came in. If there were passage right the way through, like we have, the poor animal would fall apart.

So three dimensions seems to be minimum for life.

From his book *A Briefer History of Time* (2005) from the chapter *The Forces of Nature and the Unification of Physics*:

And if a two-dimensional creature ate something it could not digest completely, it would have to bring up the remains the same way it swallowed them, because if there were a passage right through its body, it would divide the creature into two separate halves: our two-dimensional being would fall apart. Similarly it is difficult to see, how there could be any circulation of the blood in a two-dimensional creature.

On the picture below are 2D sphincters. It is not shown on the picture, but they lock and gape in the appropriate moments. Both parts always have a contact with each other, so the central nervous system can move the muscles of the sphincters coordinated way. The contact allows also the circulation of the blood, but such a creature could have two separated blood circulation systems, since both parts take part in the absorption of feed.

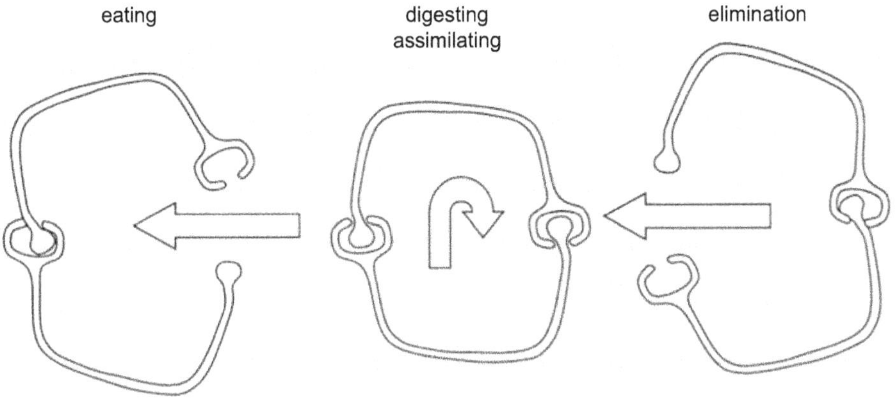

15.46 The ontological argument of Anselm of Canterbury disproven

15.46.1 Abstract

The ontological argument of Anselm of Canterbury comes from his 1078 work *Proslogion* and is disproven in this article.

15.46.2 The ontological argument of Anselm of Canterbury

From English Wikipedia from the term *Ontological argument*:

> *Since its proposal, few philosophical ideas have generate as much interest and discussion as ontological argument. Nearly all of the great minds of western philosophy have found argument worthy of their attention and criticism. The general consensus is that the argument is erroneous. However, consensus as to the exact nature of the argument's error or errors has long proved elusive to the philosophical community.*

About Bertrand Russell's criticism in the same term:

> *Bertrand Russell's, during his early Hegelian phase, accepted the argument; once exclaiming "Great God in Boots! – ontological argument is sound!" However, he later criticized the argument, asserting that "the argument does not, to a modern mind, seem very convincing, but it is easier to feel convinced that it must be fallacious than it is to find out precisely where the fallacy lies."*

And the ontological argument itself after English Wikipedia from the term *Ontological argument*:

Anselm's argument in Chapter 2 can be summarized as follows:

1. *It is a conceptual truth (or, so to speak, true by definition) that God is a being than which none greater can be imagined (that is, the greatest possible being that can be imagined).*
2. *God exists as an idea in the mind.*
3. *A beings that exits as an idea in the mind and in reality is, other things being equal, greater than a being that exists only as an idea in the mind.*
4. *Thus, if God exists only as an idea in the mind, then we can imagine something that is greater than God (that is, a greatest possible being that does exist).*
5. *But we cannot imagine something that is greater than God (for it is a contradiction to suppose that we can imagine a being greater than the greatest possible being that can be imagined.)*
6. *Therefore, God exists.*

In Chapter 3, Anselm presented a further argument in the same vein:

1. *By definition, God is a being than which none greater can be imagined.*
2. *A being that necessarily exists in reality is greater than a being that does not necessarily exist.*
3. *Thus, by definition, if God exists as an idea in the mind but does not necessarily in reality, then we can imagine something that is greater than God.*
4. *But we cannot imagine something that is greater than God.*
5. *Thus, if God exists in the mind as an idea, then necessarily exists in reality.*
6. *God exists in the mind as an idea.*
7. *Therefore, God necessarily exists in reality.*

See reference [10] for the source of this information.

In other words, the ontological "proof" is as follows:

"God" exists only as an idea in mind (A) $\Rightarrow$ There is something, that you can think of, that is greater than "God" (B) (because it additionally would exist in reality outside mind)

$$A \Rightarrow B$$

So:

$$\sim B \Rightarrow \sim A$$

So:

There is not anything, that you can think of, that is greater than "God" ($\sim B$) $\Rightarrow$ "God" does not exist only as an idea in mind ($\sim A$)

So:

There is nothing, that you can think of, that is greater than "God" ($\sim B$) $\Rightarrow$ "God" exists in reality outside mind ($\sim A$)

We assume, that there is nothing, that you can think of, that is greater than "God", so "God" exists in reality outside mind, since a true statement implicates only true statements.

15.46.3 The proof

Important note: First of all, if there is not the objective measure of greatness, then the greatest possible being does not exist. Secondly, if such a measure is infinite, then the greatest possible being also does not exist, because every being would have finite greatness, so there could always exist a being with greater greatness. Thirdly, it is worth to be noticed, that it is not so obvious, that human can imagine or even only think of the greatest possible being, due to limitations of his mind. But if we assume as Anselm, that he can, then we have, that:

It is quite easy to be disproved, because either we can think about the greatest possible thing we can think of, about which we assume, that it exists (one of its properties is, that it exists), but in reality (outside mind) it does not exist, or we cannot think about such a thing. Anselm assumes, that we can do it, in point 4 of the first proof and point 3 of the second proof in Section 15.46. So if we cannot think about such a thing, then the Anselm's argument is wrong, because it assumes the opposite of it. And if we can think about such a thing, then there is not a thing, that we can think of, that is greater due to its imagined (or thought) existence in reality, and such a thing does not exist in reality (outside mind), so conclusion in point 4 of the first proof and point 3 of the second proof is unjustified. So, again, Anselm's argument is wrong. Finally, either way Anselm's argument is wrong, so it is simply wrong.

Q.E.D.

Mistake is very simple: when you assume, that imagined (or thought) god does not exist in reality outside mind, you can automatically wrongly assume at the same time, that he is imagined (or thought) as not existing in reality outside mind, which does not have to be true, because he can be imagined (or thought) as existing in reality outside mind (when you imagine (or think), that it exists) and does not exist at the same time in realty outside mind. Anselm himself confirms that (contradicting himself at the same time) in words *"if God exists only as an idea in the mind, then we can imagine something that is greater than God (that is, a greatest possible being that does exist)"*. So according to him, if god does not exist outside mind, then you can think of a god that despite of this is conceived (or thought) as existing in reality outside mind, so why does he assume, that we did not think at the beginning about such greatest possible thing imagined (or thought) as existing outside mind but existing really only as an idea in our mind? So Anselm contradicts himself in his own proof. Notice also, that if we are not able to think of the greatest possible thing imagined (or thought) as existing, that does not exist outside mind, then Anselm's proof is still wrong, because in the above citation he assumes the opposite of it.

Simply god can be present in mind as an idea of some being existing outside mind. Then, whether he exists outside mind or not, there is not any possible thing, that is greater due to the assumption, that it exists outside mind, because this already existing in mind idea of god already uses assumption, that he exists in reality outside mind.

In other words, you can think of existing outside your mind the greatest possible thing (god) you can think of, that despite, what you think, does not exist outside your mind. Then, when you assume, that this thing exists outside your mind, there is not a possible thing you can think of, that is greater due to its imagined (or thought) existence in reality, because you already assumed, when you were thinking of that thing, that this thing, that you think of, is existing outside your mind. And in such a case you were simply wrong, because, of course, you can imagine (or think about) not existing being as existing and thus be wrong about that. Existence is just another attribute you attach to an imagined (or thought) thing. So imagining (or thinking), that something exists or does not exist, does not mean, that it automatically respectively exists or does not exist. You can imagine (or think), that something exists, when it does not exist, and you can imagine (or think), that something does not exist, when it exists.

16 Bonus

16.1 True measure – Lebesgue measure fails

16.1.1 Unambiguous measures and why Lebesgue measure fails

Lebesgue measure fails, when we need a measure of a point, and here is a proof of it.

Assume μ_L is Lebesgue measure.

We will assume, that the argument of a measure function can be a conjunction of points (a shape) as well as a set of points (a part of space).

From additive property for $a \leq b$ we have:

$$\mu_L((0,b]) = \mu_L((0,a]) + \mu_L((a,b])$$

Then for:

$$\mu_L((0,x]) = m(x)$$

We have:

$$\mu_L((a,b]) = \mu_L((0,b]) - \mu_L((0,a]) = m(b) - m(a) = \int_a^b m'(x) \, by \, x$$

$$\mu_L((a,a]) = m(a) - m(a) = 0$$

and in Lebesgue measure we have:

$$\mu_L(\{x\}) = 0$$

But then:

$$\mu_L(\{x\}) = \mu_L([x,x]) = \mu_L((x-dx,x]) = \mu_L((0,x]) - \mu_L((0,x-dx])$$
$$= \frac{m(x) - m(x-dx)}{dx} dx = m'(x)dx = 0$$

So:

$$\mu_L((0,x]) = m(x) = m(x) - m(0) = \int_0^x m'(x)dx = \int_0^x 0 = 0 \neq x$$

Q.E.D.

So what should be the correct measure of a point assuming, that $\mu((a,b]) = b - a$?

https://doi.org/10.1515/9783111441382-016

$$\mu((a,b]) = b - a = m(b) - m(a) = \int_a^b m'(x)\, by\, x$$

So:

$$m(x) = m(x) - m(0) = x - 0 = x$$

So:

$$m'(x) = 1$$

So:

$$\mu((a,b]) = b - a = \int_a^b 1\, by\, x = \int_a^b dx = \int_a^b dx\, for\, x = \int_a^b \mu(\{x\})\, for\, x$$

where:

$$\int_a^b f(x)\, for\, x = (+)\left(\coprod_a^b f(x)\, by\, x\right) = \int_a^b \frac{f(x)}{dx}\, dx = \int_a^b \frac{f(x)}{dx}\, by\, x$$

So:

$$\mu(\{x\}) = dx$$

In other words, what should not be a surprise, measure μ of a set $(a,b]$ have to be equal to sum of $\frac{b-a}{dx}$ measures of a single point. Hence:

$$b - a = \frac{b-a}{dx}\mu(\{x\})$$

From which we have the same as above:

$$\mu(\{x\}) = dx$$

But it is still a differential, so this is still standard analysis. And no matter whether we call such a measure a number or not, it can still be compared with the precision to differentials.

Analogous reasoning will prove, that n-dimensional generalization of measure (n-measure) μ of any point x in n-dimensional Euclidean space (Euclidean n-space) E for the Cartesian coordinate system is equal to:

$$\mu(\{x\}) = \mu(\{(x_1, \ldots, x_n)\}) = dx_1 * \ldots * dx_n$$

This weakness of Lebesgue measure is a consequence of too weak assumption about additive property of measure, that for $\cap_{i=1}^{\infty} S_i = \varnothing$:

$$\mu_L\left(\bigcup_{i=1}^{\infty} S_i\right) = \sum_{i=1}^{\infty} \mu_L(S_i)$$

For true n-measure μ for space E it should be as follows:

Let us define a continuum shape as the conjunction of all points of any subset of space, that cannot be expressed as the countable union of disjoint shapes, sum of the measures of which does not converge to finite measure for some degree of dimensionality. In other words, a continuum shape has a finite measure for any degree of dimensionality.

For set X of continuum shapes, for which $\cap_X s\, by\, s = \varnothing$, we have:

$$\mu\left(\bigcup_X s\, by\, s\right) = \int_X \mu(s)\, for\, s$$

To be able to calculate measure of union of sets, each of which contains one point of space, that sum to any continuum subset of space.

For any conjunctions A and B of points and for any sets A and B of points we have:

$$\mu(A - B) + \mu(B) = \mu(A \cup B)$$

$$\mu(A \cap {\sim}B) = \boldsymbol{\mu}(\boldsymbol{A} - \boldsymbol{B}) = \boldsymbol{\mu}(\boldsymbol{A} \cup \boldsymbol{B}) - \boldsymbol{\mu}(\boldsymbol{B})$$

$$\begin{aligned}\mu(A \cup B) &= \mu(B) + \mu(A \cap {\sim}B) = \mu(B) + \mu\big(\varnothing_X \cup (A \cap {\sim}B)\big)\\ &= \mu(B) + \mu\big((A \cap {\sim}A) \cup (A \cap {\sim}B)\big) = \mu(B) + \mu\big(A \cap ({\sim}A \cup {\sim}B)\big)\\ &= \mu(B) + \mu\big(A \cap {\sim}(A \cap B)\big) = \mu(B) + \mu\big(A - (A \cap B)\big)\\ &= \mu(B) + \mu\big(A \cup (A \cap B)\big) - \mu(A \cap B) =\\ &= \mu(B) + \mu\big((A \cup A) \cap (A \cup B)\big) - \mu(A \cap B) = \mu(B) + \mu(A) - \mu(A \cap B)\end{aligned}$$

So:

$$\mu(A \cup B) = \mu(A) + \mu(B) - \mu(A \cap B)$$

We can also calculate true n-measure for single point of Euclidean n-space for the Cartesian coordinate system from assumptions, that n-measure in Euclidean n-space E for the Cartesian coordinate system is translation invariant, so measure of a point from that space is uniform:

$$\big((p_1 \in E)\, and\, (p_2 \in E)\big) = \big(\mu(\{p_1\}) = \mu(\{p_2\}) = q\big)$$

And that n-measure of unit n-cube is equal to 1:

$$\mu\big((a_1, a_1 + 1] \times \ldots \times (a_n, a_n + 1]\big) = 1$$

Then:

$$\mu((a_1,a_1+1]\times\ldots\times(a_n,a_n+1]) = \int\limits_{a_1}^{a_1+1}\ldots\int\limits_{a_n}^{a_n+1}\mu(\{(x_1,\ldots x_n)\})\,for\,x_1\ldots for\,x_n$$

$$= \int\limits_{a_1}^{a_1+1}\ldots\int\limits_{a_n}^{a_n+1} q\,for\,x_1\ldots for\,x_n = \int\limits_{a_1}^{a_1+1}\ldots\int\limits_{a_n}^{a_n+1}\frac{q}{dx_1\ldots dx_n}dx_1\ldots dx_n$$

$$= \frac{q}{dx_1\ldots dx_n} = 1$$

So true n-measure μ of a n-point x is:

$$\mu(\{x\}) = q = dx_1\ldots dx_n = d\mu \neq 0$$

But:

$$\mu(\varnothing) = 0$$

And we have, that:

$$dx + dx = 2dx$$

So:

$$dx_1\ldots dx_n + dx_1\ldots dx_n = dx_1\ldots dx_{n-1}(dx_n + dx_n) = 2dx_1\ldots dx_n$$

From which we have, that:

$$\sum_{i=1}^{k} dx_1\ldots dx_n = k * dx_1\ldots dx_n$$

So we can calculate measure of any continuum shape, where we assume, that a single point in space is also continuum (in 0 dimensions), but we cannot evaluate unambiguous measure of sum of infinite countable number of continuum shapes, unless sum of their measures converges to finite value, because otherwise infinite countable set of continuum shapes can be always manipulated such a way, that we cannot even calculate average value a of element of such a sum to get $a * \aleph_0$ as a result, where $\aleph_0$ is the cardinality of infinite countable set.

In general, unambiguous n-space measures (n-measures) of subsets of n-space for given coordinate system can be sum of real value and any linear combination of differentials of appropriate sub-dimensional coordinate systems (subsystems) with only this exception, that coefficient for differential of n-point for the measure for the Cartesian coordinate system is always integer number, e.g.: $7 + 3dx_1 + 11dx_1dx_3 + 5dx_2dx_3 + 9dx_1\ldots dx_n$ and so on.

Since Euclidean measure for the Cartesian coordinate system is translation invariant, it is also rotation invariant, as a rotation is a special case of a translation of all points of a shape.

For this reason we have, that for u defined for every i as follows:

$$\int_0^1 1 = u = u(1-0) = \int_0^1 u\,dx_i = \int_0^1 \frac{1}{dx_i}\,dx_i$$

We have:

$$\mu_u(dx_1) = \mu_u(dx_2) = \cdots = \mu_u(dx_n) = u^{-1}$$

But it does not mean, that in general $dx_i = dx_j$ for $i \neq j$, because x_i and x_j in the Cartesian coordinate system are independent variables for $i \neq j$, so $x_i \neq x_j + C$.

So, for example:

$$\begin{aligned}\mu_u(7 + 3dx_1 + 11dx_1dx_3 + 5dx_2dx_3 + dx_1 \ldots dx_n) &= 7 + 3u^{-1} + 11u^{-2} + 5u^{-2} + u^{-n}\\ &= 7 + 3u^{-1} + 16u^{-2} + u^{-n}\end{aligned}$$

And this is **true measure of volume** of space, that can have even infinite number of dimensions. This is the most precise and exact unambiguous measure. Any variable t can have this measure and then $dt = u^{-1}$.

And for any two different points p_1 and p_2:

$$\mu(\{p_1\} \cap \{p_2\}) = \mu(\varnothing) = 0$$

So for any continuum shape S in n-space for the Cartesian coordinate system we have:

$$\begin{aligned}\mu(S) = \mu\left(\bigcup_S \{s\}\ by\ s\right) &= \int_S \mu(\{s\})\ for\ s = \int_E ((e \in S)?1:0)\ d\mu\ for\ e\\ &= \int_{-\infty}^{\infty} \cdots \int_{-\infty}^{\infty} \left(((x_1, \ldots, x_n) \in S)?1:0\right) dx_1 \ldots dx_n\end{aligned}$$

True uniform measure is closely related to probability (that has uniform density), that random point from unit n-cube (assuming, that the measure of an unit n-cube is equal to 1) is in given shape:

$$\mu(S) = \sum_{x_1=-\infty}^{\infty} \cdots \sum_{x_n=-\infty}^{\infty} P\left(x \in S | x \in ((x_1, x_1+1] \times \cdots \times (x_n, x_n+1])\right)$$

Existence of true measure, that in continuum shapes can consider every point separately, by virtue of additive property implicates, that Banach-Tarski paradox is impossible, because for $(A \cap B) = \varnothing$ and $\mu(A) = \mu(B) = p \neq 0$:

$$\mu(A \cup B) = \mu(A) + \mu(B) = 2p > p$$

16.1.2 Measure extended to countable sets – ambiguous measure

Of course, you can extend measure to countable sets, but then you have to remember, that for $0 < a \leq \aleph_0$:

$$a + \aleph_0 = \aleph_0$$
$$a * \aleph_0 = \aleph_0$$

and for $0 < a < \aleph_0$:

$$\aleph_0^a = \aleph_0$$

$$\aleph_0 - a = \aleph_0$$

So due to that we can no longer always calculate the ratio of one value to other value of measure, which is important, for example, for conditional probability.

And we have, that:

$$\lim_{dt \to 0} (x \pm \aleph_0 dt) = x$$

Then additive property must also be extended for countable sets.

So you can have, for example, the following measure:

$$9 + 4dt + \aleph_0 dt^2 + 7dt^2 + \aleph_0 dt^3 + 12dt^3$$

16.1.3 Example of not an ordinary continuum set

Assume, that we have a set defined as follows:

S_x = (Set of all real numbers from range $[0,1]$, that have every digit of x from position i on position $2i$ in k-base positional notation)

Since for $x \in [0,1]$ there is continuum of disjoint and equal in size continuum sets S_x, the measures of which sums to 1, we have:

$$\mu\left(\bigcup_0^1 S_x \, by\, x\right) = \int_0^1 \mu(S_x)\, for\, x = \int_0^1 \mu(S_0)\, for\, x = \int_0^1 \frac{\mu(S_0)}{dx} dx = (1-0)\frac{\mu(S_0)}{dx} = \frac{\mu(S_0)}{dx} = 1$$

So:

$$\mu(S_0) = dx$$

So $\mu(S_x) = dx$ for any $x \in [0,1]$.

16.1.4 Nonuniform measures and probability

In general, we can have measure μ of a weighted set f, that is a function, that transforms every element of the set T (space) to real value.

So we have:

$$f: T \to \left(R \cup \left\{\frac{v}{d\mu} : v \in R\right\}\right)$$

So:

$$f = \left\{((p), w) : (p \in T) \; and \; \left(w \in \left(R \cup \left\{\frac{v}{d\mu} : v \in R\right\}\right)\right)\right\}$$

And for every continuum $S \subset T$:

$$\mu(S) = \int_S \mu(s) \; for \; s = \int_S f(s) \; by \; s$$

So for n-space we have:

$$\mu(s) = f(s)\, dx_1 * \ldots * dx_n$$

$$\frac{\mu(s)}{dx_1 * \ldots * dx_n} = f(s)$$

Probability measure P is an example of such a measure, for which density function f is the weighted set and set T is the space, that is weighted by f. And probability measure fulfills additional conditions:

$$P(T) = 1$$

$f(x) \geq 0$ for $x \in T$

So $0 \leq P(S) \leq 1$ for every $S \subset T$.

As a consequence the probability of continuous random variable X being equal to single value x will not be equal to 0:

$$P(X = x) = P(x \leq X \leq x + dx) = P(X \leq x + dx) - P(X \leq x) = F(x + dx) - F(x) = F'(x)dx = f(x)dx$$

16.2 Arithmetic definition based on definitions

16.2.1 The simplest definition of arithmetic

For the definitions-based definition of new logic, set theory and theory of things see Section 5.4.

The cardinality of a set, that has one element more than some set, is, of course, greater by 1 and the cardinality of a set, that has one element less than some set, is smaller by 1. So, for example, a proof, that $1+1=2$, is not even trivial, but completely unnecessary, because we define 2 as being greater by 1 than 1. The value 2 is just the cardinality of any set, that has one element more than any single element set, and that is all.

So here is the simplest **definition of arithmetic**:

A **list** of element a is in the form: a

A list of element a and of all elements of list x is in the form: a, x

A list, that can be written down completely assuming that every element can be written down completely, is **finite**.

Any **finite set** is in the form $\{l\}$ for some finite list l.

To avoid ambiguities, if you want to write down any list l as a single thing, then you have to put it in brackets "$\langle l\rangle$", e.g.: $\{\ldots, \langle l\rangle, \ldots\}$.

The cardinality of a finite set is the only property of the set, that does not change, when we substitute all elements of the set with any other things, that do not belong to the set, and changes always when we insert new elements or remove some elements from the set.

The same can be written in the following way:

$$\bigwedge_{X \text{ is a finite set}} \bigwedge_{Y \subset X} \bigwedge_{(Z \text{ is a finite set}) \text{ and } ((Z \cap X) = \varnothing) \text{ and } (|Z| = |Y|)} (|X| = |(X - Y) \cup Z|)$$

$$\bigwedge_{X \text{ is a finite set}} \bigwedge_{(Y \subset X) \text{ and } (Y \neq \varnothing)} (|X| \neq |X - Y|)$$

$$\bigwedge_{X \text{ is a finite set}} \bigwedge_{(Y \text{ is a finite set}) \text{ and } ((Y \cap X) = \varnothing) \text{ and } (Y \neq \varnothing)} (|X| \neq |X \cup Y|)$$

The cardinality of a finite set is equal to a **natural number**.

1′ So a natural number is now defined, but it needs arithmetic:

$$0 = |\varnothing|$$

$$1 = |\{0\}| = 0 + 1$$

$$2 = |\{0,1\}| = 1 + 1$$

$$3 = |\{0,1,2\}| = 2 + 1$$

and so on.

We can give the following procedure to construct all consecutive natural numbers:

$$x_{current} = |X|$$

$$x_{next} = |X \cup \{x_{current}\}| = |\{0, 1, 2, \ldots, x_{current}\}| = x_{current} + 1$$

For any finite set A: $-|A|$ is a negative **integer number** such that:

$$-(-|A|) = |A|$$

$$-|A| + |A| = 0$$

and we can define arithmetic of integer numbers the following way for any sets A and B and any natural number n:

$$|A \cup B| = |A| + |B - (A \cap B)|$$

$$|A - B| = |A| - |A \cap B| = |A| + (-|A \cap B|) = (-|A \cap B|) + |A|$$

$$-|A \cap B| = |A - B| - |A| = |A - B| + (-|A|) = (-|A|) + |A - B|$$

$$|A \times B| = (\pm |A|) \ast (\pm |B|)$$

$$-|A \times B| = (\pm |A|) \ast (\mp |B|)$$

$$\frac{\pm |A \times B|}{\pm |B|} = |A|$$

$$\frac{\pm |A \times B|}{\mp |B|} = -|A|$$

$$(A^n \times A) = A^{n+1}$$

$$A^0 = \frac{A}{A} = 1_\times = \{\varnothing\} = \{(\Phi)\}$$

$$A \times 0_\times = 0_\times = \varnothing = \{\Phi\}$$

$$|A^n| = |A|^n$$

$$(-|A|)^n = |A|^n \ast (-1)^n$$

$$\sqrt[n]{|A^n|} = |A|$$

$$\sqrt[n]{-|A^n|} = |A| \ast (-1)^n$$

$$|A^B| = |A|^{|B|}$$

$$\sqrt[|B|]{|A^B|} = |A|$$

$$\sqrt[|B|]{-|A^B|} = |A| \ast (-1)^{|B|}$$

For more details see Sections 5.56 and 5.57:

$$|A| = |A \cup 0_\times| = |A \times 1_\times| = |A^{1\times}|$$

$$|A \times 0_\times| = |0_\times| = 0$$

$$|A^{0\times}| = |1_\times| = 1$$

$$(A \subset B) \Rightarrow (0 \leq |A| \leq |B|) = \left(\frac{0}{|B|} = \frac{|B \times 0_\times|}{|B|} = |0_\times| = 0 \leq \frac{|A|}{|B|} \leq \frac{|B|}{|B|} = \frac{|B \times 1_\times|}{|B|} = |1_\times| = 1\right)$$
$$= \left(0 \leq \frac{|A|}{|B|} \leq 1\right)$$

and so on.

16.2.2 Arithmetic defined different way

2′ But let us define arithmetic in a **different way**:

$$0 = |\varnothing|$$

$$1 = |\{0\}|$$

$$2 = |\{0,1\}|$$

Let us define an **ordered pair** for any a and b:

$$(a,b) = \Big\{\{1,\{a\}\},\{2,\{b\}\}\Big\}$$

And a **transition** f for any x and any y and any set f of pairs in the form (x,y):

$$(f\{x\} = y) = ((x,y) \in f)$$

$$\bigwedge_{X \text{ is a finite set}} \; \bigwedge_{(Y \text{ is a finite set}) \text{ and } ((X \cap Y) = \varnothing)} (|X \cup Y| = |X| + |Y|)$$

Let subtraction be the inversion of addition, then:

$$\bigwedge_{X \text{ is a finite set}} \; \bigwedge_{(Y \text{ is a finite set}) \text{ and } ((X \cap Y) = \varnothing)} (|X \cup Y| - |Y| = |X| + |Y| - |Y| = |X|)$$

From the properties of a set we have commutativity for any disjoint sets X, Y and Z:

$$|X| + |Y| = |X \cup Y| = |Y \cup X| = |Y| + |X|$$

and associativity:

$$|X| + (|Y| + |Z|) = |X| + |Y \cup Z| = |X \cup Y \cup Z| = |Y \cup X| + |Z| = (|Y| + |X|) + |Z|$$

and:

$$1 = 0 + 1 = 1$$

$$2 = 1 + 1 = 1 + 1$$

$$3 = 2 + 1 = 1 + 1 + 1$$

$$4 = 3 + 1 = 1 + 1 + 1 + 1$$

...

$$n = \overbrace{1 + \cdots + 1}^{n}$$

The set N of natural numbers:

$$N = \{1, 2, 3, \ldots\}$$

The definition of a **finite tuple** (x):

$$(x) = (x_1, x_2, \ldots, x_n) = \left\{\{1, \{x_1\}\}, \{2, \{x_2\}\}, \ldots \{n, \{x_n\}\}\right\}$$

which can be written as follows:

$$\bigwedge_{i=1,2,..,n} (\{i, \{x_i\}\} \in (x))$$

Let us define the **indexation** of tuple (x):

$$\bigwedge_{i=1,2,..,n} \left(\{i, \{(x)_i\}\} \in (x)\right)$$

For any tuple (x) and any y and any set f of pairs in the form $([\![x]\!], y)$:

$$(f(x) = y) = (([\![x]\!], y) \in f)$$

Where $[\![x]\!]$ is a tuple even for single element list x.

Such f will be called a **function**.

We also have for any tuple (x):

$$f\{(x)\} = f(x)$$

So every function can be interpreted as a transition.

A tuple can be also defined without numbers:

$$(\Phi) = \varnothing$$

$$x = (a, b, c, \ldots, d) = \{\{a\}, (b, c, \ldots, d)\} = \left\{\{a\}, \left\{\{b\}, \left\{\{c\}, \ldots \{\{d\}, \varnothing\}\right\}\right\}\right\}$$

Then:

$$head(x) = a$$
$$tail(x) = (b, c, \ldots, d)$$

and so on.

And we have:

$$x_i = x(i) = head\left(tail^{i-1}(x)\right)$$
$$length(\varnothing) = 0$$
$$length(x) = 1 + length\left(tail(x)\right)$$

The set *Z* of integer numbers; $-a$ is negative integer number for any natural number a and:

$$-(-a) = a$$
$$(-a) + a = 0$$
$$Z = \{0,\ 1,\ -1,\ 2,\ -2,\ 3,\ -3,\ \ldots\}$$

Since we can pair all elements of sets N and Z:

$$\{(0,0), (1,1), (2,\ -1), (3,2), (4,\ -2), (5,3), \ldots, (2i-1, i), (2i,\ -i), \ldots\}$$

We have the above proof, that the cardinality of the set of all integer numbers is equal to the cardinality of the set of all natural numbers.

For any positive integer numbers a and b:

$$\left((a+1) = b\right) = \left(a = (b-1)\right)$$

For any negative integer number a:

$$(a+1) = -\left((-a) - 1\right)$$
$$(a-1) = -\left((-a) + 1\right)$$

So for any integer numbers x, y and z:

$$(x > 0) = \left(x \in (N - \{0\})\right)$$
$$(x < 0) = \left((-x) \in (N - \{0\})\right)$$

$$(y > 0) \Rightarrow \left(x + y = x + \overbrace{1 + \cdots + 1}^{y} = x + \left(\overbrace{1 + \cdots + 1}^{y-1} + 1 \right) = x + (y-1) + 1 \right)$$

$$((x<0)\ and\ (y<0)) \Rightarrow \left(x+y = -((-x)+(-y))\right)$$

$$(y>0) \Rightarrow \left(x*y = \overbrace{x+\cdots+x}^{y} = \overbrace{x+\cdots+x}^{y-1} + x = x*(y-1)+x\right)$$

$$((x<0)\ and\ (y<0)) \Rightarrow \left(x*y = ((-x)*(-y))\right)$$

Then by induction:

$$x*(y+z) = x*y + x*z$$

Then:

$$\boldsymbol{x*(y*z)} = x*\left(\overbrace{y+\cdots+y}^{z}\right) = \overbrace{xy+\cdots+xy}^{z} = \boldsymbol{(x*y)*z}$$

$$\boldsymbol{x-y} = (-(-x)+(-y)) = ((-1)(-x)+(-1)(y)) = -1*(-x+y) = -\boldsymbol{(y-x)}$$

$$\begin{aligned}\boldsymbol{x*y} &= (y+(x-y))*(x+(y-x)) = (y+(x-y))*x + (y+(x-y))*(y-x)\\ &= yx + x(x-y) + y(y-x) + (x-y)*(y-x)\\ &= yx + x(x-y) - y(x-y) + (x-y) * -(x-y)\\ &= yx + (x-y)(x-y) - (x-y)(x-y) = \boldsymbol{y*x}\end{aligned}$$

So:

$$(y>0) \Rightarrow \left(x^y = \overbrace{x*\ldots*x}^{y} = \overbrace{x*\ldots*x}^{y-1} * x = x^{y-1}*x\right)$$

Then by induction:

$$x^{y+z} = x^y * x^z$$

$$((y>0)\ and\ (z>0))$$

$$\Rightarrow \left(x^{yz} = \overbrace{x*\ldots*x}^{y*z} = \overbrace{x*\ldots*x}^{\overbrace{y+\ldots+y}^{z}} = \overbrace{\overbrace{x*\ldots*x}^{y} * \overbrace{x*\ldots*x}^{y}}^{z} = (x^y)^z\right)$$

$$((y<0)\ and\ (z<0)) \Rightarrow \left(x^{yz} = x^{(-y)(-z)} = (x^{-y})^{-z}\right)$$

We will say, that $x+1$ is **greater** than x, because they are the cardinalities of respectively a proper superset of a set and the set, and $x-1$ is **smaller** than this x, because they are the cardinalities of respectively a proper subset of a set and the set. And both these properties are transitive:

$$((a<b)\ and\ (b<c)) \Rightarrow (a<c)$$

From this, by induction we have:

$$(a<b) = \bigvee_{d \in (N-\{0\})} ((a+d)=b)$$

$$(a \le b) = ((a<b)\ or\ (a=b))$$

$$(a>b) = \sim(a \le b)$$

$$(a \ge b) = \sim(a<b)$$

$$(a<0) \Rightarrow (|a| = -a)$$

$$(a \ge 0) \Rightarrow (|a| = a)$$

The existence of rational numbers come from the fact, that we can divide any segment into k equal parts, where k is any natural number greater than 1.

For any integer numbers a, b, x, y:

$$a^{-1} = \left(\frac{1}{a}\right) = (one\ of\ a\ equal\ parts\ of\ 1)$$

$$\frac{a}{b} = a * b^{-1} = (a\ of\ b\ equal\ parts\ of\ 1)$$

$$\left(\frac{b}{a}\right)^{-x} = \left(\frac{a}{b}\right)^{x} = \frac{a^x}{b^x}$$

$$\left(\frac{a}{b} * \frac{x}{y}\right) = \frac{a * x}{b * y}$$

$$\left(\frac{a}{x} + \frac{b}{x}\right) = \frac{a+b}{x}$$

$$\left(\frac{a}{x} \pm \frac{b}{y}\right) = \left(1 * \frac{a}{x} \pm 1 * \frac{b}{y}\right) = \left(\frac{y}{y} * \frac{a}{x} \pm \frac{x}{x} * \frac{b}{y}\right) = \left(\frac{y*a}{y*x} \pm \frac{x*b}{x*y}\right) = \frac{(y*a) \pm (x*b)}{x*y}$$

$$\left(\frac{\frac{a}{b}}{\frac{x}{y}}\right) = \left(\frac{a}{b} * \left(\frac{x}{y}\right)^{-1}\right)$$

The set *Q* of all rational numbers:

$$Q = \left\{0, \frac{1}{1}, -\frac{1}{1}, \frac{1}{2}, -\frac{1}{2}, \frac{2}{1}, -\frac{2}{1}, \frac{1}{3}, -\frac{1}{3}, \frac{3}{1}, -\frac{3}{1}, \frac{2}{3}, -\frac{2}{3}, \frac{3}{2}, -\frac{3}{2}, \frac{1}{4}, -\frac{1}{4}, \frac{4}{1}, -\frac{4}{1}, \ldots\right\}$$

The existence of real numbers come from the fact, that we can divide any segment into k equal parts, where k is any natural number greater than 1, and choose one of them and then divide this part into k equal parts, and again choose one of them, and so on to infinity. Every time we choose some part, the lower and the upper bounds come closer to each other and in infinity they are equal, because, if we repeat division into k equal parts infinite number of times, then we will get a $\frac{1}{k^{\infty}}$ $\left(=\frac{1}{\infty}=0\right)$ size part, so these divisions can hit any element of the whole segment. If we give measure 1 to that segment, then by these subdivisions we got a real number between 0 and 1, to which we can add a finite number of 1, to get any real number at continuum of real numbers.

If we assign numbers 0, 1, . . ., $k-1$ to the appropriate part at every step, then we will get an infinite sequence of symbols, that are called k-ary digits.

For $a_i \in \{0,1,\ldots,k-1\}$ for any i such that $i \leq n$:

$$[a_n a_{n-1} \ldots a_0.a_{-1}a_{-2}a_{-3} \ldots a_{-n} \ldots]_k$$

$$= [(k^n * a_n) + (k^{n-1} * a_{n-1}) + \cdots + (k^0 * a_0) + (a_{-1} * k^{-1})$$

$$+ (a_{-2} * k^{-2}) + (a_{-3} * k^{-3}) + \cdots + (a_i * k^i) + \cdots]_{10} = \sum_{i=-\infty}^{n} (a_i * k^i)$$

So, as you can see, all operations on real numbers are operations on rational numbers and we can do them with any precision.

Since we cannot pair all elements of the set of all integer numbers with all elements of the set of all real numbers (Cantor's diagonal argument), we have the proof, that the set of all real numbers (R) has greater cardinality, than the set of all integer numbers.

And that is, how we have defined arithmetic of all numbers.

And this arithmetic defined above this way or another is trivially true, since we can use trivial finite sets to do this arithmetic:

$$0 = |\varnothing|$$

$$1 = |\{0\}|$$

$$2 = |\{0,1\}|$$

$$3 = |\{0,1,2\}|$$

. . .

and so on.

So for every $n \in N$:

$$n = |\{0,1,2,\ldots,n-1\}|$$

The fullness of a true theory cannot implicate a statement and its negation at the same time, because a true statement implicates only true statements, so every true theory is consistent. So above arithmetic is consistent.

Since above definition of arithmetic consists only of trivially true definitions, it is consistent.

And since all possible objects (things) of arithmetic are defined above or can be derived from objects (things) defined above, its potential is exhaustive.

Q.E.D.

16.2.3 Peano arithmetic too general

The definition of natural numbers cannot be simpler. For example, Peano arithmetic is too general, because it defines natural numbers as consecutive successors of zero not specifying what they are and what is zero except the fact that they all have successors. So any natural number there is defined as $(successor^{\circ} \ldots ^{\circ}\ successor)(0)$ not specifying anything about the function *successor* except the fact that it has the same domain and codomain not counting zero, to which it gives total order. Therefore, the results and arguments can be any things, that are different from each other, that we will then call natural numbers. For example, then we will be able to call any infinite subset of natural (real too) numbers the set of all natural numbers.

In Peano arithmetic you cannot even write:

$$n = successor^{n}(0)$$

to say, that natural number n is the natural number n of recursions of function *successor*, because you did not define natural numbers, so you cannot count the number of these recursions.

Numbers are not concrete things, because, for example, the Number One yesterday had different physical attributes, than it has now (e.g., now we are saying about it, which makes a physical attribute of the Number One), but at any time every number has an existing representative and the symbols of numbers we use always refer to this one current concrete being, e.g.:

$$0^{\uparrow} = \{|\varnothing|\} = \{0\}$$

$$1^{\uparrow} = \{|\{x\}| : x \in \Omega\} = \{1\}$$

$$2^{\uparrow} = \{|\{x,y\}| : (\{x,y\} \subset \Omega)\ and\ (x \neq y)\} = \{2\}$$

$$\ldots$$

and so on.

So for every $n \in N$:

$$n^{\uparrow} = \{|\{x_1, \ldots, x_n\}| : (\{x_1, .., x_n\} \subset \Omega) \textit{ and } \textit{areDifferent}(x_1, \ldots, x_n)\} = \{n\}$$

where $\textit{areDifferent}(x_1, \ldots, x_n)$ means, that these are n different things, so these things are different from each other:

$$\textit{areDifferent}(x_1, \ldots, x_n) = \bigwedge_{1 \le i \le n} \bigwedge_{1 \le j \le n, i \ne j} (x_i \ne x_j)$$

16.2.4 Complex numbers wrongly defined

It is worth here to mention, that complex numbers are wrongly defined.

From English Wikipedia from the term *Imaginary unit*:

Imaginary unit or unit imaginary number is a solution to the quadratic equation $x^2 + 1 = 0$

From Wolfram MathWorld from the same term:

The imaginary unit $i = \sqrt{-1}$, *i.e., the square root of* -1.

It cannot be defined this way, because there is not a number, that squared gives -1. In other words, $i \ne \pm\sqrt{-1}$, because $\sqrt{-1}$ is a nobeing. So such an attempt of definition fails. And you can only define it this way:

Imaginary number has symbol "i", that is a part of symbol "$i * i$" given for number -1, where you assume, that multiplication symbol can be treated as operator, that is commutative and associative with real multiplication operator, and this operator distributes over real addition, so:

$$i * i = i^2$$

$$0 * i = 0$$

$$a * i = i * a$$

$$a + b * i = b * i + a$$

$$(a * b) * i = a * (b * i)$$

$$a * i + b * i = (a + b) * i$$

And although i has not a value, it has multiplicative inverse:

$$(x * i = 1 = -1 * i * i) \Leftrightarrow \left(x = \frac{1}{i} = -i\right)$$

Using these rules we can evaluate multiplication of complex numbers:

$$(a+b*i)(c+d*i)=a*c+a*(d*i)+(b*i)*c+(b*i)*(d*i)$$
$$=ac+ad*i+bc*i+bd*i^2=ac+bd*(-1)+(ad+bc)i$$
$$=ac-bd+(ad+bc)i$$

And using these rules we can evaluate any other operation on complex numbers.

The way, that symbol "$i*i$" is given to number -1, is as follows:

$$x^2+1=(x-x_1)(x-x_2)=x^2-(x_1+x_2)x+x_1*x_2=0$$

So:

$$\begin{cases} x_1+x_2=0 \\ x_1*x_2=1 \end{cases}$$

So:

$$x_2=-x_1$$
$$x_1*x_2=x_1*-x_1=-(x_1*x_1)=1$$
$$x_1*x_1=i*i=-1$$

And that is it. Symbol "i" is given for nonexistent root of polynomial, that multiplied by itself, if we allowed it, would give -1. This way "$i*i$" is a new symbol for number -1, but $i \neq \pm\sqrt{-1}$, because operation of square root is not allowed and not possible and has not any result for negative argument, i.e., -1. Symbol "i" does not represent any number, because any attempt to define it as a number would fail due to a contradiction with the definition of square root function, so symbol "$*$" in symbol "$i*i$" is not arithmetic multiplication at all, because arithmetic multiplication cannot be defined for arguments, that are not numbers. Only symbol "$i*i$" is a number and "i" can be defined only as a part of this symbol. That is all.

An imaginary number is not a number the same as an imaginary physical being is not a physical being. Indeed real numbers are all possible finite numbers, and we have only all subsets of the set of all real numbers.

In other words, any expression, that contains i, that cannot be eliminated by multiplication by 0 or turned back with another i from that expression to symbol of -1, has not any value. So, for example, complex number alone with nonzero imaginary part has not any value. Two such expressions are equal then and only then, when they can be transformed to each other using above rules. And some operation on two such expressions can give some value, because only then i can be eliminated by multiplication by 0, exponentiation by 0 and so on or turned back with another i into -1.

For example, Euler's equation is true, only because i is eliminated by multiplication by 0 in expansion of the expression $e^{i\pi}$ to Taylor series:

$$e^{i\pi} = \cos\pi + i * \sin\pi = -1 + i * 0 = -1$$

16.3 Cartesian geometry definition based on definitions

For the definitions-based definition of new logic, set theory and theory of things see Section 5.4.

Since we have defined arithmetic, we can define Cartesian geometry:

Cartesian space – set R^n, where every element is called a point and for every two points a, b the shortest path between them consists of the following set of points:

$$ShortestPathSet(a,b) = \{x: (x = (1-t) * a + t * b) \; and \; (t \in [0,1])\}$$

where for $t \in R$, $x \in R^n$, $y \in R^n$:

$$t * x = t * (x_1, \ldots, x_n) = (t * x_1, \ldots, t * x_n)$$

$$x \pm y = (x_1, \ldots, x_n) \pm (y_1, \ldots, y_n) = (x_1 \pm y_1, \ldots, x_n \pm y_n)$$

which means, that the space grid is not curved.

To make the right angle between dimensions (and thereby between appropriate lines of the space grid) from Pythagorean theorem we have:

$$length(ShortestPathSet(a,b))^2$$

$$= m_n(a,b)^2 = m_{n-1}((a_1, \ldots, a_{n-1}), (b_1, \ldots, b_{n-1}))^2 + (a_n - b_n)^2$$

$$m_0 = 0$$

where m_n is the real number measure in n dimensions – the measure of the number of the points between given two n dimensional points.

From this definition whole Cartesian geometry can be derived, because it defines all properties of space. Non-Euclidean geometries are not true geometries, because they simply name a curve a straight line. And spaces of such geometries are just shapes in Cartesian geometry.

16.4 How logic can help physics. Einstein was most likely right about quantum physics

Bell's theorem is one of the most important results in quantum physics, which shows that quantum mechanics is not compatible with local hidden-variable theories. The theorem was formulated by John Stewart Bell in 1964 based on the thought experiment of Einstein, Podolsky and Rosen from 1935, which suggested that quantum mechanics is incomplete and that quantum particles have some hidden properties that determine their behavior. The theorem had profound and far-reaching implications for our understanding of reality, causality and locality, but is it correct? Here is a proof, that it is incorrect:

Correlation does not imply causation, so the lack of causation does not imply the lack of correlation, because $\sim(X \Rightarrow Y) = \sim(\sim Y \Rightarrow \sim X)$ $\left(\text{in classical logic: } \sim(X \Rightarrow Y) \Leftrightarrow \sim(\sim Y \Rightarrow \sim X)\right)$.

We have:

$$\sim\left((\textit{There is a correlation}) \Rightarrow (\textit{There is a causation})\right)$$

So:

$$\sim\left((\textit{There is no causation}) \Rightarrow (\textit{There is no correlation})\right)$$

Hence the Bell's necessary condition (Bell locality of Bell's theorem) for local hidden-variable theory is wrong, because it is nothing but statistical independence (the lack of correlation), which, as you can see above and oppositely to what Bell assumed in his proof, is not implied by the assumption of causal independence (the lack of causation), which is a correct condition for local hidden-variable theory. So Bell's theorem uses a wrong assumption about local hidden-variable theory, so it is wrong.

Q.E.D.

Albert Einstein believed in local realism (which is inconsistent with Bell's theorem) to the end of his life. As you can see, he was most likely right and most likely entanglement is just a synchronization of states of entangled particles, so "to entangle particles" means most likely simply "to synchronize them". And that changes everything. For more detailed and extended proof of it see the article titled "Albert Einstein was most likely right about quantum physics" in my account (https://independent.academia.edu/ZbigniewPłotnicki) on academia.edu.

References

[1] English Wikipedia, term "Statement (logic)", point 2.

[2] Strawson, P. F. (Jul 1950). "On Referring", in Mind, Vol 59 No 235.

[3] Rouse, D. L. "Sentence, Statement and Arguments", A Practical Introduction to Formal Logic. https://people.uvawise.edu/philosophy/Logic%20Text/Contents.htm

[4] Burgess, J. P. (2009). Philosophical logic. Princeton University Press. ISBN 978-0-691-13789-6.

[5] Graham Priest. (2008). An introduction to non-classical logic: from if to is (2nd ed.). Cambridge University Press. ISBN 978-0-521-85433-7.

[6] Boole, G. (1847). Mathematical analysis of logic. George Bell: London.
——— (1856?). On the mathematical theory of logic. Boole, 1952.

[7] De Morgan, A. (1847). Formal logic. Taylor and Walton: London.

[8] Jevons, W. (1870/34). Elementary lessons in logic. Macmillan: London.
——— (1879/1958). The principles of science. Dover: New York.

[9] Venn, J. (1881/1971). Symbolic logic. Dover: New York.
——— (1889). The principles of empirical or inductive logic. London and New York.

[10] Himma, K. E. (16 November 2001). "Ontological Argument". Internet Encyclopedia of Philosophy. Chapter 4, section IV. Retrieved 2012-01-03

[11] Corcoran, J. 1998. Information-theoretic logic, in Truth in Perspective edited by C. Martínez, U. Rivas, L. Villegas-Forero, Ashgate Publishing Limited, Aldershot, England (1998) 113–135.

https://doi.org/10.1515/9783111441382-017

Index

https://doi.org/10.1515/9783111441382-018

www.ingramcontent.com/pod-product-compliance
Lightning Source LLC
Chambersburg PA
CBHW080132220326
41598CB00032B/5038
9783111440460